FIGHTER AIRCRAFT COMBAT DEBUTS

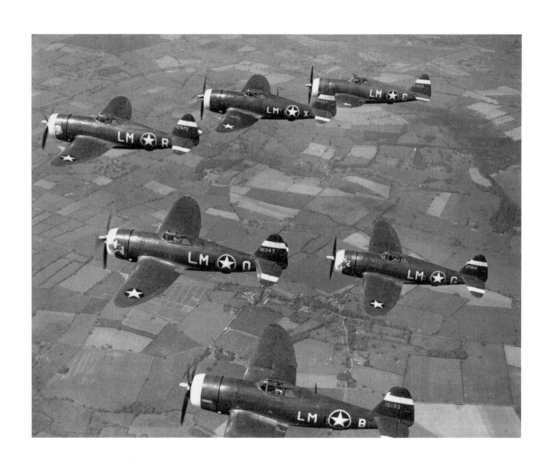

FIGHTER AIRCRAFT COMBAT DEBUTS,

1915–1945

Innovation in Air Warfare Before the Jet Age

JON GUTTMAN

WESTHOLME
Yardley

Frontis: Republic P-47C-2RE Thunderbolts of the 62nd Squadron, 56th Fighter Group, embark on a bomber escort mission from Horsham St. Faith, England, on May 25, 1943. (*US Air Force*)

Westholme Publishing, LLC
904 Edgewood Road
Yardley, Pennsylvania 19067
Visit our Web site at www.westholmepublishing.com

First Printing October 2014
10 9 8 7 6 5 4 3 2 1
ISBN: 978-1-59416-200-8
Also available as an eBook.

Printed in the United States of America.

CONTENTS

FOREWORD

Of all military aircraft, fighters hold the most mystique—perhaps because of all military aircraft, fighters are the type that can afford the least compromise. There have been numerous occasions when desperate combatants managed to achieve a surprising degree of success by improvising the most unlikely available aircraft into bombers, attack planes, and reconnaissance aircraft. There is far less room for ingenuity in the realm of fighter planes. When the goal is to seize and maintain control of the air, the confrontation is direct, with the prospect of only one out of two possible outcomes. If a pilot and his plane are better, he and his side wins; if his enemy is superior, he loses—and often dies.

Although aircraft—in the form of balloons—have appeared in battle since 1794, the concept of air superiority dates to 1914, during World War I. Since then, the development of fighter aircraft has been an ongoing seesaw battle in itself, with each new design leading to another. Like sports cars, fighter planes became sleeker and their performance greater as the competition intensified—but in contrast to peacetime competition, they also became deadlier. And with each new development, the pilots had to adjust to higher speeds and higher gs—something

they did not always do ungrudgingly. The alternative, however, was to suffer the fate of those seat-of-the-pants Italian fighter pilots who were loath to give up open cockpits, or of the Japanese who regarded dogfighting ability as the primary determinant of a fighter's worth—in both cases, these aircraft were literally left behind by newer, faster opponents.

Aside from the adrenaline rush of aerial combat itself, a great cause of excitement among fighter pilots is the arrival of a new airplane. As they admire its lines, the questions fill their minds: Will it be all that the manufacturer claims it will be? Will I be able to adjust to its idiosyncrasies? Above all else, will it give me the edge I need to win?

This book explores the first military actions for a variety of famous fighters of World War I, the conflicts of the so-called interwar years, and World War II—a thirty-year period that saw the birth of the fighter concept and its maturity on the threshold of the jet age. Comparisons of the strengths and weaknesses of fighter aircraft will probably go on as long as there are men who will fly them. This study, however, will compare them from a somewhat narrower perspective: How did they do at the very onset?

Most of the aircraft described are fairly well known to aviation historians, and a few names, such as Sopwith Camel, Fokker Triplane, Messerschmitt Me 109, Mitsubishi Zero, North American Mustang, and Supermarine Spitfire, are familiar even to the most nonaviation-minded layman. Not so well known are the circumstances of their combat debuts, in which some, such as the Zero, made their mark almost from the outset, but in which others, like the Bristol F.2A, showed rather less promise than they would ultimately realize.

While a certain amount of space must be devoted to the technical development of these famous fighters, these studies of first combats serve as a reminder that it is the human factor, with all its special quirks, that inevitably came into play when these deadly flying machines first fired their guns in anger. It was the pilot who determined how a new airplane did, and the results were not always in direct relation to the plane's capabilities. To cite a particularly striking example, the Brewster Buffalo, long vilified for its wretched performance against the Japanese Zero and *Hayabusa* fighters in the Pacific, actually saw combat for the first time over Finland nearly half a year earlier—and, thanks to the skill of the Finnish pilots, enjoyed a generous measure of success over its Soviet opponents that its American and British users would have found unbelievable.

Some of the pilots became as famous as the aircraft they flew, and some of the first men to fly the fighters recounted in this volume are exactly those whom the lay reader would expect. For example, the first aerial victory scored in a Fokker triplane was achieved by the man most popularly associated with it: Manfred von Richthofen, the Red Baron. Likewise, although Georges Guynemer may not

have gained the first aerial victory credited to a Spad 7.C1, he did score the first for the cannon-armed Spad 12.Ca1, in the development of which he had played a prominent role.

Other fighters had combat debuts that were more obscure, and sometimes, as in the case of the Supermarine Spitfire and the Vought F4U-1 Corsair, less than auspicious. Several famous types did not even enter combat with the countries that designed them, for example, the quintessentially French Spad 13.C1 scoring its first victory with a British pilot in the cockpit. Although an American did score the first aerial victory for the North American P-51 Mustang, he did so while serving in the Royal Canadian Air Force!

My original findings were published in *Fighting Firsts* in Britain in 2001. Since that time, my research for a number of subsequent book projects uncovered an astonishing wealth of new information, adding new factual details. Some revelations include combat debuts that were even earlier and under different circumstances than previously thought. There have even been a few additional fighter types, overlooked in the original book, given their due in this one. Therefore, this volume offers the fruits of more than a dozen years of fresh scholarship. Much of the newly unearthed material represents not only information from some countries from which it was previously not released, but also a proliferation of exchanges with colleagues over the Internet at unprecedented speeds. As with aviation technology, information interchange has undergone some advances in the past few decades that have radically altered the state of the art.

This begs a question that may enter the heads of some readers: Why limit things to the first thirty years of fighter development? The answer is that that period covers a classic phase of fighter origins and evolution, leading to the threshold of the jet age, when the formula underwent a new phase of refinement to accommodate an entirely new mode of propulsion, along with even higher speeds and stresses. (It had, in fact, been my intention to write a sequel covering the jet age's more notable products, which saw their own share of action over Korea, Indochina, the Middle East, and the Indian subcontinent, among others. Certainly the jet age has produced its own share of classics, worthy of their own treatment in this context. There may very well be a volume two to follow this one.)

What began for this writer, in all honesty, as a casual literary lark, soon took on the trappings of an often frustrating scavenger hunt—one, it must be added, that I could not have completed alone. Among the friends and colleagues who lent a hand to the project, I particularly wish to gratefully acknowledge the assistance of Frank W. Bailey, Dugald Cameron, Christophe Cony, Jerzy B. Cynk, Douglas Dildy, Norman L. R. Franks, Colin Heaton, Predrag Jelíc, D. Y. Louie, David Méchin, George Mellinger, Lt. Gen. Heikki Nikunen, Finnish Air Force, ret., Colin Owers, Henry Sakaida, Kari Stenman, Andrew Thomas, Greg VanWyngarden, and Paolo Varriale, as well as the late veteran pilots Robert Duncan, Svein Heglund, Robert S. Johnson, Eino Ilmari Juutilainen, and Stanislaw Skalski.

Last but by no means least, I reserve special thanks to Bruce H. Franklin of Westholme Publishing, who made the publication of this new, transatlantic edition possible, and to Laura Pfost, whose meticulous copyediting not only picked out a multitude of nits, but which frequently led to the discovery of new sources of information that produced outright changes for the better. Any errors that remain at this point are strictly the author's.

B.E.2b 487 of No.4 Squadron Royal Flying Corps at Saint-Omer represented the first generation of reconnaissance planes that became the targets of fighters in 1915. BE 487 was brought down by antiaircraft fire while bombing a railroad junction near Lille on the night of March 11, 1915. The pilot, Capt. Gilbert W. Mapplebeck, made it back to Allied territory and was awarded the Distinguished Service Order, but died in a Morane-Saulnier crash on August 24. (*Library of Congress*)

FIRST BLOOD

The Earliest Fighters, 1914–16

When Austria-Hungary declared war on Serbia on July 28, 1914, it began a chain reaction that rapidly plunged the world into a struggle which would be carried out not only on land and sea, but in the air as well.

Aircraft were then nothing new to warfare. The French had introduced observation balloons to the battlefield in 1794, and that means of intelligence gathering was subsequently employed in numerous other conflicts, such as the American Civil War, the Franco-Prussian War, and the Spanish-American War. The Italians introduced airplanes during their 1911 campaign in Libya, not only for reconnaissance but also for bombing. Airplanes were also much in evidence during the Balkan wars of 1912 and 1913. It was during World War I, however, that airmen began to take air power seriously enough to try to take control of the sky by eliminating the other side's aircraft.

Britain's Royal Flying Corps (RFC) has often taken credit for the first air-to-air victory, although that claim is not without its qualifications. The main protagonist was Lt. Hubert D. Harvey-Kelly, who at 0820 hours on August 13, 1914, had had the distinction of landing the first RFC airplane on French soil since the war began: Royal Aircraft Factory B.E.2a No. 347. Harvey-Kelly's unit, No. 2 Squadron, soon commenced reconnaissance operations, and it was during one such patrol on August 25 that three of its BEs encountered a German Rumpler Taube. Harvey-Kelly and his observer, Lt. W. H. C. Mansfield, immediately attacked the enemy plane with whatever small arms they carried, to which the German pilot responded by bringing his plane down to earth near Le Cateau. Harvey-Kelly landed nearby and saw the enemy pilot and observer running into some nearby woods. He and Mansfield gave chase but were unable to overtake them. They then returned to the Taube, and after taking some trophies from it, burned it and took off again. Theirs would not be the only victory claim that day, however. In one of three encounters in the course of it, 2nd Lt. C. W. Wilson and Lt. Cuthbert Euan Charles Rabagliati of No. 5 Squadron spotted a Taube approaching their airfield, took off in an Avro 504 and forced the German to land at Le Quesnoy, where it was captured by British troops.

A few weeks later a Russian pilot actually destroyed an enemy plane in the air, albeit by means that few of his colleagues would want to emulate. The drama began on September 7, when an Austro-Hungarian Albatros two-seater of *Fliegerkompagnie* 11, based at Zurawica, flew over the Russian XI *Korpusnoi Aviatsionny Otryad*'s (Corps Aviation Detachment's) aerodrome outside of the town of Zholkov and dropped a bomb on it. The Austrian observer, *Oberleutnant* (First Lieutenant) Friedrich *Freiherr* von Rosenthal, owned several large estates in territory that by then had been occupied by the Russians, so his action may have been inspired by a degree of personal rancor. His attack, however, was taken equally personally by one of the Russian pilots, *Shtabs-Kapitan* (Staff-Captain) Piotr Nikolayevich Nesterov.

Born in Nizhny Novgorod (now Gorky) on February 27, 1887, Nesterov had graduated from the aviation school in March 1913. Flying over Syretsk military aerodrome on September 8 of that year, he put his Nieuport IV through a complete loop in the vertical plane—the first airman to complete what was then known as a "death loop." For that achievement, Nesterov was put under ten days' arrest for "undue risk with a machine, the property of his government." Later, however, he was pardoned, promoted to *Shtabs-Kapitan* and awarded the Russian Aero Club gold medal on November 23.[1]

Appointed commander of the XI KAO in February 1914, Nesterov tried to develop a training regimen for his pilots that he described in prophetic terms:

I am perfectly convinced that it is the duty of every military aviator to be able to execute looping flights and gliding flights. These exercises must certainly be included in the training program, as they will play a great part in the aero-combats. Such a combat will resemble a fight between a hawk and a crow. The aviator who is able to give his craft the mobility and flexibility of motion of the hawk will be in a better position to seriously damage his opponent.[2]

On the morning of September 8 Rosenthal, with Moravian-born *Feldwebel* (Staff Sergeant) Franz (František) Malina as his pilot, reappeared over Zholkov, spurring Nesterov to put his theories into practice. Flying a Morane-Saulnier G two-seat monoplane, he attacked the Lohner with a pistol, but accomplished nothing. A second sortie and a second encounter produced an identical result. When Rosenthal made a third appearance that day, Nesterov jumped into the cockpit of Morane-Saulnier G No. 281, and when one of his lieutenants came up to offer his Browning pistol, he replied, "That's all right; I shall manage without it." Taking off and gaining altitude, Nesterov then dived on the Albatros and rammed it. He may have intended to recover from the deliberate collision, but some of the Albatros's wing fabric wrapped around his propeller, and although the Morane-Saulnier tore free as the two planes spun down, Nesterov, who had neglected to fasten his seat belt, was thrown from it. His body was found thirty or forty feet from the wreckage that held the remains of Rosenthal and Malina. For his sacrificial victory, Nesterov was buried in Askold's Grave, a resting place for heroes in Kiev, and posthumously awarded the Order of Saint George, Fourth Class, on July 22, 1915. His status as a pioneer of military aviation was also honored later by the Soviet Union, which on December 3, 1951, renamed Zhovkva in Lvov Province to Nesterov, and renamed the Zholkovsky region as the Nesterov region.[3]

It might be argued that the first serious attempt to attain air superiority was mounted by the Japanese when, after evoking a 1902 treaty with Britain to enter the war on the Allied side, they blockaded the German concession port of Tsingtao on August 27. Among the ninety-five ships involved in the campaign was the "hydroplane mother ship" *Wakamiya Maru*, carrying three two-place and one three-place Maurice Farman M.F.7 floatplanes—the first airplane-carrying vessel to see wartime use. Joining those naval units was the Japanese army's Provisional Air Corps, consisting of four M.F.7s and a Nieuport 6M monoplane. Facing them was the entire German air arm in China: a Rumpler Taube flown by a navy pilot, *Linienschiffsleutnant* (Lieutenant Commander) Günther Plüschow.

On September 5, one of the three-seater Farmans, piloted by 1st Lt. Hideo Wada, dropped several fourteen-pound and forty-five-pound artillery shells fitted with tail fins on one of the city's strongpoints, Fort Bismarck. The next day, in a specific effort to eliminate the Germans' only air asset, two Japanese planes dropped three bombs on Plüshow's hangar but failed to hit their target.

The Germans responded by organizing a warning and air defense system, while Plüschow erected a new hangar for his plane under a rocky overhang at the other side of the airfield. He also devised a decoy Taube out of bamboo, tin, canvas, and bicycle wheels, which he periodically moved around the hangar to reinforce the illusion that it was a real airplane. The ruse worked—in several subsequent raids the Japanese bombed the fake Taube, only to experience puzzled frustration thereafter at the sight of Plüschow scouting their lines again.

On September 28, Plüschow privately noted an incident in which he attacked a

A Maurice Farman floatplane attached to the Japanese seaplane tender *Wakamiya Maru* taxis for takeoff. The tender's aircraft tried—unsuccessfully—to eliminate the sole German Rumpler Taube operating from the besieged German concession of Tsingtao in 1914. (*Imperial Japanese Navy photograph*)

Farman with his Mauser pistol, firing thirty 9mm rounds until he saw it spin down to crash. The Japanese did record a Flying Lieutenant Shigematsu being killed in a Maurice Farman sometime in the course of the siege, but it may never be ascertained whether he was the victim of Plüschow, who in any case did not officially report what might otherwise have gone down as history's first air-to-air shootdown. With nine Japanese planes gunning for the "sky spy," as they called him, Plüschow had specific orders from Governor Alfred von Meyer-Waldeck to avoid any contact with them rather than risk Tsingtao's sole air asset. To report his combat in violation of that order would therefore have invited court-martial.

After a few attempts by Plüschow to bomb Japanese targets—including *Wakamiya Maru*—with large coffee canisters filled with twenty-five sticks of dynamite, horseshoe nails, and detonators, the Japanese redoubled their efforts to eliminate the pesky Taube. Mounting a machine gun on one of the M.F.7s only handicapped its already sluggish performance, and Plüschow easily evaded attempts to drop hand grenades on him. The Japanese naval floatplanes carried out forty-nine sorties over the German airfield, dropping 199 bombs, to achieve eight confirmed and sixteen probable hits, but the only damage they inflicted was on Plüschow's dummy plane.

Ultimately, the Japanese never did eliminate Plüschow's Taube before Tsingtao itself fell. On November 5, Meyer-Waldeck ordered him to leave the doomed city with war diaries and official documents. The next morning he took off and flew 150 miles to Haichow, where he crash-landed in a muddy field. After selling the plane's engine and his Mauser pistol and ammunition to a local mandarin, he burned the Taube. Tsingtao surrendered on November 7, but Plüschow, after escaping Chinese internment and undertaking a remarkable trek that included capture and escape from the British, eventually made his way back to Germany.

The first confirmed aerial shootdown occurred on October 5, 1914, when a French Voisin 3LA two-seat pusher of *Escadrille* V.24 crewed by *Sergent* (Sergeant First Class) Joseph Frantz and *Sapeur* (Combat Engineer Private) Louis Quénault attacked an Aviatik B.I two-seater of *Feldflieger-Abteilung* 18 over Jonchery-sur-Vesle, near Reims. Manning an 8mm Hotchkiss M1909 machine gun on a front-mounted tripod, Quénault expended two fifty-eight-round magazines without success. With the Germans then firing at him with rifles, Quénault produced one of his own and managed to hit the enemy pilot, *Sergeant* Wilhelm Schlichting, and the Aviatik crashed, also killing the observer, *Obltn.* Fritz von Zangen. After that, the airmen of both sides began experimenting in earnest with means of more effectively clearing the skies of their counterparts.

That first instance notwithstanding, the most effective weapon for accomplishing the task was the machine gun, but the most efficient way of using it was at first contingent upon how to fire it without shooting one's propeller off. Under the circumstances, there were initially two schools of thought regarding fighting aircraft—large flying fortresses with anything from one to three gunner's positions; and single-seat scouts, in which the pilot aimed his gun by aiming the plane itself at the target. Ultimately, the latter came to be recognized as the best fighter, but the problem of clearing the propeller arc remained to be solved.

At first, Britain literally found a way around the dilemma. In 1912 Vickers Ltd.'s newly created aviation branch received an order from the Royal Navy for a "fighting biplane" armed with a machine gun, and after devising a number of experimental variations came up with the F.B.5, which had a one-hundred-horsepower Gnome monosoupape rotary engine behind the two-man crew's nacelle. The motor drove a pusher propeller amid a latticework of struts and wires that connected the wing to the tail surfaces, which allowed the front observer to man a .303-inch Lewis machine gun. Dubbed the "Gunbus," the Vickers offering attracted production contracts for both the Royal Naval Air Service (RNAS) and the RFC on August 14, 1914, although it would not reach the front in force for more than eleven months.

On February 5, 1915, the first F.B.5, No. 1621, arrived in France. Flown by 2nd Lt. M. R. Chidson, it was initially assigned to No. 2 Squadron and then reassigned to No. 16 Squadron five days later. On February 28, however, Chidson and his observer, 2nd Lt. D. C. W. Sanders, were forced to land in German lines, where they were taken prisoner, and Britain's first fighter fell virtually intact into enemy hands.

In spite of that unfortunate debut, more F.B.5s reached the front and were allocated to various squadrons to defend the RFC's reconnaissance planes and harass the enemy's. On July 25, the first unit entirely equipped with F.B.5s, No. 11 Squadron, arrived at Vert Galant aerodrome.

Britain's first specialized fighter squadron had come none too soon, since Fokker E.I monoplanes, armed with synchronized machine guns, were just starting their own rampage among the Allied reconnaissance planes. Wasting no time to accept the challenge, on July 28 Capt. Lionel Wilmot Brabazon Rees and Flt. Sgt. James McKinley Hargreaves, in F.B.5 1649, engaged one of the vaunted "Eindeckers" and were credited with driving it down in enemy lines.

An aggressive Welshman with a dozen years of army service, and a year in the RFC under his belt, Rees continued to assert himself over the next few months. On August 31, he and his Scottish observer, Hargreaves, destroyed an LVG C.II near Achiet-le-Grand and drove down an Ago C.I on September 21, followed by an Albatros the next day. On September 30, they forced an Albatros down in British lines at Gommecourt, the crew, *Leutnant der Reserve* (Lieutenant of the Reserves) Fritz Kölpin and *Obltn.* Ernst Leonhardi of *Fl.-Abt.* 23, dying of their wounds. Rees was awarded the Military Cross and Hargreaves the Distinguished Conduct Medal for their exploits, which also made them the second Britons to attain what would come to be called ace status with five accredited aerial victories.

On August 31, Rees and another observer, Flight Sergeant Raymond, downed an LVG. In September, No. 11 Squadron moved from Vert Galant to Villers-Bretonneux, from which aerodrome a second F.B.5 pilot would earn a

different distinction. On November 7, Lt. Gilbert Stuart Martin Insall was flying F.B.5 5074 with 1st Class Air Mechanic T. H. Donald as his observer when they attacked an Aviatik near Achiet-le-Grand, and after a merry chase forced it to land southeast of Arras. Ignoring ground fire—including shots fired by the enemy aircrew—Insall descended to drop a small incendiary bomb, which set the Aviatik afire. On the way home they strafed the German trenches, but return fire holed their fuel tank.

Insall landed in a wood five hundred yards inside Allied lines, where he and Donald stood by their plane while subjected to enemy artillery fire. That night, working by torches and other lights, the duo repaired their plane, and the next dawn they took off and returned to their aerodrome.

On December 23, Insall was gazetted for the Victoria Cross and Donald for the DCM. Neither man was there to receive them, however, because nine days earlier they had been brought down wounded and taken prisoner, either by ground fire or by a two-seater they had engaged, crewed by ace-to-be *Hauptmann* (Captain) Martin Zander and *Leutnant* (Second Lieutenant) Gerche of Royal Bavarian *Fl.-Abt.* 9. After two failed attempts, Insall and two companions escaped from Ströhen prison camp on August 28, 1917, and reached the Dutch border nine nights later. Awarded the Military Cross for that feat, in mid-1918 Captain Insall returned to service commanding A Flight of No. 51 (Home Defence) Squadron at Bekesbourne. Remaining in Royal Air Force service until July 30, 1945, Gilbert Insall died at Bawtry on February 17, 1972.

Although reliable and well liked by its crews, the F.B.5 was only capable of seventy miles per hour and was soon outclassed by single-engine opposition, so it was replaced by another two-seat pusher design, the Royal Aircraft Factory "Farman Experimental" F.E.2b. The impression the Gunbus left, however, was as striking as it had been brief. For the next two years the Germans became accustomed to referring to every pusher fighter they encountered, whether single or two seat, as a "Vickers."

The first successful single-seat fighter was a crude improvisation devised on a French Morane-Saulnier L two-seat parasol reconnaissance plane. The idea was to mount steel wedges on the propeller, so any bullets that struck it would most likely be deflected aside. One of the early advocates of the system, *Sgt.* Eugène Gilbert of *Escadrille* MS.23, abandoned it after two of his friends were killed by ricocheting bullets during ground testing. Another prewar pilot in neighboring MS.26, Roland Garros, also discussed the idea with Raymond Saulnier, who devised steel deflectors that could be bolted onto the propeller blades. Garros and his mechanic, Jules Hue, improved on Saulnier's design by narrowing the width of the propeller blades at the point where the deflectors were attached. Garros then installed the modified airscrew on a Morane-Saulnier L armed with a forward-firing Hotchkiss machine gun and, attaching himself to MS.23 at Saint-Pol aerodrome, went up looking for trouble.

Garros was on a bombing mission to Ostend on the morning of April 1, 1915, when he encountered a lone Albatros two-seater and immediately attacked. *Gefreiter* (Private) August Spacholz and *Ltn.* Walter Grosskopf of *Fl.-Abt.* 40 thus had the dubious distinction of being the first aircrew to be shot down by a single-seat fighter. Over the next two weeks, Garros was the terror of the Western

Front, attacking any German aircraft he could find. He brought down an Aviatik on April 15 and another Albatros three days later, but shortly after scoring his third victory, his luck ran out and he was forced down in enemy lines—either due to engine trouble, or a single rifle bullet through his fuel line by a German soldier, depending on whose story one believed.

The Germans were delighted to have captured France's hero of the hour, and even more pleased to have gotten their hands on the secret of his success. They soon discovered flaws in the system, however. While French copper-jacketed ammunition bounced off the deflectors, German steel-jacketed bullets tended to shatter the wedges.

One of the aircraft designers whom the German High Command had asked to adapt, or even improve upon, Garros's deflectors, Anthony Fokker, had a better idea. Since 1913, Franz Schneider of the *Luft-Verkehrs Gesellschaft* (LVG) had held a patent for using a series of cams and rods attached to the trigger bar, to interrupt the machine gun's fire whenever the propeller was in its way. Schneider's idea had been strictly theoretical up to that time, but Fokker put it into practice, adapting it to the 7.92 mm Parabellum 08/14 machine gun and to his M.5K single-seat scout. Similar in appearance to the Morane-Saulnier H shoulder-wing monoplane, the Fokker M.5K differed in having an airframe of steel tubing rather than wood and was powered by an eighty-horsepower Oberursel U.0 seven-cylinder rotary engine.

The German High Command ordered thirty of Fokker's *Eindecker* (monoplane) scouts armed with the synchronized machine gun armament, which were initially designated M.5L/MGs but subsequently standardized as E.Is. Several were assigned to Döberitz for pilot training,

One of the Fokker E.I prototypes, M5K/MG 5/15, was used by *Ltn*. Kurt Wintgens to score the first, albeit unofficial, Eindecker victory on July 1, 1915. Although the Germans did not confirm it, French records indicate that Wintgens had indeed drawn blood that day. (*P.M. Grosz*)

while the rest were farmed out to front-line units, to be used by their most experienced pilots.

Among the fighter's early recipients was *Ltn*. Kurt Wintgens, a bespectacled twenty-year-old army officer's son from Neustadt. After some time as an observer, Wintgens began training as a pilot and demonstrated such innate skill that he was assigned to fly a Fokker with *Feldflieger-Abteilung* 67 and then with *Fl.-Abt*. 6. It was with the latter unit that he claimed a Morane-Saulnier parasol monoplane east of Lunéville at 1800 hours on July 1, 1915—a success that, had it been confirmed, would have been the first German fighter victory, for it probably resulted in the wounding of *Capitaine* (Captain) Paul du Peuty and *Lieutenant* (First Lieutenant) Louis de Boutiny of MS.48 that day. Wintgens claimed another Morane-Saulnier parasol on July 4, again without confirmation. He then transferred to *Fl.-Abt*. 48 on July 5, although he stated in a letter to a friend that he had been granted what amounted to a roving commission. Finally, on July 15, Wintgens downed a Voisin over Schlucht and went on to score two more victories by early 1916, when he came down with influenza. Returning to action that spring, he resumed his scoring on May 20 with a Nieuport 12 of N.68 whose pilot, *Maréchal-des-Logis* (Sergeant) Léon Beauchamps, was taken prisoner, but whose observer, *Sous-Lieutenant* (Second Lieutenant) Philippe Debacker, was killed. By June 30 he had brought his total up to eight, for which, on the next day, he became the fourth German fighter pilot to be awarded the *Orden Pour le Mérite*. Wintgens's official score reached nineteen when he was shot down over Villers-Carbonnel by *Lt*. Alfred Heurt-eaux of *Escadrille* N.3 on September 25, 1916.

Another early recipient of Fokker Eindeckers was *Fl.-Abt*. 62, the ranks of which included a twenty-four-year-old *Leutnant* from Dresden named Max Immelmann. Immelmann had been flying LVG two-seaters with the unit since March 1915, and was overjoyed when his flying section was assigned two E.Is for escort and hunting duties. Another pilot who got to fly the new type was *Ltn*. Oswald Boelcke, who had already proven his mettle on July 4, when he and his observer, *Obltn*. Heinz von Wühlisch, flying LVG C.I 162/15, destroyed a Morane-Saulnier parasol over Valenciennes, killing *Lts*. Maurice Tétu and Georges de la Rochefoucauld of MS.15.

Although the two budding fighters were close friends, the only thing they held in common was an aggressive spirit and a shared belief that the Fokker E.I represented the future of aerial warfare. Immelmann was self-centered, arrogant, and unpopular; one of his instructors accused him of having "a truly childish temperament."[4] Boelcke, though almost a year younger than Immelmann, was more mature in attitude and was the more experienced pilot. Although Boelcke preferred to think of himself as being as much of a loner as Immelmann, he proved to be a natural leader and mentor for a future generation of German aces. Boelcke himself once wrote, "You can win the men's confidence if you associate with them naturally and do not try to play the high and mighty superior."[5] While Boelcke thought nothing of taking German nurses up on "joy rides" (for which he was censured by his superiors), and who courted a young French girl living near his base, Immelmann seemed to have only one woman in his life—his mother.

As the "old hand" in the flying section, Boelcke got the first chance to fly the Eindecker and attacked a French two-seater in June 1915. He was still unfamiliar with the E.I's characteristics, however, and when he dived and loosed a long burst at the French machine, his gun jammed. A nearby German two-seater crew swore to have seen the French plane go down, but nobody saw it crash and it went unconfirmed.

On August 1, flying crews of *Fl.-Abt.* 62 were sleeping off the previous evening's drinking binge when they were shaken awake by bombs falling on their aerodrome. B.E.2cs of No. 2 Squadron RFC were staging a surprise raid on Douai. Immelmann scrambled into whatever flying garb was available and took off in

Fokker E.I 3/15, followed soon after by Boelcke in E.I 1/15. The BEs, too stable for their own good, made easy targets for the Fokkers, and Immelmann quickly brought one down, landed next to it and took its pilot, Lt. William Reid, prisoner. Boelcke also lined up a target, only to suffer another gun jam. After pounding on the mechanism of his LMG 08/14, he gave up and returned to Douai. For his feat, Immelmann was awarded the Iron Cross, First Class.

Boelcke finally got his chance on the evening of August 19, when he downed a Morane-Saulnier biplane. Soon, the formidable pair would wreak such havoc on Allied reconnaissance planes as to achieve a measure of local air superiority. "They treat my single-seater with a holy respect," Boelcke wrote. "They bolt as quick as they can."[6]

Disobeying orders to stay within his own lines, Boelcke began to hunt in Allied territory. When he was almost shot down by an Allied plane while attacking another, however, he realized the flaw in lone-wolf tactics. He formulated the idea of two Fokkers working as a team, with a wingman flying slightly above and to the side, to guard the leader's tail. He and Immelmann soon put that idea into practice. In spite of a friendly rivalry that developed between them, they worked quite effectively as a team, with encouraging results. Their scores were tied at six on January 12, 1916, when both were awarded the *Pour le Mérite*. On June 18, however, Immelmann—then with fifteen victories to his credit—was killed when his Fokker E.III, 246/16, suffered structural failure while attacking a Royal Aircraft Factory F.E.2b pusher of No. 25 Squadron, crewed by Capt. J. R. McCubbin and Cpl. J. H. Waller. The British credited him to McCubbin and Waller, but the Germans attributed his

Sergent Eugène Gilbert stands beside his Morane-Saulnier N MS388 *Le Vengeur*, armed with a Hotchkiss machine gun and deflectors on the propeller blades as used by his friend Roland Garros to achieve the first single-seat fighter victories in April 1915. (*Service Historique de Défense—Air*)

loss to a malfunction of the machine gun's synchronization system, resulting in his shooting his own propeller off. Immelmann's name lives on in a maneuver he developed to regain altitude, involving a half-loop with a half-roll at the top.

On the same day Garros fell into German hands, April 18, 1915, *Sgt.* Eugène Gilbert was being transferred to a newly formed *escadrille*, MS.49. Upon learning of his friend's capture, Gilbert embarked on a personal campaign that he best summed up with the words he emblazoned on the side of his plane: "*Le Vengeur*" ("The Avenger"). Gilbert's machine was a Morane-Saulnier N, a racing and aerobatic single-seat monoplane that Garros and he had flown before the war. The Model N featured extra stringers—which gave the fabric-covered fuselage a rounded, streamlined shape—and a large cone-like propeller spinner. The deflectors that Garros had employed on his modified Morane-Saulnier L were ultimately meant for the smaller, nimbler single-seater, and Gilbert set out to continue what his friend had started.

Gilbert's quest for revenge began with an indecisive combat on June 6, although he may have killed the enemy plane's observer, *Ltn.* Fritz Rössler of *Fl.-Abt.* 34. On the following day, he was credited with driving an enemy plane down in German lines near Saint-Amarin, which was added to three previous victories he had scored in two-seaters. Gilbert claimed a two-seater in flames on June 11, and although it was never confirmed, he may again have killed the observer, *Ltn.* Joachim von Maltzahn of *Fl.-Abt.* 48, and wounded his pilot, *Vizefeldwebel* (Sergeant Major) Rudolf Weingarten. Gilbert achieved ace status on June 17, when he shot down another Aviatik two-seater of *Fl.-Abt.* 48 northeast of Saint-Amarin, killing the pilot, *Vzfw.* Hugo Grabitz, and wounding his observer, *Ltn.* Karl Schwartzkopff, but Gilbert's own plane was badly shot up in the fight.

Gilbert's fighting career as "*Le Vengeur*" ended on June 27, when his plane suffered engine trouble during a bombing raid and he was compelled to force land in Rheinfelden, Switzerland. After two attempts, he finally escaped from Swiss internment, but he was killed in an accident at Villacoublay on May 17, 1918.

A somewhat modified version of Gilbert's Morane-Saulnier, the Model Nm (the last letter signifying "*militaire*"), was put into production by the French and saw some use in the summer of 1915, but the deflectors were never a satisfactory solution to the problem of firing a machine gun through the propeller arc. The fighters were never allotted to squadrons in more than twos or threes and some early aces—such as Jean Navarre, Georges Pelletier d'Oisy, and Jean Chaput—scored a few victories in them before moving on to the more practical Nieuport scouts.

Although the Morane-Saulnier N did not last long in French service, it did see more extensive use in the RFC, equipping No. 60 Squadron in the summer of 1916. By that time, the deflector system and the plane on which it was mounted were completely outdated, and the unit suffered heavy casualties. Four examples of the Morane-Saulnier I—with 110-horsepower Rhône 9J rotaries in place of the original 80-horsepower 9B rotary; and Alkan-Hamy interrupter gear, a Fokker-like system developed by *Sergent-Mécanicien* (Sergeant Mechanic) Robert Alkan of N.12, in place of the deflectors—were also delivered to 60 Squadron, while others served in the Imperial Russian Air Service. The installation of a heavier and more powerful engine only made a tricky airplane even harder to control, and one of 60 Squadron's pilots, Lt. William M. Fry, tellingly described it as the only plane he ever flew which gave the constant impression that it was doing its sincere best to kill him. Enlarging the wings on the Morane-Saulnier V did little to alleviate the problem. The last of the Morane-Saulniers were withdrawn in October, 60 Squadron reequipping with other scouts of French design, Nieuport 16s and 17s.

By then, the Morane-Saulniers had been in service much too long, but they left behind a stigma in the minds of the RFC's senior officers that would also persist far longer than it should have—an almost pathological distrust of, and prejudice against, monoplanes.

While Morane-Saulnier sought to fire a machine gun through a spinning propeller, Nieuport explored its own way around the problem. The progenitor of a line of Nieuport fighters, the XB—the "B" signifying that it was a biplane version of the Type X monoplane—was developed by Gustave Delage in 1915. Unlike most biplanes, the Nieuport's single-spar lower wing was of much narrower chord than the upper, to which it was braced by V-shaped interplane struts, giving the pilot and the observer, who sat in front of him, a greatly improved downward view. Introduced into military service as the Nieuport 10 and powered by an eighty-horsepower Clerget or Le Rhone 9B rotary engine, the two-seater was soon being flown by French, British, Belgian, and later, Italian, aircrews.

When the matter of armament began to be addressed, Nieuport put a central aperture in the upper wing center section, allowing the observer to stand up and fire a rifle or carbine over the top of the propeller arc. Inevitably, Allied pilots improvised means of mounting a lightweight Lewis machine gun above the upper wing, and then the observer's cockpit was faired over, turning the plane into a single-seat fighter. Nieuport eventually developed a standard mount which allowed the Lewis gun to be pulled down by the pilot so that he could reload it.

One of the earliest to reequip with the Nieuport 10 was MS.3, a reconnaissance unit that had shown considerable aggressiveness, even while flying Morane-Saulnier L parasols. The unit's first air-to-air victory had been scored on July 3, 1915, when its commander, *Capitaine* Antonin Brocard, attacked an Albatros two-seater of *Fl.-Abt.* 2 over Dreslincourt and, though armed only with a carbine, brought it down, killing the observer. Then, on July 19, *Caporal* (Corporal) Georges Marie Ludovic Jules Guynemer went up in a Morane-Saulnier L with an improvised machine gun mount in the rear cockpit, manned by his mechanic, *Soldat* (Private) Jean Guerder, and the pair shot down an Aviatik between the lines, killing *Unteroffizier* (Corporal) August Ströbel and *Ltn.* Werner Johannes of *Fl.-Abt.* 26. Both Guynemer and Guerder were awarded the *Médaille Militaire* for their feat. *Escadrille* MS.3 began receiving Nieuports in July, and on August 16 it moved from Vauciennes to Breuil-le-Sec. Brocard shot down an enemy plane north of Senlis on August 28.

MS.3 was fully reequipped with the sesquiplanes and officially redesignated N.3 on September 20. Its next success occurred on December 5, when *Sergent* Guynemer took off in a modified single-seat Nieuport 10 with an infantry Lewis gun, complete with stock and mounted above the wing, and brought down an Aviatik over Bois de Carré. Guynemer struck again on December 8, shooting down an LVG between Roye and Nesle and killing *Vzfw.* Kurt Diesendahl and *Ltn.* Hans Reitter of *Fl.-Abt.* 27. On December 14 Guynemer teamed up with a two-seater Nieuport 10, crewed by *Adjudant* (Warrant Officer) André Bucquet and *Lt.* Louis Pandevant, in shooting down a Fokker *Eindecker* over Hervilly.

The French were not alone in making aggressive use of their Nieuports. At 1011 hours on the morning of September 12, a

Sergent Georges Guynemer of Escadrille N.3 beside his Nieuport 10 N328 *Vieux Charles III* with 110-horsepower Rhône engine, Lewis machine gun mounted on the upper wing, and modified into a single-seater with headrest. Flying this plane on June 22, 1916, he teamed up with *Sgt*. André Chainat to shoot down an LVG at Rosière-en-Santerre for his ninth victory, *Leutnante* Erwin Bredow and Peter Tulff von Tscheppe und Weidenbach of *Feldflieger-Abteilung* 42 being killed. (*Service Historique de Défense—Air*)

Nieuport 10 of the 2*ème Escadrille Belge* attacked an Aviatik two-seater and sent it down to crash at Oud-Stuivekenskerke. The Nieuport's pilot was *Sous-Lt.* Jan Olieslagers, a prewar champion motorcycle racer and aviator, known as the Antwerp Devil. His success was confirmed as the first Belgian aerial victory to be scored in a single-seater, as well as the first of an eventual wartime total of six for Olieslagers.

Contemporary with the Nieuport X was the much smaller Nieuport XI, a monoplane that was also redesigned to use the XB's sesquiplane arrangement. Originally designated the BB-XI, and later the military designation of 11.C1 (the suffix signifying *chasse*, or fighter, single-seat), this smaller and aesthetically pleasing progeny of the Nieuport 10 was popularly known as the *Bébé* (baby), a word play that sprang naturally from its first designator. It was also called the "Type 13" in reference to its 13.3 square meter wing area.

The prototype Nieuport XI first flew in the summer of 1915 and entered production that autumn. Britain's RNAS, which had already made substantial orders for Nieuport 10s, also ordered 11s and seems to have received its first *Bébé* in November 1915. The French, preferring to stockpile their fighters for maximum effect, delivered the first Nieuport 11s to *escadrille* MS 31 on January 5, 1916, and by the critical month of February there were 90 serving in various sectors.

As Nieuport 11s began reaching the front, *Commandant* (Major) Jean Baptiste Marie Charles, baron de Tricornot, marquis de Rose, then in charge of the air assets of the V*e Armée*, ordered N.12, which had already exchanged most of its Morane-Saulniers for Nieuport 10s, to be completely reequipped with *Bébés*. This made it the first single-seat fighter squadron in history.

Early in 1916, as single-seat *Bébés* began to replace the Nieuport 10s in N.3, Guynemer received Nieuport 11 N836, on the fuselage of which he applied the legend "*Le Vieux Charles*," a reference to *Sgt.* Charles Bonnard, a well-liked member of old MS.3 who had transferred to the Macedonian front. Guynemer was flying that plane on February 3 when he encountered and attacked an LVG near Roye. "I did not open fire until I was at 20 meters," Guynemer later wrote. "Almost at once my adversary tumbled into a tailspin. I dived after him, continuing to fire my weapon. I plainly saw him fall in his own lines. That was all right. No doubt about him. I had my fifth. I was really in luck, for less than ten minutes later another plane, sharing the same lot, spun downward with the same grace, taking fire as it fell through the clouds."[7]

German records only record one fatality from the combat: *Ltn.* Heinrich Zwenger, an observer of *Fl.-Abt.* 27,

killed between Roye and Chaulnes. Both LVGs were confirmed, however, to which double victory Guynemer added a seventh success on February 5 with an LVG downed at Herbecourt, again killing the observer, *Ltn.* Rudolf Lumblatt of *Fl.-Abt.* 9.

Another early Nieuport fighter unit was C.65, which was originally formed at Lyon-Bron on August 2, 1915, as an "*escadrille provisoire de chasse*," with two Nieuport 11s, three Nieuport 12 two-seaters, and three Caudron G.4s. The unit's first success, an enemy plane forced to land on October 16, involved a two-seater crew, and not until February 21 was it redesignated a full-fledged fighter squadron as N.65. Long before that, however, N.65's first success in the *Bébé* Nieuport was achieved by a hero about as far removed from Guynemer in temperament as could be imagined: Charles Nungesser.

While Georges Guynemer was a sickly twenty-year-old boy driven by a single-minded devotion to his country, Charles Eugène Jules Marie Nungesser was an athletic man of the world who had raced cars, boxed, and learned to fly while in Argentina before the conflict. During the war's early days, Nungesser had served with distinction with the 2*ème Régiment des Hussards*, earning the *Médaille Militaire* before transferring into aviation in November 1914, and also his military pilot's brevet on March 17, 1915. Assigned to VB.106, he flew fifty-three bombing missions in a Voisin 3, the front nacelle of which he personalized with a black skull and crossbones. In the early morning hours of July 31, *Adjudant* Nungesser and his mechanic, *Soldat* Roger Pochon, went up in a new Voisin armed with a Hotchkiss machine gun—an unauthorized flight since Nungesser was supposed to be on standby duty that night—but as fortune would have it, five Albatros two-seaters staged a raid on Nancy that night, and the Voisin crew was able to catch one and send it down to crash. For deserting his post, Nungesser was confined to his quarters for eight days. For downing the German plane, he received the *Croix de Guerre*. His commander subsequently sent him for training in Nieuports, in which he could put his hell-raising attitude to more productive use.

Such was Nungesser's background when he joined N.65 in November 1915 and received a Nieuport 11. He was evidently delighted with the new fighter, for while flying it to N.65's aerodrome at Malzéville on November 26, he buzzed the city of Nancy, flying around the church steeples, looping over the town square, and zooming down the main street as low as thirty feet. Upon landing at Malzéville, he discovered that a telephone call from the Nancy town elders had preceded him, as he got a dressing down from his commander, *Capitaine* Louis Gonnet-Thomas, who caustically remarked that he should be scaring the enemy, and not his fellow Frenchmen.

Nungesser obeyed in his own way— after refueling his Nieuport, he crossed the lines and gave an even wilder aerobatic display at a German aerodrome. The Germans were too astonished to fire at him, and upon his return Nungesser reported to Gonnet-Thomas: "It is done, *mon capitaine!*"[8]

Nungesser was rewarded with another eight days of house arrest, but on November 28 he was permitted to take his plane up to practice gunnery on ground targets. He had barely taken off, however, when he spotted two Albatros two-seaters crossing the lines near Nomeny. Climbing to eight thousand feet

Sergent Charles Nungesser of N.65 before his Nieuport 16 N880, marked with his initial on the after fuselage, in the spring of 1916. Nungesser later adopted a personal trademark black heart with death's head, candles, and a coffin for all of his planes and survived the war as France's number three ace with a least forty-three victories. (*Service Historique de Défense—Air*)

and placing the sun at his back, Nungesser attacked. One of the Germans fled but the other fought, and fought well. Nungesser made four passes at the plane, each time closing to a hundred feet and using up a drum of ammunition without effect. Then, after placing his last drum on his Lewis gun, he went directly at his opponent, ignoring the gunner's return fire as he closed to thirty feet before emptying the magazine into the Albatros. The pilot was hit and the Albatros went into a dive, but Nungesser reported a scene that took much of the triumph out of his second victory:

The observer, still alive, clung desperately to the mounting ring to which his machine gun was attached. Suddenly the mounting ripped loose from the fuselage and was flung into space, taking with it the helpless crewman. He clawed frantically at the air, his body working convulsively like a man on a trapeze. I had a quick glimpse of his face before he tumbled away through the clouds . . . it was a mask of horror.[9]

The observer whose fall Nungesser witnessed was *Ltn.* Wilhelm von Kalkreuth of *Brief A.M.*, whose body was found at Nomeny. His pilot, *Vzfw.* August Blank, crashed to his death at Mailly. Nungesser had trouble eating and sleeping for some time after that, but he eventually got over it. Gonnet-Thomas helped—he commuted the remaining six days of his arrest and recommended that he be made a *Chevalier de la Légion d'Honneur*, an honor that Nungesser received on December 4. In spite of numerous crashes, Nungesser survived the war as France's third-ranking ace with forty-three victories to his credit—only to vanish during an attempt to fly nonstop across the Atlantic Ocean in May 1927.

Another early scorer in the Nieuport 11 was an impatient Canadian in the Royal Naval Air Service. Born the son of a barrister in Winnipeg, Manitoba, on August 11, 1886, Redford Henry Mulock had been a lieutenant in the 13th Canadian Field Artillery in 1911, but after arriving at the Western Front, he transferred to the RNAS in January 1915. While patrolling the English coast in an Avro 504 on the night of May 16–17, he attacked the German airship LZ38 over the Thames, but his gun jammed, allowing the crew to drop ballast and escape.

As a member of A Squadron, 1 Wing RNAS, at Saint-Pol-sur-Mer, Flight Sub-Lieutenant Mulock used one of the unit's new Nieuport 11s to drive a two-seater down "out of control" near the Houthulst Forest on December 30. Flying Nieuport 3977 on January 24, 1916, he forced a two-seater to land near Westende and claimed another "OOC" near the town of Nieuport two days later.

Matching Mulock's aggressive zeal, Flt. Sub-Lt. Noel Keeble in Nieuport 3178 forced a floatplane down in the sea on January 25. Others in the naval squadron would follow suit.

Although the nature of his accredited victories are somewhat suspect in relation to whether the enemy planes were actually destroyed, "Red" Mulock (a nickname derived from his first name, not his hair color) ended up with five, making him the first Canadian ace. He went on to distinguish himself more for his leadership, however, commanding No. 3 Squadron RNAS and, at the end of the war, a wing of four-engine Handley-Page V/1500s slated to bomb Berlin had the armistice not gone into effect.

By the end of 1915, fifteen French *escadrilles* were equipped with Nieuport 10s and 11s. By February there were ninety Nieuport 11s serving in various sectors, ready to be unleashed to take the sky back from the Fokkers.

Their arrival in quantity came at a critically opportune time, just as the Germans launched their offensive in the Verdun sector on February 21, 1916. With German chief of staff Gen. Erich von Falkenhayn's all-out effort to break the stalemate on the Western Front—first by taking the French forts in the area and, failing to do that, by drawing the French army into a protracted war of attrition aimed at bleeding it white—came the first serious effort to seize control of the air. To do this, the Germans began with standing patrols all over the front, a waste of resources that *Hptmn.* Oswald Boelcke, upon being transferred to that sector in mid-March, replaced by having his small detachment of Fokker Eindeckers, or *Kampf-Einsitzer Kommando* (KEK), stationed at Sivry, just eleven kilometers from the front, and who was informed of Allied air activity by telephone from the lines. Boelcke's tactic allowed him and his pilots to respond quickly where they were needed, saving time, fuel, and wear on their Fokkers' rotary engines. Within days the Germans were causing such disruption in the French gathering of air intelligence that *Général* Henri Philippe Pétain felt compelled to issue a significant order to his aggressive chief of air operations: "Rose, sweep the sky for me! I'm blind!"[10]

He certainly did not have to tell de Rose twice. By March 15, the major had responded by concentrating his scout *escadrilles* around the Verdun-Bar-le-Duc sector in provisional combat groups (or *groupes de combat provisoires*). One consisted of N.65 and N.67 supporting Pétain's IIe *Armée* from Bar-le-Duc, with N.23 serving as a reserve at Vadelaincourt. N.15 and N.69, also at Bar-le-Duc, were tasked with driving off German aircraft in the Xe *Armée* sector. N.57 soon joined the effort from Lemmes. Instead of conducting barrage patrols in certain assigned sectors, all of these units were to seek out and destroy any enemy plane they saw, a proactive shade of difference that was welcomed by the budding fighter pilots.

Early in 1916 the Nieuport 11 was joined by a more powerful derivative, the Nieuport 16. This was essentially the same airframe, though powered by a 110-horsepower Rhône 9J that gave it somewhat higher performance, but it also made it nose heavy and gave it a higher wing loading (8.46 pounds per square foot compared to the 11's 7.4 pounds per square foot). Many pilots considered it a handful, but a number of pilots adapted to it successfully enough for it to dominate the skies over Verdun alongside its more tractable forebear.

Gustave Delage's solution to the Nieuport 16's handling problems was to redesign the airframe with more wing area—14.75 square meters compared to the *Bébé*'s 13.3. The new plane also got a synchronized .30-caliber Vickers machine gun, using Alkan-Hamy interrupter gear. The horseshoe-shaped cowling of earlier Nieuports was replaced by a circular one faired smoothly to the fuselage sides, and was further streamlined by a *cône de pénétration* attached to the engine shaft rather than to the propeller so that it remained stationary instead of spinning. Transparent Cellon panels were installed in the upper wing center section to give the pilot a better view upward.

Officially designated the 17 but often referred to as the "fifteen-meter" Nieuport, the new scout proved far more amenable to the 110-horsepower Rhône 9J, as well as to the 120-horsepower 9Jb, than to its smaller predecessors. A later version, using a fuselage rounded with stringers called the 17bis, used the 130-horsepower Clerget 9B or 9Z engine. The French encountered some problems with cracks in the cowling, which sometimes necessitated cutting the bottom away, or outright replacement with a 16's cowl until a more flexible mounting was perfected. The *cône de pénétration* was eventually abandoned.

Arriving at the front in April 1916, the first Nieuport 17s are alleged to have been delivered to *Escadrille* N.57 in May, although it is not certain when they flew their first mission. *Sous-Lieutenant* Jean Firmin Louis Robert was probably flying one of the *Bébé* Nieuports when he downed a Fokker Eindecker for the unit's first victory on April 25. *Sergent* Louis Coudouret, however, was probably in a 17 when he started the next month's scoring on May 4, by sending an LVG two-seater crashing near Hermeville for his first of an eventual six victories. The squadron acquired an experienced member on May 7 when *Sous-Lt.* Jean Chaput transferred over, after having already been credited with three victories with *Escadrilles* C.28 and N.31. Nine days later, Chaput showed his mettle by attacking an Aviatik over Esparges and severely damaging it. *Lieutenant* André Bastien forced an enemy plane down in its own lines the next day, and *Adj.* Léon François Acher did the same on May 19, although he was badly wounded in the course of the fight. *Lieutenant* Charles Dumas sent an Aviatik down to crash near Eparges on May 21, and on the following day *Lieutenants* André Charles Dubois de Gennes and Chaput each flamed German balloons, the latter's victory bringing his total to five and acedom. The Nieuport 17's first month ended on a blue note, however, when the recently promoted *Lieutenant* Robert failed to return from a mission on the twenty-fourth.

In the months to come, the Nieuport 17 became one of the war's most ubiquitous fighters. In addition to the French, the British used it extensively, although they usually replaced the synchronized Vickers with a Lewis gun above the wing on their own Foster mounting. The Duks factory built Nieuports for Russian use, and Nieuport-Macchi built them for the Italians. The Belgian 1*e* and 5*e Escadrilles* were equipped with Nieuport 17s in 1917. American volunteers in N.124 and other French *escadrilles* flew them in combat, and later, US Army Air Service pilots trained in them.

Even German aces such as Hermann Pfeiffer and Paul Bäumer were known to fly captured Nieuport 17s at times, as did Austrian ace of aces Godwin Brumowski. More significantly, the Nieuports so impressed the Germans that the *Inspektion der Fliegertruppen* (Inspectorate of Aviation Troops, or *Idflieg*) would order the fighter, or at least its sesquiplane configuration, copied to produce a fighter of com-

Nieuport 17s newly delivered to N.103 show the stationary cônes de pénétration that were soon abandoned to improve engine cooling, as well as Lewis machine guns mounted above the wing due to early problems with the plane's standard synchronized Vickers guns. (*Service Historique de Défense—Air*)

parable ability. This would turn out to be a grave mistake.

For the balance of the bloody Verdun campaign, the efficiency of the Fokkers' interrupter gear and the superior performance and more concentrated numbers of the Nieuport scouts virtually canceled one another out. Although confrontations between the two fighters were comparatively infrequent, both took a substantial toll on the other side's reconnaissance planes, leaving a bloody stalemate above Verdun proportionate to the more epic mutual slaughter continuing on the ground.

After being credited with his nineteenth victory on June 27, Oswald Boelcke was ordered out of his base at Sivry for an inspection tour of the Balkans and Turkey by a *Fliegertruppe*, which since Immelmann's death was unwilling to risk the life of its highest-scoring fighter pilot. Characteristically, Boelcke used the downtime to put on paper all he had learned from his combat experience thus far, and formulated a list of fundamentals for his colleagues that came to be known as the "Dicta Boelcke:"[11]

1. Seek advantage before attacking. If possible, keep the sun at your back.

2. Having begun an attack, always follow through.

3. Only fire at short range, and only when your opponent is positively in your sights.

4. Never lose sight of your opponent, and do not be fooled by his tricks.

5. In every attack it is important to approach your opponent from behind.

6. If your opponent attacks from above, do not try to evade, but fly to meet him.

7. When over enemy territory, never forget your path home.

8. For the *Staffel*—attack on principle in flights of four or six. When single combat ensues, take care that many do not go for one opponent.

Boelcke would soon have occasion to put his dicta into practice. Although subject to refinement in future as aircraft themselves advanced, many of his basic principles remain relevant to fighter pilots today.

While the first fighters were appearing over the Western Front, Russia was producing its first—and, as it turned out, only—indigenous fighter plane, courtesy of Igor I. Sikorsky, the same man who had designed the world's first four-engine airplane, the *Bolshoi Baltitsky*, in 1913. His follow-up to the multiengine giant, the *Il'ya Muromets* or IM, was a long-range reconnaissance plane with secondary bombing capability that was by December 1914 operating with an *Eskadra Vozdushnykh Korablei* (Flying Ship Squadron) or EVK, based at Yablonna, north of Warsaw. At the same time, Sikorsky had designed several single-engine monoplane and biplane scouts for the Russian navy, and in October 1914 he began work on the S.16, a biplane somewhat influenced by the Sopwith

Tabloid—but employing ailerons rather than wing warping—with a wingspan of only twenty-seven feet six inches and seating a crew of two within a single cockpit. Production delays, improvements by Sikorsky's design team, and problems obtaining suitable engines held up delivery of fifteen production, or "serial," S.16ser aircraft until January 1916, and by the time they were delivered to the EVK to serve as trainers and scouts, a new role had been envisioned for them. One of Sikorsky's design team, *Leitenant Flota* (Naval First Lieutenant) Georgy I. Lavrov, had patented a simple interrupter mechanism for a forward-firing Vickers machine gun on December 1, 1915, and this was installed on the S.16ser on March 21, 1916. Thus armed, the S.16ser would escort and defend the IMs from enemy aircraft.

At that point, IRAS pilots had learned of Russia's first armed scout and were petitioning for its use beyond the EVK. In response, Grand Archduke Aleksandr Mikhailovich, a strong advocate of air power and a founding father of the IRAS, arranged for the delivery of S.16s to the corps air squadrons attached to the Seventh and Eleventh Armies. The first, powered by a Rhône engine, was delivered to the XXXIII KAO on March 17, and ten days later *Kapitan* (Captain) Konstantin K. Vakulovsky used it to intercept and drive off a German scout.

Three more S.16sers were delivered to the 7th *Aviatsionny Otryad Istrebitelei* (Fighter Aviation Detachment) at Kiev, led by *Podporuchik* (Sub-Lieutenant) Ivan A. Orlov, and two to the 12th AOI at Pskov under Max G. von Lerche. Sent to the front in April, the 7th settled in the Galician town of Khmelevka, and on April 15 one of its crews, *Podporuchik* Orlov and *Kornet* (Ensign) Longin Lipsky, intercepted and drove off an Austro-Hungarian reconnaissance plane. Two days later, however, Orlov discovered what would become an ongoing problem when the Lavrov interrupter gear malfunctioned, forcing him to disengage. There would be many more interceptions, but no real damage was done until April 27, when *Kornet* Yuri Vladimirovich Gilsher of the 7th AOI, flying with *Praporshchik* (Warrant Officer) Vladimir Kvasnikov as his observer in S.16ser No. 201, attacked an Austro-Hungarian two-seater near Burkanov and after firing 120 rounds saw the enemy plane descend emitting a trail of black smoke. Aviatik B.III 33.02 came down three miles behind Russian lines with a damaged engine, and Russian cavalrymen captured its crewmen, *Zugsführer* (Lance Corporal) Alexander Galbavy and *Kadett der Reserve* (Cadet in the Reserves) Julius Horak of *Flik* 20, before they could burn their plane.

That first victory for a Russian-built fighter was also the last. While Gilsher was returning from a second patrol that same day, his ailerons jammed and he spun down twelve hundred meters. S.16 No. 201 was demolished, and though its sturdy airframe saved the lives of its crew, Gilsher's left foot was crushed and had to be amputated. He would learn to fly with a prosthetic, however, and after returning to the 7th AOI, he scored five more victories before being shot down and killed on July 7, 1917.[12]

By the middle of May, all IRAS S.16s had been withdrawn and returned to the EVK. Analysis of its short combat record concluded that the Lavrov interrupter gear was defective, and the S.16's Russian-built versions of the Gnome rotary engine did not produce enough power for it to sustain its attacks on enemy two-seaters, since its average speed, even without the observer along, was seventy-three miles per hour and its

ceiling ten thousand feet. From then the IRAS fighter force would be dominated by imported French fighters such as the Morane-Saulnier N and I and the Spad SA, while its factories built the Nieuport 10, 11, 17, and 21 under license.

While the French and Germans were addressing the matter of developing an effective single-seat fighter, the British were doing so in their own way, with one airman in particular playing an instrumental role. In October 1914, the same month that Frantz and Quénault scored their aerial victory, No. 6 Squadron RFC arrived in France, its flying personnel including a twenty-three-year-old officer from a distinguished military family named Lance George Hawker.

When his unit traded in its ungainly Henry Farman pushers for somewhat faster B.E.2cs, Hawker began complementing his reconnaissance patrols with more aggressive activities, such as a bombing attack on Zeppelin sheds, for which he received the Distinguished Service Order.

In 1915, No. 6 Squadron began to receive F.E.2b pushers—powered by a rear-mounted 120-horsepower, six-cylinder Beardmore engine, and armed with a .303-inch Lewis machine gun in a gunner's pit in front of the pilot—to escort the BEs. The squadron also acquired a single-seat Bristol Scout C, serial number 1609, for short-range frontline reconnaissance flights. Designed by Frank Sowter Barnwell for the British & Colonial Aeroplane Company, the Bristol Scout was originally powered by an 80-horsepower Gnome Lambda nine-cylinder rotary engine and flew for the first time in February 1914. Its wing, which had a span of only twenty-two feet, employed ailerons, rather than the older

Bristol Scout C 1611, the second such plane on which Captain Lance G. Hawker of No. 6 Squadron, RFC, installed his oblique Lewis gun mount, in which he downed two Albatros two-seaters on June 25, 1915, for which he became the first single-seater pilot to be awarded the Victoria Cross. (*Author*)

but still prevalent wing warping, for lateral control. Compact and clean for its time, the Scout embodied the biplane fighter configuration that would remain essentially standard for more than twenty years to follow.

Captain Hawker had the privilege of ferrying the single-seater in from Saint-Omer on June 3. On June 7 he wrote: "I have a beautiful little toy, a new Bristol Scout that goes at 80 and climbs at five or six hundred feet a minute. I'm having a machine gun fitted to see how they like it."[13]

The gun to which Hawker referred was a lightweight .303-inch Lewis, but he faced a problem that the FE pusher crews did not have to deal with—finding a way to fix such a weapon to a single-seat tractor airplane so that he could fire it without shooting his own propeller off. Hawker devised a solution, which Air Mechanic Ernest J. Elton duly fabricated: mounting the gun so that it could fire forward, downward, and outward at an angle. The arrangement avoided the propeller arc but posed a challenge as far as aiming was concerned.

Nevertheless, Hawker sallied forth on June 21, 1915, and attacked a DFW two-seater over Poelcapelle. It was officially credited to him as "brought down out of control," although there is no evidence of any aircraft or crew losses in German records. On the following day, Hawker overturned his "toy" during a forced landing, and it had to be taken to No. 1 Aircraft Park for repairs. It was replaced by another Bristol C, No. 1611, on which Hawker promptly installed his Lewis gun mounting and resumed his hunting.[14]

In the course of three sorties on June 25, Hawker attacked three German aircraft. During his second combat, he forced an Albatros two-seater of *Fl.-Abt.* 3 to land near Passchendaele at about 1845 hours, then succeeded in shooting down another in flames southeast of Zillebeke fifteen minutes later, killing *Obltn.* Alfred Uberlacker and *Hptmn.* Hans Roser, also of *Fl.-Abt.* 3. Hawker was credited with a double victory and subsequently awarded the Victoria Cross—the first man to receive Britain's highest decoration for air-to-air combat. Hardly one to rest on his laurels, he flew more patrols in the Bristol until 1611 was withdrawn to the air park on October 22. After that, he continued to seek combat in the F.E.2bs, supplementing the gunner's firepower with his own Lee-Enfield rifle.

Flying F.E.2b 4227 with Lt. A. Payze as his observer on August 2, Hawker attacked a German two-seater and forced it to land at Wulverghem. Hawker was credited with a second "double" on August 11, when he and his gunner, Lt. N. Clifton, destroyed an Aviatik two-seater at Houthem at 1545 hours and downed an attacking Fokker Eindecker in the area of Lille-Roubaix at 1915 that evening. Returning to Bristol Scout 1611 on September 7, Hawker shot down an enemy biplane over Bixschoote for his seventh victory. Hawker was now not only Britain's pioneer single-seat scout pilot but also its first ace.

Posted back to England late in 1915, Captain Hawker was sent to Hounslow on September 28 to take command of the RFC's first single-seat fighter unit, No. 24 Squadron. The squadron's equipment, the de Havilland D.H.2, represented yet another early method for a machine gun-armed single-seater to circumvent the propeller. After leaving the Royal Aircraft Factory to join the Aircraft Manufacturing Company in June 1914, Geoffrey de Havilland had designed a two-seat pusher fighter of similar configuration to the RAF's F.E.2, the Airco D.H.1. In the early summer of 1915, however, he introduced a more compact single-seat version, the D.H.2, which was sent to France for operational evaluation on July 26.

The new scout's frontline debut could scarcely have been less promising. Its one-hundred-horsepower Gnome monosoupape rotary engine had to be replaced immediately upon its arrival in France. It was attached to No. 5 Squadron at Saint-Omer, but disappeared over the lines on August 9. A German plane, piloted by future naval ace *Ltn. zur See* (Ensign) Theodor Osterkamp of the II. *Marine Fliegerabteilung*, subsequently dropped a message stating that the plane's pilot, Capt. Robert Maxwell-Pike, had died of wounds, but otherwise they did not indicate whether the plane had been shot down or had crashed after another engine failure.

As it turned out, Pike had attacked an Albatros two-seater crewed by *Obltn. zur See in der Reserve* (Naval Reserve Lieutenant) Alfred Ritscher and *Ltn. zur See* Heinrich Maas, the latter of whom had struck the British pilot in the head with a single bullet. Pike came down east of Ypres before he died. A German photograph showed that the plane had flipped over on its back upon landing, but was otherwise quite intact. In spite of their being presented with an almost perfect advance look at Britain's latest fighter, however, for more than a year thereafter, the Germans tended to misidentify D.H.2s as "Vickers scouts" in their combat reports with curious regularity.

In spite of the disastrous loss of the prototype, Airco put the D.H.2 into production, and the first example was delivered to No. 18 Squadron on January 9, 1916, while another went to 11 Squadron two days later. Back in Britain, No. 24 Squadron got its first D.H.2 on January 10, and by the time the newly promoted Major Hawker took his unit to Bertangles, France, on February 10, its strength was up to twelve D.H.2s.

In its original form, the D.H.2 carried a Lewis gun with a forty-seven-round magazine within a fairing on the port side of the pilot's nacelle. The production version placed the gun centrally within the front of the nacelle on a vertical shaft that gave it a considerable range of movement. In practice, Hawker judged what he called the "wobbly mounting," which required the pilot to fly the plane with one hand and aim his weapon with the other, to be utterly useless, and devised a clamp to fix

In a disastrous debut, the first prototype de Havilland DH.2 4732 was attached to No. 5 Squadron, RFC, and promptly fell into German hands on August 9, 1915. In spite of that, what they misidentified as the "Vickers scout" would come as a nasty surprise to them when unleashed in three-squadron strength in July 1916. (*Aaron Weaver*)

ly opened the squadron's account when he shot down an enemy plane over Bapaume. On April 25, three of 24's D.H.2s were escorting a B.E.2c of No. 15 Squadron when they were attacked by a flight of Fokker Eindeckers. The D.H.2s turned on the enemy, drove down one of the monoplanes and chased off the rest. That action sealed 24 Squadron's confidence in the D.H.2, and from then on its pilots attacked all German aircraft they encountered without hesitation.

The Fokker bogey had been laid to rest and the Allies had retaken the sky. But they would not hold it for long.

the gun in a forward-firing position. Higher authorities prohibited such practice, but Hawker ultimately came up with a compromise—a spring clip that would theoretically allow the pilot to release the gun for flexible use at will. In practice, Hawker's men almost never unclipped their guns when they went into battle.

Another problem encountered in the D.H.2 was a tendency to fall into spins, which caused a number of training accidents. At that time, little was known about how one could recover from a spin, but Hawker understood the concept of centralizing the controls and repeatedly put his plane into a spin and pulled out of it to show his pilots that it could be done. He also dealt with the numbing cold that had to be endured in an open nacelle with no engine to shield and warm the pilot, by designing fur-lined flying clothes, including the hip-high "fug boots," which would become a popular item among British airmen throughout the war.

Once its personnel had gained sufficient experience and confidence in the D.H.2, No. 24 had its first engagement with the Germans on March 19. On April 2, 2nd Lt. David Marie Tidmarsh official-

Two

Biplanes, Boelcke, and the Baron

HALBERSTADT AND ALBATROS SCOUTS, 1916–17

By the time the Battle of the Somme commenced on July 1, 1916, Allied fighters had virtually retaken command of the sky. *Major* (Major) Wilhelm Siegert, commander of the *Idflieg* (*Inspektion der Fliegertruppen*, or Inspectorate of Aviation Troops), wrote in outspoken retrospect: "The start of the Somme battle unfortunately coincided with the low point in the technical development of our aircraft. The unquestioned air supremacy we had enjoyed in early 1916 by virtue of our Fokker monoplane fighters shifted over to the enemy's Nieuport, Vickers and Sopwith aircraft in March and April. Our monthly aircraft output did not even allow a squadron to be equipped with a common type. For example, *Flieger-Abteilung* 23 had a complement of five different aircraft types."[1]

By October 1916, the aerial balance of power began to shift again. In large degree, Siegert attributed the German resurgence to the "enterprise of Boelcke and his 'school' in conjunction with the new Halberstadt D.III fighter."[2]

Originally formed on April 9, 1912, as the Deutsche Bristol-Werke Flugzeug GmbH and founded to build Bristol air-

craft under license, the Halberstädter Flugzeugwerke GmbH produced an original design, the B.I biplane, in 1914. A product of technical director Karl Theis and Swiss-born Hans Burkhard, before the latter left the firm to join Gotha, the B.I was powered by an eighty-horsepower Oberursel U.0 rotary engine but its successor, the B.II, used a hundred-horsepower Mercedes D.I inline engine. Both were used as trainers, although their controls were noted to be more sensitive than most such aircraft. Those flying characteristics were no doubt on Theis's mind in early 1916, when letters from Boelcke, Immelmann, Wintgens, Otto Parshau, and other aces expressed an urgent need for a successor to the Fokker Eindeckers. Anthony Fokker responded in May 1916 with the D.I, a biplane powered by a Mercedes engine that was otherwise similar in structure to the E.III, and which proved to be only a marginal improvement. Theis designed a more compact, sturdier single-seat version of the B.II, with a forward stagger to the double-bay wings and a hundred-horsepower Mercedes D.I engine. On February 26, 1916, static testing of the Halberstadt

D.I, as the fighter version was called, produced a safety factor of 5.78, which contrasted markedly with the fragility shown
by the Fokker D.I, D.II, and D.III
biplanes. An initial production contract
for twelve aircraft was placed in March,
and the new fighter was officially flight-
tested and accepted in May.

It is generally believed that the first
Halberstadt D.I was intended for Max
Immelmann, but he was killed four days
before it arrived at Douai on June 22,
1916. Boelcke visited the aerodrome
later that month and wrote:

> One evening I flew the new Halberstadt
> biplane, the first appearance of this
> machine at the Front. Because it had a
> slight resemblance to the British BE, I
> was able to completely surprise an
> *Englander*. Undetected I got within 50
> meters and shot his jacket full. But since
> I was too fast and did not have the
> machine in hand like my Fokker, I had to
> dive under the *Englander*. He turned
> around at once and began to descend. I
> went after him, the belt jammed and I
> had to turn away. By the time I had cor-
> rected the problem, the enemy was
> gone.[3]

A few days after that promising first
flight, Boelcke was ordered to take leave
to Turkey, so he did not have time to
become better acquainted with the
Halberstadt. By June 29, however, there
were eight Halberstadt D.Is, now powered by 120-horsepower Mercedes
engines, at the front. Two were delivered
to *Fl.-Abt.* 32 by *Ltne.* Gustav Leffers and
Franz Diemer, who flew them with the
unit's fighter detachment, called *Abwehr
Kommando Nord*. Leffers was probably flying the Halberstadt when he shot down
an F.E.2b of No. 11 Squadron RFC over
Miraumont on July 9, although Allied
witnesses misidentified it as a Fokker.

Halberstadt D.Is and D.IIs proliferated
in the months to follow, and became
important transitionary fighters for the
Germans—not only as a technological
advancement in itself, but as equipment
for new, specialized fighter squadrons
(*Jagdstaffeln*) that began to form in August
1916. *Abwehr Kommando Nord* became the
nucleus for *Jagdstaffel*, or *Jasta* 1, under
the command of *Hptmn.* Martin Zander
on August 22, although its initial standardized aircraft complement was comprised of less desirable Fokker D.Is and
D.IIs, and even an E.IV that *Ltn.* Kurt
Wintgens insisted on keeping because it
had twin machine guns. Boelcke, recalled
from Turkey on August 11, formed *Jasta* 2
on August 27, its initial equipment
including two Fokker biplanes and the
first frontline example of another new
fighter type: the Albatros D.I.

Built in the summer of 1916, the Albatros
D.I was based on a racing plane developed
just before the war began by Robert
Thelen, supervisor of the Albatros design
committee. Although its single-bay, twin-
spar wing structure was standard for the
time, the D.I featured a streamlined plywood fuselage with a neatly cowled 160-
horsepower Mercedes D.III engine and a
spinner over the propeller. Taking advantage of the more powerful engine, Thelen
built the new fighter to carry not one but
two synchronized 7.92mm LMG 08/14
machine guns, an arrangement that more
than doubled the rate of fire. Although it
was not as maneuverable as most of its
Allied opponents, German airmen soon
decided that they could live with the
superior speed and firepower of the
Albatros. Early D.I fliers, however, complained of the way the upper wing and the
trestle-type center-section struts blocked
their upward vision. Thelen responded by

From left, *Obltn.* Ernst *Freiherr* von Althaus, *Ltn.d.Res.* Alfred Lenz, *Hptmn.* Oswald Boelcke, and an officer named Pietsch are photographed before a Halberstadt biplane in late July 1916. (*Lance Bronnenkant*)

lowering the upper wing and supporting it with N-shaped cabane struts splayed outward, designating the modified result the D.II. He also later replaced the drag-producing "ear"-type Windhoff radiators on the fuselage sides of the D.I and early D.IIs with a Teeves und Braun radiator installed flush within the center section of the upper wing.

Nimble and viceless though the Halberstadt D.II had been, the Albatros D.I made an even more dramatic impression on its pilots. One of the earliest members of *Jasta* 2, *Ltn.* Erwin Böhme, wrote of the Albatros fighters: "Their climb rate and maneuverability are astonishing—it is as if they were living, feeling beings that understand what their master wishes. With them, one can dare and achieve everything."[4]

Boelcke scored *Jasta* 2's first victory while flying a Fokker D.III, bringing down a D.H.2 of No. 32 Squadron north of Thiepval on September 2, and cordially showing its uninjured pilot, Capt.

Robert E. Wilson, around his aerodrome at Bertincourt before sending him off to the prisoner of war camp. Boelcke continued to fly Fokker and Halberstadt scouts over the next two weeks, adding six more Allied aircraft to his score, until September 16, when five Albatros D.Is and one D.II arrived. Later that same afternoon, *Ltn.* Otto Walter Höhne flew a patrol in one of the new D.Is and brought down an F.E.2b of No. 11 Squadron at Manacourt at 1800 hours. The crew, 2nd Lt. A. L. Pinkerton and Lt. J. W. Sanders, were taken prisoner.

On the following day, Boelcke inaugurated an innovation more significant than the Albatros—a team effort at gaining local air superiority. At 1100 hours, he led five of his men toward British lines and spotted fourteen British aircraft heading for Marcoing railway station. Boelcke led his formation to intercept them, but the British reached their target first and were bombing the station when Boelcke's flight arrived. While Boelcke

Leutnant Manfred *Freiherr* von Richthofen (left, in sweater) and a *Staffel* mate scan the western sky for incoming Allied "customers" before his Albatros D.II 491/16 and three Albatros D.Is at *Jagdstaffel* 2's aerodrome at Lagnicourt in November 1916. (*Greg Van Wyngarden*)

held back in order to be ready to help any inexperienced pilot who got into trouble, five Albatros scouts dived on the bombers and their escorts, broke up the formations, and then proceeded to go after lone targets.

One of Boelcke's young disciples, *Ltn.* Manfred *Freiherr* von Richthofen, pursued an F.E.2b, which took evasive action, and led him on a merry chase until he was able to get a good shot into the engine compartment. The FE finally came down at the German aerodrome at Flesquières, and Richthofen landed nearby to inspect the first of the eighty aircraft that he was destined to bring down. The observer, twenty-one-year-old 2nd Lt. Thomas Rees of No. 11 Squadron, was dead; the pilot, nineteen-year-old 2nd Lt. Lionel B. F. Morris, died shortly after being taken to the hospital at Cambrai.

Another F.E.2b team from No. 11 Squadron, 2nd Lt. T. P. L. Molloy and Sgt. G. J. Morton, were somewhat more fortunate than Rees and Morris, being brought down alive south of Thescault by *Ltn.d.Res.* Hans Reimann for his second victory. Boelcke himself downed another of 11 Squadron's FEs over Equancourt, Capt. D. Gray and Lt. L. B. Helder also becoming POWs. Finally, *Leutnant*

Böhme, who felt that one could do anything in an Albatros, opened his account by bringing down a Sopwith 1 1/2-Strutter of No. 70 Squadron, its pilot, 2nd Lt. Oswald Nixon, being killed and the wounded observer, Lt. Ronald Wood, taken prisoner.

That night, a German army tradition was broken and a new air force tradition established when enlisted men were permitted to join the officers in the squadron *Kasino* to celebrate. Amid the festivities, Boelcke pinned the Iron Cross, First Class, on Böhme. Richthofen, who had already received that decoration, wrote a letter that evening to a jeweler in Berlin, ordering a plain silver cup, two inches high and one inch wide, on which was to be inscribed: "1. Vickers 2. 17.9.16."[5]

The Albatros D.II, combined with the adoption of the "Boelcke Dicta," the first formal fighter doctrine to be listed for universal scrutiny, resulted in another reversal of fortune over the Western Front in the autumn of 1916. *Jasta* 2 would be the vanguard of a general resurgence of German air power, during which Boelcke would bring his personal score up to forty before being killed in a midair collision with Böhme on October 28, 1916. Böhme managed to bring his stricken plane down, but blamed himself for the death of his beloved mentor, and Richthofen only narrowly dissuaded him from committing suicide. Death caught up with Böhme of its own accord on November 29, 1917, when he fell victim to an Armstrong Whitworth F.K.8 crew of No. 10 Squadron, but not before he had accounted for a total of twenty-four Allied aircraft and received the *Orden Pour le Mérite*.

The New Year brought a new fighter to the growing number of *Jagdstaffeln* as the first Albatros D.IIIs arrived. Inspired by the maneuverability and improved downward vision of the Nieuport 11 and 17, the Albatros design team tried to achieve the best of both worlds by applying a sesquiplane wing arrangement to the D.II. The result, which featured two wings with a long curving rake at the tips, the lower of which was of considerably reduced chord, certainly looked more graceful than the squared-off wings of the D.II. However much they may have enhanced the appearance, maneuverability, and climb rate of the D.III, though, the single-spar lower wing in the sesquiplane arrangement displayed a disturbing tendency to twist about on its axis during a prolonged dive.

Manfred von Richthofen had sixteen victories to his credit when the first D.IIIs reached *Jasta* 2 on January 7, 1917. Before he could try the new plane out, however, he was reassigned to *Jasta* 11, a fighter squadron attached to the 6. *Armee*, which had accomplished nothing since it became operational on October 11, 1916. Richthofen was to take over command from *Obltn.* Rudolf Emil Lang, who in turn would take command of a new unit, Royal Württemberg *Jasta* 28.

Richthofen received the *Orden Pour le Mérite* on January 12, and eight days later he arrived at *Jasta* 11's aerodrome at La Brayelle, northwest of Douai. At that point Richthofen, deciding that camouflage did nothing to hide his plane, went the opposite route and had it completely overpainted in red. When his men protested that he would be too vulnerable to being singled out by the enemy, Richthofen suggested that all of them adopt some red as a unit marking, and add an additional color for personal identification.

Richthofen inaugurated his new all-red Albatros D.III to combat on January 23, shooting down an F.E.8 single-seat pusher fighter of No. 40 Squadron over Lens and killing its Australian pilot, 2nd Lt. John Hay, for *Jasta* 11's first and his seventeenth victory. On the following day, he brought down an F.E.2b west of Vimy, the wounded crew of which, Capt. Oscar Greig and Lt. John E. MacLennan of No. 25 Squadron, were taken prisoner. Richthofen again landed near his victims, but it was not solely for the purpose of having a chat—in the course of the combat, his wing had cracked during a dive. Richthofen's confidence in the new D.III was seriously shaken, and he reverted to the Halberstadt D.II for much of his flying until the end of March 1917, scoring as many as eleven victories in the older but more reliable biplane. Eventually, Albatros was able to strengthen and brace the wing cellule enough to make the D.III satisfactory, and in the hands of Richthofen and the aggressive pilots he inspired, *Jasta* 11 made it the terror of the Western Front in April 1917.

By the spring of 1917, the fundamentals of air-to-air combat were in place, but few of its pioneers were around to see the fruition of their efforts. Garros was a prisoner of war, while Immelmann, Boelcke, and Wintgens were dead. Nor did Maj. Lance Hawker live to see the arrival of newer, more efficient fighters at No. 24 Squadron. He was leading his men as usual on November 23, 1916, when he engaged in an epic eight-minute duel with an Albatros D.II of *Jasta* 2. At last, running low on fuel, Hawker tried to make a break for home but was shot down and killed, as victory number eleven for Manfred von Richthofen.

Hubert Harvey-Kelly, victor of the RFC's first aerial combat, had a somewhat longer career than Hawker's, eventually rising to the ranks of major and commander of No. 19 Squadron before he too became a victim of the new form of warfare he had helped pioneer. He was flying a Spad VII as he took off from Vert Galant aerodrome at 1000 hours on the morning of April 29, 1917, accompanied by 2nd Lts. Richard Applin and William N. Hamilton, to patrol the area around Douai. There they encountered eight Albatros D.IIIs, five of which were described as silvery gray and three of which were red. The bright-colored Germans were from *Jasta* 11, led by Hawker's killer, *Hauptmann* von Richthofen, who by then was gaining notoriety as the Red Baron.

Six Sopwith Triplanes of No. 1 Squadron, Royal Naval Air Service, were also in the area, so Harvey-Kelly confidently attacked the Germans. The Triplanes indeed joined in and claimed three Albatros scouts as "out of control" in the course of the twenty-minute melee. Things turned out disastrously different for the Spads, however, as Applin was sent crashing to his death in a swamp near Lecluse by Manfred von Richthofen, who claimed him as his forty-ninth kill—and the first of four that he would gain before the day was over. Hamilton was more for-

With its faulty wing cellules bolstered, the Albatros D.III sesquiplane reached the apex of its career during what the British called "Bloody April" of 1917—no more so than with *Jasta* 11 under Manfred von Richthofen, whose all-red plane can be seen second from foreground in this lineup. (*Greg Van Wyngarden*)

tunate, being brought down and taken prisoner, the thirteenth victory for the Baron's younger brother, *Ltn*. Lothar von Richthofen. Harvey-Kelly crashed after a duel with another of *Jasta* 11's Albatros pilots, dying of head wounds in a German hospital three days later. He was the tenth victim of *Ltn*. Kurt Wolff.[6]

Three

A Deadlier Breed

FIGHTER DEVELOPMENT, 1916–17

By the end of 1916, the essential fighter configuration had been defined: a single-seater with a forward-firing machine gun, usually synchronized to fire through the propeller. For the rest of World War I, aircraft manufacturers of the warring powers were engaged in a constant struggle to improve the breed, seeking to gain that edge in speed, climb, or maneuverability which would bring them control of the sky. Neither the Allies nor the Central Powers were able to attain that goal for long before the other side came up with a new type which would equal or surpass the performance of the current world-beater.

An airplane's worth often depends on its power plant, and the water-cooled, eight-cylinder marvel that Swiss engineer Marc Birkigt created in 1915, the 150-horsepower Hispano-Suiza 8A, was to inspire several great Allied fighters. The first had its origins in the dissolution of the French *Société Provisoire des Aéroplanes Deperdussin* and its resurrection in August 1914 as the *Société anonyme pour l'Aviation et ses Dérivés*, retaining the company acronym of Spad. In its new incarnation, Spad also retained Deperdussin's talented designer, Louis Béchereau.

Among Béchereau's first wartime designs was the Spad A2, which sought to solve the problem of firing a machine gun past the propeller arc by placing a gunner in a pulpit held by means of struts in front of the propeller and its 80-horsepower Rhône rotary engine. Although placed into limited production, the Spad A2 and its 110-horsepower successor, the A4, were too terrifying—the front gunner having little chance of survival in the event of a nose over upon landing—to last long. The basic airframe, however, was sound. On June 4, 1915, Béchereau applied for a patent for the plane's single-bay wing cellule, which featured intermediate struts of narrow chord, to which the bracing wires were attached at the midpoint. That arrangement added strength and, by reducing vibration in the wires, reduced drag as well.

Béchereau's next fighter, the Spad G, used the same airframe and rotary engine as the A2 but dispensed with the gunner, relying instead on mounting a remotely controlled machine gun in a sizeable housing in front of the propeller. That, too, was a failure, so then Béchereau adapted the airframe to two new developments: the 150-horsepower Hispano-Suiza 8A engine and a synchronized .303-inch Vickers machine gun. At first called the Spad HS, the prototype had a large conical spinner in front of a circular radiator. The spinner was soon abandoned but the radiator was retained, although it would undergo several changes in configuration to address an early spate of cooling problems.

First flown in April 1916, the Spad HS was an instant success, and the French air service placed its first order for 268 of the new Spad 7.C1, as it was officially designated—or Spad VII, as it was more commonly called—on May 10, 1916. The Royal Flying Corps and the Royal Naval Air Service were also quick to order the new fighter.

The first pilot to receive a Spad VII was reportedly Lt. Armand Pinsard of N.26, who used S122 to force an LVG to land between Combles and Rancourt on August 23, 1916. That was not confirmed, but on September 7 he drove an enemy plane down near Pertain, where British artillery demolished it, and scored a second victory near Lechelle on November 1. Later in the month he left N.26 to take command of N.78, and with that unit he would add another fifteen to his total of twenty-seven victories.

Another early Spad VII recipient was *Sous-Lt.* Georges Guynemer, the leading ace of *Escadrille* N.3 "*Les Cigognes*," who scored his fifteenth victory in Spad S115 over Hyencourt on September 4, 1916, his victims being the Aviatik C.II crew of *Leutnante* Hans Steiner and Otto Fresenius of *Kampfstaffel* 37.

On September 23 Guynemer downed two Fokkers, plus a third unconfirmed. As he returned over the lines at three thousand meters altitude, however, nervous French antiaircraft gunners struck his unfamiliar-looking plane with a 75mm

Sous-Lieutenant Georges Guynemer of N.3 flies his new Spad VII S115 in September 1916. He scored three confirmed victories in the plane before being shot down in error by a French antiaircraft gun crew; his survival of the resultant crash ultimately sold him on the robust fighter. (*Service Historique de Défense—Air*)

shell. Guynemer spun down but managed to regain enough control to pull up and crash-land in a shell hole, emerging with a cut knee and a slight concussion. As for the plane, he wrote his father: "Only the fuselage was left, but it was intact. The Spad is solid, with another [plane] I would now be thinner than this piece of paper."[1]

Production of the Spad VII got off to a slow start, but eventually some 3,500 were built and served in numerous air arms around the world. Aces from France, Britain, Russia, Belgium, Italy, and the United States all scored victories in Spad VIIs at one time or another in their careers, as did André Bosson and Jacques Roques, two Swiss volunteers who became aces in French service. Problems with the later twin-gunned Spad XIII's geared Hispano-Suiza engine made it necessary to retain the Spad VII in frontline service until the end of the war, and French fighter pilots were still training in Spad VIIs as late as 1928.

By the time the RFC received its first Spad VII to evaluate on September 9, 1916, the Royal Aircraft Factory at Farnborough was engaged in designing its own Hispano-Suiza-powered fighter. In fact, RAF engineer Frank W. Goodden had had two ideas in the works as early as June. One, designated the F.E.10 (Fighter Experimental No. 10), was a rather impractical-looking contraption with the pilot and his machine gun perched in front of the propeller in a nacelle braced to the undercarriage and upper wing, in a manner similar to that used by the gunner of the unsuccessful Spad A2. The other, designated S.E.5 (Scouting Experimental No. 5), was similar in overall layout, but with the engine in front and the pilot seated aft of the wings—essentially a smaller, more compactly proportioned single-seat version of the B.E.2c. Not surprisingly, the conventional design was selected for further development, but the F.E.10's vertical tail surfaces were retained in lieu of the smaller fin and rudder originally proposed for the S.E.5.

When the first twenty-one French-made Hispano-Suiza 8A engines were delivered to the RFC on September 20, most were slated for installation in British-built Spad VIIs, but two were used to power the first and second S.E.5 prototypes, A'4561 and A'4562. On November 28, the Royal Aircraft Factory received its first example of the new geared 200-horsepower Hispano-Suiza 8B, which it subsequently installed in the third prototype, A'4563—thereby creating the first S.E.5a.

S.E.5 A'4562 broke up during a test flight on January 28, 1917, killing Major Goodden. Simple modifications corrected the plane's structural problems, however, and the first production S.E.5 A'4845 cleared its final inspection on

March 2, 1917. The first production batch of S.E.5s did not make a promising impression on their pilots, who complained of poor lateral control—a shortcoming that was alleviated somewhat (but never entirely) by shortening the wingspan and reducing the rake of the wingtips. Early production S.E.5s also featured an overhead gravity tank, a large half-canopy which pilots came to call the "greenhouse," and a mechanism that could raise or lower the pilot's seat. The mixed armament consisted of one .303-inch Vickers machine gun in the fuselage, synchronized by means of CC fire control timing gear; a new hydraulic system developed by Major George E. Colley and Romanian physicist George Constantinesco—the earliest examples of which frequently failed; and a .303-inch Lewis gun mounted above the upper wing.

The first S.E.5s were assigned to No. 56 Squadron under Maj. Richard Graham Blomfield, a new unit that nevertheless had the benefit of a handpicked cadre of experienced pilots. The most famous of the "old hands" was the twenty-year-old leader of A Flight, Capt. Albert Ball, an eccentric but brilliantly aggressive loner whose exploits as a Nieuport pilot in No. 60 Squadron were already legendary in the RFC. Ball, whose score then stood at thirty-two victories, had high expectations for the S.E.5, but after giving the first prototype a ten-minute test flight on November 23, 1916, he remarked with bitter regret that the new scout had "turned out a dud."[2]

On April 7, thirteen S.E.5s of No. 56 Squadron landed at Vert Galant aerodrome, joining the Spad VIIs of No. 19 Squadron and the Sopwith Pups of No. 66 Squadron. The unit and its new fighters reached the front at the start of the Battle of Arras, a British offensive that was in fact meant merely as a diversion for the larger French push which would be launched along the Aisne River on April 16.

Ball had made no secret of his dislike for the SE, and when Maj. Gen. Hugh Montague Trenchard visited the sector, Ball flew to Le Hameau and entreated him to replace the new fighters with Nieuports. Trenchard lent him a sympathetic ear, and Ball was convinced that the SEs would be replaced, but he had already taken the liberty of modifying his personal S.E.5 A'4850 while at London Colney. He replaced the "greenhouse" with a small Avro windscreen, which reduced drag and gave the pilot better access to the upper Lewis gun. He also removed the adjustable armored seat and replaced it with a board until a simpler, lower seat could be installed. The Lewis machine gun mounting was altered, with the slide lengthened by two inches to make it easier for the pilot to replace the ammunition drums. Ball discarded his synchronized Vickers gun and replaced it with an obliquely mounted downward-firing Lewis. He also removed the petrol and water gravity tanks from the upper wing and installed long Spad-type exhaust pipes. Ball noted that his alterations resulted in a considerable improvement in performance, but he still regarded the SE to be "a rotten machine."[3]

Not all of Ball's modifications met with the RFC's approval—the obliquely mounted downward-firing Lewis gun was not a good idea—but many of them were made by his squadron mates to their own planes, and went on to be officially adopted for production aircraft, and the S.E.5 was the better for it. The undercarriage wheels were also moved farther forward, and the external overwing tank was replaced by internally fitted fuel and water gravity tanks behind the leading

Captain Albert Ball in S.E.5 A'4850 at London Colney in March 1917. Although unimpressed with it, he affected a number of custom touches that helped turn it into a far more effective fighter. (*Imperial War Museum*)

edge of the upper wing center section, which was also strengthened and covered with plywood to withstand the recoil of the Lewis gun.

Snow and bad weather delayed test flying in the modified aircraft until April 13. Ball was told that afternoon that Trenchard had authorized him a Nieuport for his personal use, although he still had to fly S.E.5 A'4850 on squadron patrols. Ball was delighted, and that evening Capt. Henry Meintjes flew Nieuport Scout No. B1522 in from Candas, while Ball brought in his modified S.E.5.

The first operational S.E.5 patrol was dispatched at 1018 hours on April 22, the pilots' enthusiasm being somewhat tempered by orders that they were on no account to cross the front lines. Ball led A Flight in A'4850, but one of his five planes dropped out when Lt. Gerald Maxwell developed engine trouble due to failure of the oil to circulate properly. At an altitude of 11,000 feet, Ball spotted an Albatros C.III over Adinfer and attacked. Although he fired three drums of ammunition into the enemy plane at a range as close as 150 feet, the German managed to escape in a steep dive over the lines, at which point Ball had to break off the chase. Another enemy plane was seen at noon, but it was too far away to engage. Meintjes led C Flight on another patrol that afternoon, but it was equally unproductive.

Captain Cyril M. Crowe led the squadron's first offensive patrol the next morning, but again the British encountered no enemy activity. Ball, however, had taken off on his own at 0600 hours, flying his Nieuport between Douai and Cambrai with the expectation of encountering a German en route to or from one of those aerodromes. Indeed, two Albatros two-seaters appeared 8,000 feet over Cambrai, and Ball carried out his usual tactic—a dive and then a pullout underneath his quarry, at which point he pulled up his wing-mounted Lewis gun and fired at the underside. The first German eluded him, but Ball slipped under the second, fired half a drum of Lewis into it and then pursued his diving prey until it crashed into the side of the road between Tilloy and Abancourt. Thus, No. 56 Squadron's first official victory was not scored in the unit's assigned aircraft.

Ball found and attacked another Albatros a few minutes later, but its pilot was an experienced hand and throttled back when Ball dived to attack, causing him to overshoot. With the tables thus turned, the German put fifteen bullets in the spars of the Nieuport's bottom wing. Ball dived away and landed safely at 0845, but his Nieuport was out of action until a new lower wing could be installed. Ball therefore flew his second patrol of the day in his unloved S.E.5.

Taking off at 1045 and climbing to 12,000 feet, Ball spotted an Albatros C.III over Adinfer, dived, pulled up and opened fire—only to suffer a gun jam. Apparently forgetting that he had another Lewis gun in his plane, Ball landed at Le Hameau aerodrome, rectified his jam and took off again. At 1145, he sighted five Albatros

scouts over Sévigny and again dived to the attack, firing 150 rounds into one opponent, which fell out of control and burst into flames before reaching the ground. The other four Germans put some rounds into Ball's plane, but he used the SE's superior diving speed to escape. Three-quarters of an hour later, Ball encountered yet another Albatros C.III north of Cambrai, dived underneath and fired half a drum of Lewis into it. The German pilot, *Vizefeldwebel* Ebert of *Flieger-Abteilung* 7, retired in a steep dive, made a good landing and then helped his observer, *Leutnant* Berger, who had suffered a severe neck wound in Ball's attack.

Meintjes led a five-plane patrol up at 1315 that afternoon, but half an hour later Lt. William B. Melville turned back with engine trouble, and his S.E.5 A'4852 overturned while landing and was wrecked. The rest of the flight chased a German two-seater south of Lens but failed to bring it down. In the final sortie of the day, Ball led Lts. Clarence R. W. Knight and John Owen Leach in search of enemy balloons, but they returned empty-handed at 1735.

So ended the S.E.5's first day. Only the redoubtable Ball had shot anything down, yet squadron morale was high. Disappointing though they had been at the onset, the modified S.E.5s had not performed badly. Even Ball came to appreciate it and used it to add eleven victories to his final total of forty-four before being killed in it shortly after a costly run-in with the Red Baron's Circus on May 7, 1917. The replacement of the S.E.5's 150-horsepower Hispano-Suiza with a more powerful 200-horsepower model, along with further refinements, produced the S.E.5a, the first of which began arriving at No. 56 Squadron in June. Fast, rugged, and almost viceless, the S.E.5a became a mainstay of the RFC, and later of the RAF, over the Western Front right up to the end of the war. The first unit to employ it, "Fighting Fifty-Six," was also the most successful, being credited with 401 victories by the end of the war and producing numerous famous aces, two of whom—Ball and James Thomas Byford McCudden—were awarded the Victoria Cross.

In the same month that the S.E.5 saw combat for the first time, another British fighter was having a far less auspicious debut. Although it, like the S.E.5, would evolve into one of the great fighters of the war, the first Bristol F.2As to cross the lines added to the notoriety of "Bloody April"—and to that of the Flying Circus.

In 1916, the RFC began seeking a replacement for the intolerably vulnerable B.E.2c. The Royal Aircraft Factory responded with the R.E.8, while Frank Barnwell at Bristols designed the R.2A, a two-seater that had the fuselage raised above the lower wing by struts in order to improve the pilot's view over the upper wing. A revised version of the Bristol design with unequal-span wings and a 150-horsepower Hispano-Suiza engine was designated the R.2B, but it was quickly superseded by another variant, with equal-span wings and a new Rolls-Royce Falcon I engine. At that point, the Bristol design was being regarded as a reconnaissance fighter, and by the time the first prototype, A3303, flew on September 9, 1916, it had been redesignated as the F.2A. With its side-mounted "ear" radiators replaced by a circular radiator within a deeper nose cowling, A3303 underwent its official tests between October 16 and 18, using both a two- and four-bladed propeller. The RFC was already impressed with the new plane's

potential, and testing only confirmed its decision to order fifty production examples. The first operational F.2As were delivered to No. 48 Squadron in December, armed with a forward-firing .303-inch Vickers gun synchronized by CC gear, and a .303-inch Lewis gun on a Scarff ring in the observer's cockpit. After training at Rendcomb, No. 48 Squadron was deployed to Bellevue, France, in March 1917, its aircraft being held until the start of the Battle of Arras in hopes of their achieving a degree of surprise over the enemy.

As with the S.E.5s of No. 56 Squadron, the Bristol crews of No. 48 Squadron were led by a cadre of veteran flight and deputy flight leaders. Captain Arthur T. Cull had seen considerable action, and Capt. Alan M. Wilkinson was already credited with ten victories flying D.H.2s with No. 24 Squadron. Captain William Leefe Robinson, who had been awarded the Victoria Cross for destroying Schütte-Lanz airship SL11 over north London on the night of September 2–3, 1916, was the most famous of 48 Squadron's flight commanders, but he also had the least experience in the more intense fighting environment that had developed over the Western Front.

As misfortune would dictate, Robinson was slated to lead six F.2As on their first operational sortie, an offensive patrol to Douai at 1000 hours on the morning of April 5—a job for which, seen with the benefit of hindsight, he was in woefully over his head.

"Robinson did not appreciate the fact that the Bristol could be used as an offensive weapon by the pilot, and that it was not necessary to provide one another with protection," Wilkinson later remarked. "We were not bothered about Robinson's VC, only that he seemed to dislike any opinion other than his own in matters [of] which, I might say, he had no experience."[4]

Largely ignoring the higher speed and greater maneuverability that the Bristol offered over two-seat reconnaissance aircraft, Robinson adopted the tactic of closing up the flight when attacked, so that the rear gunners could concentrate their fire. On top of that, prior to the mission he had reportedly learned that the lubricating oil in the machine guns froze at high altitude. His solution was to tell the gunnery officer, "If the guns are freezing up through the freezing of the oil, stop oiling the guns and therefore there will be nothing to freeze."[5]

By the time Robinson's flight neared Arras at 1100, German ground observers had reported its presence to the fighter unit at Brayelle, and soon five Albatros D.IIIs of *Jasta* 11 were taking off to intercept them, led by *Obltn*. Manfred von Richthofen.

As the Germans closed in, the Bristols closed up, their gunners waiting for the enemy to come within range. Then the Albatros scouts bore in fast, twin machine guns blazing, and almost immediately a Bristol fell out of formation and was forced down near Lewarde by Richthofen himself. Its wounded crew, 2nd Lts. Arthur N. Leckler and Herbert D. K. George, were taken prisoner, but George died of his wounds shortly afterward.

While Richthofen hastened to rejoin the chase, *Vzfw.* Sebastian Festner drove down Robinson and his observer, 2nd Lt. Edward D. Warburton, both of whom became prisoners of war. Catching up with another Bristol near Cuincy, Richthofen downed Lts. Alfred T. Adams and Donald J. Stewart, who were also wounded and taken prisoner, their only consolation lying in both having survived their run-in with the Baron. A fourth Bristol was brought down by *Ltn.* Georg

Simon, resulting in another wounded aircrew, Lt. Horace A. Cooper and 2nd Lt. Alan Boldison, falling into German captivity.

The remaining two F.2As limped home full of holes. One was crewed by Lt. P. Pike and 2nd Lt. Hugh Bradford Griffith, who had been an observer in F.E.2bs and F.E.2.ds with Nos. 20 and 11 Squadrons before transferring to 48 Squadron on March 17, 1917. Griffith reported that his plane became separated from the formation during the combat, and in a lone rearguard action he had driven down an assailant at 1050. He also said he saw Robinson and Warburton drive an opponent down out of control before they fell. The two claims were officially credited as No. 48 Squadron's first victories, but the sad truth was summed up by Richthofen: "After the attack, which was similar to a cavalry charge, the enemy squadron lay demolished on the ground. Not a single one of us was even wounded."[6]

The Baron also noted that his latest adversary "appears to be quick and rather handy. A powerful motor, V-shaped, 12 cylinder; its name could not be recognizable."[7] The Bristol's "black day" was far from over, as a second patrol, aggressively led by Captain Wilkinson, encountered three Albatroses near Douai, two of which Wilkinson and his twenty-four-year-old observer, Lt. Laurence W. Allen, drove down. The patrol claimed another enemy "driven down" before it returned safely. While leading the day's third patrol, Irish-born Capt. David M. Tidmarsh, a former infantryman who had scored three victories in D.H.2s with No. 24 Squadron before joining 48, engaged a two-seater east of Douai, only to see it escape into the clouds. Second Lieutenant Oswald W. Berry attacked another enemy plane southeast of Douai and drove it down in a steep spiral, while Lt. George

N. Brockhurst attacked and drove down a red scout. Only Wilkinson and Allen's victory was credited as "out of control," but those second and third sorties did much to restore 48 Squadron's shaken confidence in its new machines.

Patrolling continued unabated, with 48 Squadron's fortunes improving as its crews adjusted their tactics to their new mounts. And if losing VC recipient Robinson had been a blow to British morale, that blow was redressed on April 8, when a flight of F.2As, flying an offensive patrol between Arras, Lens, and Vitry, were attacked by Albatros D.IIIs of *Jasta* 4. Skillful mutual support resulted in claims for six Germans driven down, of which one was later collectively credited as "out of control" to Captain Tidmarsh and 2nd Lt. C. B. Holland, and 2nd Lts. Oswald Berry and F. B. Goodison, while a second was credited to the teams of 2nd Lts. George Brockhurst and C. B. Boughton, Robert E. Adney and Leslie G. Lovell, and A. J. Riley and L. G. Hall. This time one of their claims was genuine, as D.III 1958/16 broke up before crashing between Vitry-Sailly, killing *Jasta* 4's commander, *Ltn.* Wilhelm Frankl, a recipient of the *Orden Pour le Mérite* who had just scored his twentieth victory the day before. Frankl's demise may well have been caused, or at least abetted, by structural failure of his single-spar lower wing, a problem that had been plaguing the Albatros D.III since its debut in late January. In any case, his loss was a demoralizing one for the Germans—and far from the last they would suffer at the hands of Bristol Fighters.

In spite of their dubious initial showing, another two hundred Bristol Fighters were ordered for the RFC, modified with longer span tail planes and a slope to the upper fuselage longerons. The first hundred fifty F.2Bs, as the altered Bristols were called, used Hispano-Suiza engines but the next fifty were powered by the 220-horsepower Rolls-Royce Falcon II, with radiator shutters to help control the engine's temperature. Even while Bristol Fighters were being delivered to Nos. 11, 20, 22, 48, 62, and 88 Squadrons, pilots were realizing that the new plane's speed and maneuverability would be more effectively brought into play if it were flown aggressively like a scout with a sting in the tail, rather than as a typical two-seater. They changed their tactics accordingly, and the Bristol F.2B went on to acquire an outstanding fighting record.

Having provided the British forces with the excellent Pup and Triplane scouts in 1916, Sopwith did not produce a new single-seat design until December 1916—six months after the Triplane—and structurally, at least, it differed little from its predecessors. The F.1 biplane had a shorter, deeper fuselage than the Pup, with the rotary engine, cockpit, and guns concentrated closer up front. To facilitate production, Sopwith eliminated dihedral on the one-piece upper wing and compensated by doubling the dihedral of the lower one. The most significant improvement offered in the F.1 was the replacement of the single .303-inch Vickers machine gun of the Pup and Triplane with a twin-gun installation partially covered by fairings that looked somewhat like a hump from the side, leading to the plane being christened the Camel.

The first F.1 was powered by a 110-horsepower Clerget 9Z rotary engine when Harry Hawker got in its cockpit and, as he put it, "bounced into the air" from Brooklands aerodrome on December 16, 1916.[8] Subsequent prototypes were powered by several other engine types. The first production batch from Sopwith used either the 130-horse-

One of the earliest Sopwith F1 Camels to enter service with No. 4 Squadron, Royal Naval Air Service at Bray Dunes in June 1917, N6346 was usually flown by Flight Cmdr. G.M.T. Rouse. (*Leslie Rogers via Aaron Weaver*)

power Clerget 9B or the 150-horsepower Bentley B.R.1.

Similar though it may have been in construction, the Camel's altered configuration gave it flying characteristics that differed radically from the docile Pup and the manageable Triplane. The torque of the rotary engine, combined with the concentration of weight up front, endowed it with breathtaking maneuverability, but it also made it murderously unforgiving of pilot error. Few, if any, German aircraft could match a Camel in a right-hand turn, but failure to compensate for the torque during takeoff or landing could send a careless pilot careening into the ground. Pilots who mastered the Camel's idiosyncrasies swore by it; those who did not swore *at* it—provided they survived to do so.

The first unit to use the Camel in combat was No. 4 Squadron, RNAS, which began replacing its Pups with the new type in the first week of June 1917. On June 4 Flt. Cmdr. Alexander M. Shook, patrolling in N6347, attacked an enemy plane fifteen miles off Nieuport. The German dived steeply and escaped into a dense sea haze. Shook was up again the next day when he sighted fifteen German aircraft between Nieuport and Ostende, and immediately attacked. The Canadian pilot was more successful this time, sending an enemy scout down to crash on the beach and shooting a second down out of control. The first combat loss occurred on June 13, when Flt. Sub-Lt. Langley F. W. Smith, an American volunteer from Chicago, Illinois, who had already been credited with eight victories flying Pups, was killed in N6362—according to some witnesses, his Camel broke up while he was stunting above the German aerodrome at Neumünster. On June 25, Shook and Flt. Sub-Lt. Arnold J. Chadwick attacked a German two-seater and its escort of scouts. In the course of the melee, Chadwick managed to shoot down the two-seater in flames.

A prelude of somewhat different things to come occurred on the early morning of July 4, when five of Naval 4's Camels encountered sixteen Gotha bombers flying at an altitude of between 12,000 and 15,000 feet, thirty miles northwest of Ostende. Shook attacked one of the

Gothas and last saw it diving and trailing black smoke. He then attacked a second bomber until his guns jammed. Shook was subsequently awarded the Distinguished Service Cross for his part in the action, but his fellow pilots were equally keen to come to grips with the twin-engine giants. Flight Sub-Lieutenant Sydney E. Ellis dived into the center of the bomber formation and fired three hundred rounds into one of the Gothas, which stalled and fell away erratically, with brown smoke spewing from the rear gunner's cockpit. Flight Sub-Lieutenant Albert J. Enstone claimed to have damaged one Gotha in the fight and forced a second down low over the Netherlands.

The Camels of Naval 4, joined by those of Naval 3, accounted for several more German aircraft over the next few days. On July 10, however, Flt. Sub-Lt. Eric W. Busby was killed in action, and two days later Naval 4 was reminded of the Camel's unforgiving nature when Ellis spun into the ground and was killed instantly. Nevertheless, overall it had been a satisfactory first month for the touchy but nimble debutante that would eventually fly with the RNAS, the RFC, and the Royal Air Force over every battle-front, and it would be credited with more aerial victories than any single fighter type of World War I, and endure as one of the most famous aircraft of the war—and of all time.

Two other British fighters of less conventional configuration than that of the S.E.5, Bristol and Camel were doomed to less successful careers. Late in 1916, Geoffrey de Havilland built a tractor-engine successor to his D.H.2, powered by a 110-horsepower Rhône 9J radial and armed with a synchronized Vickers gun that could be elevated as high as sixty

degrees in order to attack opponents from below. The most unusual feature of the D.H.5, however, was the placement of the pilot forward of a back-staggered upper wing, which was intended to give him as good a forward view as a D.H.2 pilot had had. Although its 100-miles-per-hour speed at 10,000 feet was a distinct improvement over the D.H.2's 77 miles per hour, the D.H.5 was slower and climbed less rapidly than the Sopwith Pup, and its service ceiling of 14,300 feet was less than the Pup's 17,500 feet. In spite of that, 400 D.H.5s were ordered on January 15, 1917.

The first D.H.5s reached No. 24 Squadron on May 1, although the unit still had a few of its old and hopelessly obsolescent D.H.2s on strength when the Battle of Messines began on June 7. The D.H.5's first success came on May 25, when 2nd Lt. Stanley Cockerell, a D.H.2 veteran, shot down an Albatros D.III over Ligny for his sixth victory. Only nineteen other claims were to be made by the squadron, however, before it was reequipped with S.E.5as, and in only two cases were the enemy aircraft seen to crash.

D.H.5s also equipped Nos. 32, 41, 64, and 68 (Australian) Squadrons but successes were few, and more frequently the D.H.5 pilots experienced the frustration of enemy aircraft escaping by outrunning or outclimbing them. The ruggedly built D.H.5s performed yeoman service as ground strafers during the Battle of Cambrai in November 1917, but attrition to ground fire and enemy fighters was inevitably high. Even while No. 64 Squadron was bringing its new D.H.5s to the front in October 1917, No. 41 Squadron was beginning to replace its D.H.5s with S.E.5as. Later, No. 24 Squadron's historian wrote: "On December 25th the Squadron received the best of all Xmas presents—a new machine—and both pilots and mechanics heaved a sigh of thankfulness to heaven. The new machine was the S.E.5, with 200 h.p. Hispano-Suiza engine." By January 1918, the last D.H.5s had been withdrawn from the front.[9]

Among the most advanced single-seat designs to enter British service during World War I was also the most underutilized. Designed by Frank Barnwell—creator of the redoubtable F.2B—the Bristol M.1A was a monoplane fighter that first flew on July 14, 1916. Its appearance could not have been more untimely, for at that time the Morane-Saulnier Nm monoplane was taking part in the Battle of the Somme with No. 60 Squadron, making a wretched name for itself and for monoplanes in general. By mid-October the RFC's commander, Gen. Hugh Trenchard, was canceling all further purchase orders for Morane-Saulnier scouts and was left with a near-pathological antipathy for all things with single wings.

That was unfortunate, because the Bristol had none of the Morane-Saulnier's faults. Although its thin wings were wire-braced to the fuselage underside and to a rounded pylon above the cockpit, they proved to be sturdy enough to withstand the stresses of combat and had ailerons rather than wing warping like the earlier French monoplane. The Bristol's tail surfaces had horizontal and vertical stabilizers, making it far less tricky to fly than the Morane-Saulnier, and the synchronization mechanism for its single Vickers machine gun was more reliable than that of the Morane-Saulnier I and V—to say nothing of the terrifying propeller-mounted metal deflectors on the Morane-Saulnier Nm. The most remarkable feature of the Bristol monoplane was the performance it achieved with a mere 110-horsepower Clerget 9Z rotary engine—130 miles per

hour and the ability to climb to 10,000 feet in 8.5 seconds, as compared to the 109.4-miles-per-hour top speed of a Morane-Saulnier I, which took twice as long to reach 10,000 feet. Later, test pilot Oliver Stewart rated the Bristol M.1's maneuverability as superior to that of the S.E.5a and the Sopwith Snipe, the latter of which was a 1918 design.

Bristol took steps to remedy the prototype's principal shortcoming—a complete lack of downward visibility—by removing a section of fabric from the right wing root of the revised M.1B, which also featured a pyramidal bracing pylon above the cockpit. The first M.1B, A5139, went to France, and on the twenty-third it was flown by three veteran pilots, including Lt. D'Urban V. Armstrong and Capt. Roderic Hill, both of whom had flown Morane-Saulnier monoplanes in No. 60 Squadron. Hill and the third pilot, Capt. Alan M. Lowery of No. 70 Squadron, wrote favorable reports about the Bristol, but Trenchard had made up his mind about the new monoplane before he even read them, citing its limited downward view as the official excuse to reject it out of hand. Frontline fighter pilots who had heard about the Bristol and who hoped to see it issued to their squadrons were flabbergasted. "It is true that the downward and forward view was slightly restricted," Oliver Stewart wrote, "but there was not a fighting pilot of experience who would not have exchanged that view for the speed and climb of the Bristol Monoplane with alacrity and enthusiasm."[10]

Although no Bristol monoplane ever fought over the Western Front, in June 1917 three M.1Bs—A5140, A5141, and A5142—were shipped to Palestine in response to desperate pleas from Brig. Gen. William G. H. Salmond for modern fighters to support large-scale operations that were about to commence there. They arrived in August and were assigned to the newly forming fighter unit, No. 111 Squadron at Deir el-Belah, where they flew alongside Vickers FB.19 Bullets, Bristol F.2Bs, D.H.2s, and a Bristol Scout D. Soon after joining the squadron, A5142 was modified, with both wing root panels uncovered and the machine gun repositioned from the left to a central position. Later, it was apparently transferred to Heliopolis, Egypt, to serve as a trainer with a camera gun with its spinner removed to improve engine cooling.

Lieutenant William S. Lighthall, a Canadian pilot in No. 111 Squadron, recorded his impressions of the M.1B in desert service: "These were far ahead of their time aerodynamically, but were fragile and difficult to get in and out of due to a low triangle of struts over the pilot's cockpit and the close proximity of the Vickers gun which had a tendency to break the pilot's nose in a bad landing."[11]

The first success for 111 Squadron was achieved on October 8 by a Bristol Fighter team, 2nd Lt. Robert C. Steele and Lt. John J. Lloyd Williams, who brought down an Albatros D.III with the pilot, *Obltn.* Gustav Adolf Dittmar of *Fl.-Abt.* 300, taken prisoner. The German plane was subsequently repaired and given several evaluation flights by Capt. Arthur H. Peck. Steele and Lloyd Williams downed a second Albatros D.III on October 15, and the team of Peck and Lloyd Williams had accounted for three other enemy planes by November 8. In regard to the Bristol monoplanes, the squadron history noted that they "had some usefulness in that they made it a little more difficult for the enemy airmen to reconnoiter, except from great heights, but their very limited endurance prevented them from being used to escort the

long-distance strategical reconnaissance aeroplanes." Major General W. Sefton Brancker, who took command of the Middle East Brigade in October 1917, wrote: "The Bristol Monoplanes and Vickers Bullets are not very much good except to frighten the Hun; they always seem to lose the enemy as soon as he starts manoeuvring."[12]

M.1B A5141, in fact, became No. 111 Squadron's first operational loss, although its pilot, 2nd Lt. Edgar Percival, returned on November 16. His squadron mate, Lighthall, wrote of the incident:

> Edgar Percival, later a well-known aircraft designer and owner of Percival Aircraft Company, had a forced landing in a Bristol Monoplane far behind the Turk lines in the Judean hills. He only saved his nose by ducking very low and took the bump on the top of his head. He crawled out of the mess of the struts and wires, cowed some hostile Arabs with his Very pistol, and made them guide him through the Turkish lines and home to our squadron at Megdil.[13]

However good it might have been had it fought over the Western Front, the Bristol M.1B's debut over Palestine was distinguished primarily by Percival's desert adventure. By March 1918, more fighters with greater range and better adaptability to a desert environment, in particular the S.E.5a, were available to 111 Squadron, and the last of its monoplanes were withdrawn. On August 3, 1917, the War Office ordered 125 Bristol M.1Cs, which had panels cut out of both wing roots and were powered by 110-horsepower Rhône engines, but they, too, were destined to see action only in sideshows. On March 2, 1918, No. 72 Squadron arrived at Basra, equipped with D.H.4s, S.E.5as,

Spads, Martinsydes, and eight M.1Cs. The Bristol monoplanes got off to a unique start in Mesopotamia when two of them gave an aerobatic display before some Kurds, resulting in the entire tribe switching its allegiance from the Turks to the British. Other than that, 72 Squadron's M.1Cs gave good service in the ground-attack role.

A handful of M.1Cs were also shipped to Nos. 17 and 47 Squadrons in Macedonia in early January 1918. On April 1, 1918, while the Royal Flying Corps and Royal Naval Air Service were being amalgamated into the Royal Air Force, A Flight of No. 17 Squadron and B Flight of No. 47 Squadron were being amalgamated into a new unit, No. 150 Squadron. Although specifically a fighter squadron, its equipment was far from homogeneous, consisting of S.E.5as, Bristol M.1Cs, and Nieuports, to which would later be added some Camels. It was with No. 150 Squadron that the Bristol M.1C finally got the opportunity to show its worth in air-to-air combat on April 25, when Lt. Arthur Eyquem de Montaigne Jarvis—then still officially on No. 17 Squadron's roster—fired 150 rounds at a DFW C.V over Yenimah and last saw it descending in a steep dive toward Rupel Pass. Later that same morning, Jarvis in M.1C C4913 and Lt. Acheson G. Goulding in an S.E.5a attacked another DFW near Nihor and, in spite of an attempt by two Albatros D.IIIs to intervene, sent it down vertically over Angista, apparently on fire. On the next day, Jarvis in his M.1C and Lt. J. J. Boyd-Harvey in a Nieuport attacked a DFW over Prosenik and sent it crashing into Hristos Gully. The M.1Cs were quite successful over Macedonia, and one of their pilots, Capt. Frederick D. Travers, even managed to claim five victories while flying one, his wartime total being nine. Nevertheless, it

is doubtful that more than thirty-five M.1Bs and M.1Cs saw any combat use, the rest being used as trainers. The Bristol, however, would not be the only promising monoplane fighter design destined for a disappointing career before the war was over.

While the Camel was able to coexist with the S.E.5 and Bristol F.2B in the RFC, France's general satisfaction with the Spad VII undermined the prospects of numerous other creditable fighter designs finding acceptance. Nieuport produced some refined variants on the Nieuport 17—the 17bis, 23, 24, 25, and 27—but they served primarily as a supplement until enough Spad VIIs or XIIIs were available to replace them. By late 1917 the sesquiplane, with its structural limitations, clearly had no future, and Nieuport's engineers devoted themselves to developing a biplane, the 28.

Another noteworthy French fighter design, long believed to be among the doomed Spad competitors, was in fact designed with an entirely different client in mind. Lacking a suitable indigenous fighter design when it entered the war, Italy had used Nieuports, most of which were built under license by Giulio Macchi, who had steadfastly resisted demands for his factory to build bombers. In the autumn of 1916 *Capitano* (Captain) Ermanno Beltramo, a member of the Italian Military Aviation Mission in Paris, contacted René Hanriot and his chief engineer, Emile Eugène Dupont. Years later Beltramo wrote, "Of my own initiative I asked Mr. Hanriot if he were willing to build, without obligation, a single-seat fighter following the concepts which I would suggest in order to create a type better suited than the Nieuport to the fighter needs of our theatre of operations."[14] By November, Dupont had creat-

Hanriot HD.1 No.519 was a French-built plane assigned to the 76a *Squadriglia* and bearing the personal motif of *Tenente* Giuseppe Retinò when photographed at Casoni aerodrome in February 1918. (*Paolo Varriale*)

ed a compact single-seater powered by a 120-horsepower Rhône 9J engine, the HD.1, which underwent flight testing at Villacoublay on December 20.

Structurally conventional with the usual wood and canvas (but not including the aluminum cowl panels), the HD.1 was notable primarily for its W-shaped cabane strut arrangement, similar to that of the Sopwith 1 1/2-Strutter which Hanriot was then license-building for the French air service, and also for the greater span and dihedral of its upper wing in relation to the lower. In spite of its angular fuselage and wings, the HD.1 was well proportioned and stands alongside the Sopwith Pup as one of the war's most aesthetically pleasing aircraft designs.

By February 1917, the Hanriot was being test-flown in Italy, where it was praised by *Maggior Generale* (Major General) Giovanni Battista Marieni of the General Direction of Aeronautics for its superiority to the Nieuport 17 in speed, handling, visibility, and "automatic stability which prevents it from spinning and overturning." France ordered 50 HD.1s, while Macchi negotiated an initial order to license-produce 200 in June 1917. Italy ended up purchasing a total of 360

HD.1s from Hanriot, and of the 1,700 it ordered from Macchi, 831 would be delivered by the end of the war, with another 70 completed by February 1919.[15]

The first Hanriot-built HD.1s arrived at the 76a *Squadriglia* at Bergamo in July 1917, replacing its mixed bag of Nieuports and Spads but deliveries from France were slow, and by October 26 only the 78a *Squadriglia* had likewise converted—just as the Austro-Hungarians, with German assistance, were achieving a decisive breakthrough at Caporetto. In consequence, both squadrons had to destroy sixteen of their new fighters and retreat to Istrana. At that point, Macchi was producing its own Hd.1s (to use its designation for them), which it delivered to the 82a *Squadriglia* in early November, followed by the 70a and 80a by the end of the month and the 72a in December.

The first HD.1 victory was scored by *Sergente* (Sergeant) Alessandro Contardini of the 82a on November 6, and the first loss was *Aspirante* (Acting Second Lieutenant) Vittorio Aquilino of the 78a, shot down near Maser on the fourteenth. He may have been the victim of a *Leutnant der Reserve* Bussmann of *Jasta* 1, one of three German *Jagdstaffeln* then assigned to the Italian Front, who claimed a "Sopwith" over Il Montello that was not confirmed to him.

By mid-1918, the Hanriot HD.1 was the principal Italian fighter, with Spads and Nieuports playing a supplementary role in the squadrons. Among its many successful exponents was Italy's second-ranking ace, Silvio Scaroni, who scored at least nineteen of his twenty-six victories in HD.1s.

In June 1917, the exiled *Aviation Militaire Belge* also ordered 125 HD.1s from the Hanriot firm, the first of which arrived on August 22, and which was assigned to the 1*e Escadrille* at Les Möeres. That unit's pilots had been so content with their Nieuport 17s that they had already declined an offer to replace them with Spad VIIs. They likewise rejected the first HD.1 out of hand, starting with the squadron's leading ace, André de Meulemeester, followed by Jan Olieslagers, and so on down the line to the most junior member, Willy Coppens. Coppens was still flying the squadron's last Nieuport 16, the poorly balanced and treacherous flight characteristics of which he hated with a passion, so he was more than game to give the HD.1 a try. His serendipity upon discovering how well the Hanriot handled soon converted Coppens's squadron mates, and once they overcame some early problems with gun jams and inaccurate sights, the pilots of the 1*e Escadrille* did quite well with the fighter that France rejected. Coppens eventually became the HD.1's greatest user—and Belgium's leading ace—with a total of thirty-seven aerial victories, thirty-five of which were balloons.

In addition to the improved performance it gave to the S.E.5a, the geared 200-horsepower Hispano-Suiza 8B engine led

to the development of another famous French fighter, the Spad XIII, and to one of the war's most heavily armed single-seat fighters, the Spad XII. The conception of both fighters was also heavily influenced by the Spad VII's most enthusiastic proponent and France's leading ace at the time, Georges Guynemer. In December 1916, *Lieutenant* Guynemer wrote a letter to Spad's chief designer, Louis Béchereau:

> The 150-hp Spad is not a match for the Halberstadt. Although the Halberstadt is probably no faster, it climbs better, consequently it has the overall advantage. More speed is needed; possibly the airscrew might be improved.[16]

Hispano-Suiza came up with one answer to that problem: by increasing the compression ratio of its original engine to produce the 180-horsepower 8Ab. On June 11, 1916, the company successfully bench-tested a new geared V-8 engine, the 8B, which could produce 208 horsepower at 2,000 revolutions per minute at ground level. Béchereau tested the new engine in the Spad VII but soon concluded that a somewhat larger, more robust airframe would be necessary to accommodate it. In addition to its size, the Spad 13.C1, which was ordered into production in February 1917, had rounded wingtips, forward-staggered cabane struts with a frontal bracing wire, and twin .30-caliber Vickers machine guns. *Sous-Lieutenant* René Dorme, one of the many aces in Guynemer's *Escadrille* Spa.3, flight-tested one of the new Spad XIIIs at Buc on April 4, 1917, and at least one was undergoing evaluation in a frontline *escadrille* on April 26. In spite of a maximum speed of 124 miles per hour and a climb rate of 13,000 feet in 11 minutes, problems with the Hispano-Suiza 8B's spur reduction gear were to delay the

Spad XIII's introduction at the front and would handicap it for months thereafter. In November 1917, for example, units were reporting that their Spad XIIIs were grounded two days out of three because of engine malfunctions.

Notwithstanding its frontline evaluation by the French, the first Spad XIII shootdown was credited to the RFC. The British had enthusiastically adopted the Spad VII in the summer of 1916 and were equally keen to acquire its successor. On April 26, 1917, General Trenchard instructed Capt. William James C. K. Cochran-Patrick of No. 23 Squadron to go to La Bonne Maison aerodrome and inspect—but not fly—the Spad XIII. On May 1, Cochran-Patrick and accompanying British officers drafted a report on the new fighter which included locally recorded performance figures of 120 miles per hour at 4,000 meters (13,120 feet), a climb to that altitude in 11 minutes, and a climb to 7,000 meters (22,965 feet) in 28 minutes. "The machine is said to be less handy near the ground," the report stated, "but considerably handier at a height than the 150-hp Spad."[17]

At the end of May 1917, the British Aviation Supplies Depot received Spad XIII S498. Given the British serial number B3479 and tested at Candas, it exceeded expectations, reaching a speed of 140 miles per hour at 15,000 feet and climbing to that altitude in 16 minutes and 18 seconds. On June 9, B3479 was sent to No. 19 Squadron for frontline evaluation, and while flying it on June 13, Capt. Frederick Sowrey drove down a German two-seater for his third victory. Lieutenant G. S. Buck destroyed an Albatros the next day, and Fred Sowrey became an ace in B3479 on July 21, when he drove an Albatros D.III down out of control northeast of Ypres. Soon, howev-

er, the British would experience the same teething troubles with the geared Hispano-Suiza engine that were delaying the Spad XIII's combat debut in the French *escadrilles*.

Another suggestion that Guynemer had made to Béchereau at the end of 1916 had been for a fighter capable of mounting a cannon. Since the geared Hispano-Suiza engine raised the propeller above the cylinder heads, Birkigt was indeed able to arrange a 37mm Puteaux cannon with a shortened barrel to fire through a hollow propeller shaft. Taking matters from there, Béchereau designed an enlarged version of the Spad VII, which was designated the 12Ca.1 (for Type 12, cannon-armed single-seater).

In spite of its unusual armament, the Spad XII did not look like a freak—in fact, with its lack of bulged fairings on the cowling and slightly forward-staggered wings, it was one of the most elegant-looking Spads ever built. Inside the cockpit, however, the cannon breech protruded between the pilot's legs, necessitating Deperdussin-type elevator and aileron controls on either side of the pilot, instead of a central control column. When *Capitaine* Guynemer flew the Spad XII in the spring of 1917, he recorded a maximum speed of 137 miles per hour at ground level and a maximum ceiling of 23,000 feet. The main shortcoming that Guynemer saw in what he called his "*avion magique*" was the fact that the 37mm cannon was a single-shot weapon which had to be reloaded by hand. However, the Spad XII also had a single synchronized .30-caliber Vickers machine gun, which could be used to help sight the cannon on a target before it was fired, or to help the pilot fight his way out of trouble after the cannon had been fired.

The first operational Spad XII, S382, was delivered to Guynemer at N.3 in July

Captaine Georges Guynemer's cannon-armed Spad XII S382 at Saint-Pol-sur-Mer in July 1917. Although he scored four victories in the "*avion magique*," it was a handful, or, rather, two handfuls-to fly. (*Service Historique de Défense—Air*)

1917. During the new fighter's first frontline sortie on July 5, Guynemer attacked a DFW C.V two-seater, but the enemy gunner managed to inflict such damage that Guynemer was compelled to disengage and return his Spad XII for nearly three weeks of repairs. It returned to the squadron on July 20, and on the twenty-seventh Guynemer used it to shoot down an Albatros fighter between Langemarck and Roulers with eight machine gun rounds and one cannon shell, killing *Ltn.d.Res.* Fritz Vossen of *Jasta* 33. He downed a DFW C.V over Westroosebeke the next day using two shells and thirty bullets, but again his victim's return fire caused enough damage to send S382 back to the repair shop until August 15.

Guynemer scored a double victory on August 17—one German two-seater with machine gun fire alone, and a second with the cannon—but an inconclusive fight with an aggressive two-seater crew the next day put the cannon Spad temporarily out of action once more. In spite of the cannon fighter's mixed fortunes,

Guynemer's performance resulted in Spad getting a contract for three hundred aircraft. It is unlikely that anywhere near that number were built, however, for the few cannon-armed Spad XIIs that saw service in 1917 and 1918 proved to be a handful for any but the most skilled pilots, such as Guynemer and the consummate hunter who would eventually surpass him as France's ace of aces, René Fonck.

While Guynemer waited for his cannon fighter to be repaired, he received Spad XIII S504, armed with twin Vickers machine guns with 380 rounds each. He wasted no time in blooding that new machine on August 20 when he used it to shoot down a DFW C.V over Poperinghe. Guynemer's first Spad XIII victory was also his fifty-third—and last.

On August 24, Guynemer visited the Spad factory at Buc to inspect the repairs to his Spad XII and recommend new improvements. Typically for Guynemer, who seldom rested, he was in a dangerously nervous state, and on the twenty-eighth he commented to a friend, "I shall

Guynemer's Spad XIII S504, in which he scored his 53rd and final victory, and in which he was killed on September 11, 1917. (*Service Historique de Défense—Air*)

not survive." Back at N.3, he took off with *Sous-Lt*. Benjamin Bozon-Verduraz on September 8, but bad weather forced them to abort their patrol. With the weather still unsuitable for flying on the ninth, Guynemer attended Sunday mass at Saint-Pol-sur-Mer. He took off in S504 on September 10, but had to land at the Belgian aerodrome at Les Möeres when the plane's water pump control became stuck. After returning to Saint-Pol, Guynemer borrowed *Lt*. Albert Deullin's Spad for another patrol and attacked a large number of German aircraft, but took four bullets in his plane and was forced to land with a disabled air pump. Returning by automobile, he took off in another borrowed Spad, but this time the fuel overflowed due to a loosened carburetor cover, and Guynemer was again compelled to land when the engine caught fire.[18]

Visibly annoyed by the day's bad luck, Guynemer ordered his mechanics to have S504 working and ready to fly at 0800 the next morning, but fog delayed takeoff until 0325 on September 11. He was to have been accompanied by Bozon-Verduraz and *Sgt*. Louis Risacher, but the latter's engine was having trouble starting, and Guynemer impatiently took off with Bozon-Verduraz only.

The haze disappeared at 5,000 feet altitude, and at 12,000 feet the Frenchmen spotted a DFW C.V northeast of Ypres. After making a desultory wave to Bozon-Verduraz, Guynemer dived on the German from above and behind. Bozon-Verduraz followed him, fired, missed, and then pulled up when he noticed eight German aircraft approaching. The enemy formation turned away, behaving as if its pilots had not noticed either Spad, so Bozon-Verduraz went back to find Guynemer, only to see no trace of him or the DFW. Following another indecisive combat over Poperinghe, Bozon-Verduraz returned to Saint-Pol-sur-Mer at 1020, his first words being, "Has he landed yet?"[19]

Guynemer never returned, and it would be another month before anything was learned of his fate. The Germans announced that *Ltn*. Kurt Wissemann of

Jasta 3 had killed the great ace for his fifth victory, but by then Wissemann himself was not available for comment, having been killed on September 28 by two S.E.5as of No. 56 Squadron flown by two of Albert Ball's disciples—Capt. Geoffrey Hilton Bowman and Lt. Reginald T. C. Hoidge. A sergeant of the German 413th Regiment certified that he had witnessed the Spad's crash and identified Guynemer's body, noting that the ace had died of a head wound, although one of his fingers had also been shot off and a leg was broken. An Allied artillery barrage drove the Germans back before they could recover or properly bury the body, which thus vanished amid the chaos that was the Western Front.

A more plausible cause of Guynemer's demise would be Rumpler C.IV 1463/17 from *Fl.-Abt.(A)* 224w, crewed by *Flieger* (Private) Georg Seibert and *Ltn.d.Res.* Max Psaar, who may have shot down his attacking Spad before being themselves shot down and killed soon afterward near Oudekapelle, seventeen kilometers north of Poelkapelle, by *Lt.* Maurice Medaets of the Belgian 1e *Escadrille*. Whatever the tragic circumstances of his death, Georges Guynemer went into French legend as its second-ranking ace who, a popular myth insisted, had flown so high he could never come down. The Spad XIII went on to greater things, although attempts to improve its lateral flying characteristics by squaring off the wingtips with added pocket extensions proved to be another of Spad's more frightful ideas. The final production version, built with squared wingtips from the outset, became one of the most famous (if not one of the best) fighters of World War I.

As Allied fighters improved and proliferated, their adversaries began to feel their control of the air, even on their side of the lines, slip away. In June 1917 the Germans introduced a more refined version of the Albatros D.III, the D.V, but its sesquiplane wing layout proved to be even more vulnerable to stress and failure than that of its predecessor. A strengthened version, the D.Va, was heavier, but that handicap on its performance was offset somewhat by the development of a higher compression variant of the Mercedes engine, the D.IIIa. In spite of Albatros's continued commitment to quantity production, however, the *Luftstreitkräfte* needed a new fighter design, or its fighter units would lose their tenuous hold on the sky.

One potential solution appeared in the late spring of 1917, when the prototype Pfalz D.III made its first flight. Designed by chief engineer Rudolph Gehringer and his team at the Bavarian Pfalz Flugzeugwerke GmbH, the D.III had been built in response to the same general appeal from the *Idflieg* for a fighter to match the Nieuport 11 and 17, which had led to the development of the Albatros D.III and D.V. Unlike the Albatros scouts and the many Nieuport copies produced by other German firms, the Pfalz was not really a sesquiplane, but a biplane with a smaller lower wing braced with U-shaped cabane and interplane struts. The fuselage, constructed in two halves from three-ply veneer strips, incorporated a compound wing root, along with integral fin and tail plane surfaces, based on Pfalz's previous experience building the unsuccessful Roland D.II under license.

Accepted and put into production in mid-1917, the sharklike Pfalz D.III was one of the most rakishly handsome aircraft of its time. When the first three arrived at the front at the end of August, however, the reaction was less than enthusiastic. *Leutnant* Werner Voss, commander of *Jasta* 10, received one of the first machines and flew it on at least a few operational missions but he is said to have preferred the superior rate of climb and maneuverability of his Albatros D.V, compared to the newer plane's stronger wing structure and better diving characteristics. Similar complaints of relatively sluggish handling characteristics and performance indicative of an unfavorable power-to-weight ratio undermined the Pfalz's popularity as it arrived in quantity at *Jastas* 4 and 10, as well as at Bavarian *Jastas* 16, 32, 34, and 35, during the fall of 1917. Another unappreciated feature was the enclosing of the D.III's twin machine guns within its fuselage, a drag-reducing measure that nevertheless made it impossible for the pilot to reach them if they jammed in combat.

At about the same time he received the disappointing Pfalz, Voss got something more to his dogfighting taste, in the form of the nimble Fokker F.I triplane. Another early Pfalz recipient, *Ltn.* Oskar *Freiherr* von Boenigk, a member of *Jasta* 4 who was flying D.III 1396/17 in September, may have been among the first to demonstrate that, in the right hands, the Bavarian shark had teeth. On September 3, he downed a Camel over Houthem and also claimed a Pup that could not be confirmed. He was credited with another Sopwith over Poelcapelle on September 9, bringing his personal score to five before being posted to take command of *Jasta* 21 on October 21. Other German aces would have their share of successes in the Pfalz, which later that year would get an altered lower wing with rounder tips, enlarged horizontal tail surfaces, guns raised above the fuselage where the pilot could clear jams, and the high-compression Mercedes D.IIIa engine, resulting in the Pfalz D.IIIa. Produced in greater numbers than the D.III, the improved

Typifying the "workhorse" role with which it was saddled, Pfalz D.III 4114/17 served in *Kampfeinsitzerstaffel* 8, an interceptor unit in Flanders, attached to Jagdgruppe Nord along with *Jasta* Boelcke 17, 20, and 28 in the autumn of 1917. (*Author*)

D.IIIa became one of the principal German fighters in the spring of 1918; but its performance was at best barely the equal of its Allied opposition, and its pilots desperately yearned for something significantly better.

Germans were not the only airmen in the Central Powers who longed for a fighter to match those of the Allies in 1917. Like Italy, Austria-Hungary was unsuccessful in developing an effective indigenous scout design in the early war years, its pilots using German designs such as the Fokker E.III and Brandenburg D.I, the latter of which had been created by a design bureau headed by Ernst Heinkel. A biplane braced by a unique junction of eight interplane struts into a starlike arrangement and featuring a deep fuselage whose rear upper decking tapered aft into a vertical stabilizer, the Brandenburg D.I, or KD (*Kampf-Doppeldecker*), entered production for the *Kaiserliche und Königliche Luftfahrtruppen* (Imperial and Royal Aviation Troops, or LFT) in May 1916. *Oberst* (Colonel) Emil Uzelac, commander of the LFT, personally test-flew the new fighter on November 7, 1916—and promptly crashed, suffering a concussion. Nevertheless, the Brandenburg D.I (series 65), fitted with a

Type II VK canister holding an MG 08/15 machine gun and ammunition above the upper wing, was committed to the front, and on 3 December *Obltn.* Godwin Brumowski of *Flik (Fliegerkompagnie)* 12, flying D.I 65.53, teamed up with *Linienschiffsleutnant* Gottfried Banfield of the Trieste naval air station and *Zugsführer* Karl Cislaghi of *Flik* 28 to bring down a Caproni Ca.1 bomber over Mavinje.

Only two other victories would be scored by Brandenburg-built D.Is. One of its pilots, *Zugsführer* Julius Arigi of *Fluggeschwader* 1, added a vertical stabilizer and an enlarged rudder, which improved the KD's flight characteristics somewhat, and which were incorporated in the seventy-two D.Is (series 28) built by the Phönix firm. Phönix-built D.Is took a reasonable toll on Italian aircraft, virtually in spite of themselves, thanks to the skill of their pilots. Still, by June 1917 *Flik* 41/J, a crack fighter squadron organized along the lines of Richthofen's *Jasta* 11 by its commander, *Hauptmann* Brumowski, was reporting that "the KD is absolutely useless . . . the best pilots (and only they can fly the type) are shackled, ruin their nerves and perish in crashes over the airfield, without their expert skill achieving anything."[20]

At about that time, Austrian versions of the Albatros D.II and D.III, license-built by the Oesterreichische Flugzeugwerke A.G. (Oeffag) and powered by 185-horsepower Austro-Daimler engines, were entering service. Later versions of the D.III—constructed with greater care and a better grade of wood than the German Albatros scouts, and powered first by 200-horsepower, and later by 225-horsepower, Austro-Daimler engines—would be the most successful fighters to serve in the LFT, and the best of the Albatros sesquiplane series. Nevertheless, neither the Phönix-built

Brandenburg D.I nor the Oeffag-Albatros D.III was an indigenous fighter, but was instead a refinement of foreign design.

The first true Austrian-designed fighter was almost rejected out of hand before it reached the drawing board. On June 8, 1916, Julius von Berg of the Öster-reichisch-Ungarische Flugzeugfabrik "Avia-tik" GmbH proposed a fighter to use a 160-horsepower Daimler engine and a similar single-seat reconnaissance design to be powered by a 120-horsepower Daimler. Both were rejected by the LFT's *Fliegerarsenal* (*Flars*) because it wanted Aviatik to concentrate on manufacturing Knoller CII two-seaters rather than pursue its own projects. When *Oberst* Uzelac heard of Berg's intentions in August, however, he intervened, and Berg was allowed to proceed.

Just two months later, on October 16, Aviatik fighter prototype serial 30.14 was ready for its test flight at Aspern. Combining a flat-sided wooden fuselage with Knoller-style wings featuring a reflex curvature, the 30.14 underwent a somewhat wobbly first flight, and then as its test pilot, *Fw.* Ferdinand Könschel, throttled down for a landing, it suddenly plunged into the ground, killing him. In spite of that discouraging start, Aviatik engineer Julius Kolin was convinced that the center of pressure shift had been responsible for the crash and set about redesigning the wing. Several follow-up prototypes were built and tested until success was finally achieved with 30.19, which first flew on January 24, 1917. Test-flying the machine on March 31, *Flars* pilot *Obltn.* Oszkar Fekete praised its "fabulous climb and enormous maneuverability." The new fighter was subsequently ordered into production as the Aviatik D.I.[21]

On May 15, 1917, 30.19, retrofitted with an improved one-piece upper wing, was sent along with the first production Aviatik D.I, 38.01, to *Fluggeschwader* 1 on the Isonzo front for combat evaluation. *Hauptmann* Karl Sabeditsch took 30.19 up for its first frontline mission on May 18, with an unsynchronized machine gun mounted on the upper wing. On the twenty-third he managed to shoot down a Caproni Ca.1. In July, Aviatik 30.19 was so badly damaged that it had to be written off, but it had demonstrated the D.I's worth. The Aviatik D.I, also called the Berg D.I, was produced by a number of subcontractors and appeared in equally numerous forms, using a succession of increasingly powerful engines, a variety of radiator configurations, and different forms of armament ranging from single guns firing over the upper wing to synchronized twin weapons buried in the cowling, and finally to the twin guns in front of the cockpit, which is the setup that pilots favored. Although extraordinarily nimble for an airplane powered by an inline engine, the Aviatik D.I occasionally suffered wing failures due to shoddy construction, depending on the manufacturer, and in spite of the ubiquity it achieved by the end of 1917, Austria's first fighter never achieved a significant level of ascendancy over any of its Allied counterparts.

Having already improved the Brandenburg D.I, Phönix set about creating a replacement for it and unveiled two prototypes in June 1917. One, bearing the serial number 20.14, introduced a wireless variation on the Nieuport sesquiplane wing cellule, designed by Leo Kirste, which combined a V-interplane strut with a third strut that braced the lower wing to the middle of a somewhat elongated Brandenburg D.I fuselage. The 20.15, designed by Edmund Sparmann, utilized a more conventional wire-braced wing cellule based on his observations of German aircraft—and on the pros and cons of the Nieuport-style sesquiplane layout as employed on the Albatros D.III and D.V. Test flights showed the 20.15 wing to be more robust than that of the 20.14, in addition to which it provided a superior climb rate and flight characteristics that were vastly superior to those of the Brandenburg D.I. The 20.15's speed was not significantly higher than that of the Brandenburg, but that was remediable by installing a more powerful engine than the 185-horsepower Daimler—which it got in the form of a 200-horsepower Hiero. Production of the Phönix D.I began in August 1917, with armament consisting of two synchronized machine guns mounted on either side of the engine.

Austria-Hungary had a more poorly developed industrial capacity than Italy, and the arrival of Phönix fighters to the front was considerably delayed. In November 1917, *Oberst* Uzelac learned that fifty D.Is were in storage, still awaiting the installation of machine guns and the new Zaparka interrupter gear. Again taking matters into his own hands, he arranged to have the aircraft delivered to frontline units, where their armorers would install the armament as it arrived. By December, Phönix D.Is were entering the inventories of *Fliegerkompagnien* 4/D, 15/D, 17/D, 48/D, 54/D, and 66/D, where they were to serve as escorts to those units' reconnaissance aircraft, and at fighter *Fliks* 14/J, 30/J, 60/J, 61/J, and 63/J.

Activated in November 1917 and dispatched to Italy the following month, *Flik* 60/J was based at Grigno in the mountainous Val Sugana, about sixty miles north-northwest of Venice—a most unforgiving environment for taking off and landing. Fully equipped with Phönix D.Is of series 128, 228, and 328, the unit's initial flying personnel consisted of the commander, *Obltn.* Frank Linke-Crawford, and six enlisted pilots. Linke had already been credited with thirteen victories while flying Brandenburg D.Is and Oeffag-Albatros D.IIIs in *Flik* 41/J under the tutelage of the formidable Brumowski, and it may have been he who blooded the Phönix in fine style. On January 10, 1918, he took off in D.I 228.16, accompanied by *Stabsfeldwebel* (Master Sergeant) Kurt Gruber, another veteran with five victories to his credit, in D.I 228.24. Over Valstagna, they encountered what Linke described as a "Sopwith two-seater" accompanied by escort fighters and attacked. In short order, they sent the two-seater—which was most likely a SAML of the 115*a Squadriglia*, crewed by *Sgt.* Ubaldo Lenzi and *Sottotenente* (Second Lieutenant) Stefano Achenza—down just within the Italian side of the lines, followed shortly afterwards by one of its Nieuport escorts. It is possible that the Nieuport was from the 79*a Squadriglia*, another of whose members, *Sgt.* Antonio Reali, claimed an unidentified enemy fighter over Primolano, although it was not confirmed; Reali would eventually be credited with eleven out of the twenty-nine victories he claimed in the course of the war. Since Austria-Hungary credited shared victories in the French manner, two were added to both Linke-Crawford's and Gruber's personal accounts. Both men would add an individually scored victory to their tallies before the month was over.

Opinions varied regarding the Phönix D.I's merits. During a flight comparison in September 1917, it was judged to be superior in speed and climb rate to the Oeffag-Albatros D.III and to have better flight characteristics than the Aviatik D.I.

Phönix D.I 228.45, shown during acceptance trials at Aspern in the winter of 1917, later served as a photoreconnaissance fighter with Fliegerkompagnie 37/P in 1918. (*Österreichisches Luftfahrt Archiv*)

A German report in October stated that the Phönix "possesses totally amazing qualities, especially the quickness of maneuver and stability when throttled down. The pilot can stall the aircraft virtually on the spot and drop several hundred meters without losing control." In February 1918, however, *Flik* 60/J reported that the "D.I is not favored by pilots because the speed and climb are inferior to the Nieuport, Spad, and Sopwith fighters." *Flik* 30/J complained that it was too slow and "almost too stable for quick combat maneuvers."[22] Whatever its faults, the Phönix D.I, like the S.E.5a, the Spad XIII, and the Pfalz D.IIIa, acquired a reputation for sturdiness and for the ability to escape trouble in a high-speed dive without fear of its wings coming off, as was the case with the Nieuport 17, Albatros D.V, or Aviatik D.I. The Phönix D.I and its successors—the D.II, D.IIa, and D.III—also did much to provide Austria-Hungary's small but spirited LFT with a competitive fighter of its own design.

Four

The Triplane Craze

SOPWITH TRIPLANE AND FOKKER DR.I

World War I aviation was dominated by the biplane. Monoplanes were potentially faster, but they could not be practical until engineers could design a reliable means of bracing their wings. With a biplane, on the other hand, the two wings provided the matrix for the wooden spars and struts and the bracing wires that held the whole arrangement together.

The potential of the cantilever wing structure was first proved when Professor Hugo Junkers's all-metal J1 monoplane first took to the air in December 1915, but that and other cantilever airframes would not work on fighter planes until they could be made light enough to be propelled by engines producing only 80 to 160 horsepower. Eventually a fighter with wings of wooden cantilever structure did take to the air, and it proved itself a match for any biplane in the sky. Ironically, however, the Fokker Dr.I was not a monoplane but more a step in the opposite direction—a triplane!

The process that led to what amounted to a brief "triplane craze" from which the Fokker Dr.I emerged began with a series of aircraft from the British firm of Thomas O. M. Sopwith and his chief

designer, Herbert Smith. The first was a handsome single-bay two-seat biplane with a 100-horsepower Clerget 9Z rotary engine, which was passed by the Sopwith experimental department on December 12, 1915—the same day that the Junkers J1 first flew. This was somewhat ironic, since the Sopwith was of conventional wood and wire construction, but its layout was advanced for its time, with a synchronized Vickers machine gun for the pilot and a flexibly mounted Lewis machine gun for the observer. It also had an adjustable tail plane, the incidence of which could be modified by means of a handwheel in the pilot's cockpit. The 1 1/2-Strutter, as it was dubbed because of its W-shaped cabane struts, was the first true two-seat fighter when it entered production in February 1916.

Although the 1 1/2-Strutter had originally been ordered by the Admiralty, the need for air superiority for the coming British Somme offensive resulted in twenty-seven production aircraft being transferred from the Royal Naval Air Service to the Royal Flying Corps in the summer of 1916. Formed to use the new type on April 13, No. 70 Squadron flew its first combat mission on the opening day of the Somme offensive, when four 1 1/2-Strutters took off at 0600 hours on July 1, 1916. They met no aerial opposition and returned with useful information on German troops' movements and rail activity around Cambrai. On the following day the same four Sopwiths, accompanied by Martinsyde G.100 Elephants of No. 27 Squadron, were reconnoitering Cambrai and Douai when they were attacked by six German aircraft. One Sopwith that was lagging behind the group, piloted by 2nd Lt. Jack Manley, was picked out for special enemy attention; but he and his observer, 2nd Lt. Robert C. Oakes, managed to fend off their attackers, and all the aircraft returned safely to Fienvillers aerodrome.

The first casualty occurred five days later when Capt. Guy L. Cruikshank, with his nineteen-year-old cousin 2nd Lt. Andrew J. T. Cruikshank acting as his observer, carried out another reconnaissance to Cambrai and became separated from the two accompanying aircraft. Cruikshank's plane was attacked by three Fokker Eindeckers and had to fight a twenty-two-minute running battle, during which he made it to Allied lines, but with his plane damaged and his observer seriously wounded. Captain Cruikshank landed at Saint-Omer in the hope of finding better medical facilities for his cousin, but Lieutenant Cruikshank died of his wounds later the same day.

Two more Sopwiths were lost to antiaircraft fire on July 7, and another was lost on the nineteenth, Lt. Henry R. Hele-Shaw and his observer, nineteen-year-old Robert Oakes, being shot down and killed by *Ltn.* Kurt Wintgens of *Kampfeinsitzer-Kommando Vaux*. Five of the 1 1/2-Strutters attacked a Roland C.II on July 20, but the only result was that Manley was wounded in the face and

Establishing the British two-seater formula of a synchronized Vickers machine gun forward and a Lewis on a Scarff ring mount aft, with center section panels removed to give the pilot a better upward view, Sopwith 1 1/2-Strutter A'1924 served in No. 70 Squadron RFC until written off in a crash on October 20, 1916. (*RAF Museum*)

shoulder and had to be hospitalized. The squadron finally scored its first victory as four Sopwiths were escorting No. 27 Squadron's Martinsydes on a bombing mission to Hervilly on July 29, when its aircraft were credited with driving an enemy plane down out of control.

Good though it was, the Sopwith 1 1/2-Strutter did not remain an effective fighter for long. By the time a second unit, No. 45 Squadron, arrived in France on October 15, German biplane scouts such as the Halberstadt D.II and Albatros D.II were outperforming it in both speed and maneuverability. From then on its versatile airframe was employed in other tasks—as a reconnaissance plane and as a bomber.

Sopwith's next fighter design originated with a small, lightweight, single-seat, single-bay biplane, powered by a fifty-horsepower Gnome rotary engine, which had been designed in the autumn of 1915 by test pilot Harry Hawker, and which was generally referred to as "Hawker's Runabout." The frail machine produced such remarkable performance, including a maximum speed of 84.6 miles per hour, that the Sopwith staff began working on a smaller, more compact and robust version, to be powered by an eighty-horsepower Rhône engine for military purposes. The prototype that emerged from their efforts was cleared on February 9, 1916, and subsequent testing by the Admiralty resulted in a glowing report. The plane had a maximum speed of 110 miles per hour at 6,500 feet and climbed to 10,000 feet in 12 minutes and 29 seconds; but what really distinguished it was its ease in handling, combined with outstanding maneuverability, as well as the excellent visibility afforded from the cockpit—all wrapped up in one of the most nicely balanced and aesthetically pleasing aircraft of the war. The RNAS put in its first order for the little fighter in April 1916, and when Gen. Hugh Trenchard read a copy of the Admiralty's report on it, he penciled in a concise comment of his own: "Let's get a squadron of these."[1]

Armed with a single Vickers machine gun using Sopwith-Kauper interrupter gear, the plane was officially designated the Sopwith Scout, but was more popularly known by another name. Allegedly, it was then-Col. Sefton Brancker of the RFC who, upon seeing the new fighter alongside its larger two-seater forebear, exclaimed: "Good God! Your 1 1/2-Strutter has had a pup."[2] Whatever its origin, the sobriquet "Pup" rapidly proliferated in spite of all official efforts to discourage its use.

The first Pups arrived at No. 1 Wing RNAS, based at Dunkirk, in July 1916. The single-seater's first success may have occurred on September 24, when Flt. Sub-Lt. Stanley James Goble sent an LVG two-seater down out of control near Ghistelles. This was the third victory for the Australian-born Goble, who had

The Sopwith Pup flew well regardless of which engine was installed—in the case of the plane shown, a 100-horsepower Gnome monosoupape—but its single machine gun proved a weakness in 1917. (*RAF Museum*)

claimed two earlier ones while flying Nieuports and would survive the war with a total of ten.

On the twenty-fifth, Flt. Sub-Lt. Edward Rochfort Grange scored the type's second success, as described in a letter to the author sixty-five years later:

> My first confirmed victory was a large seaplane shot down in the sea about five miles off the coast at Ostend. I saw [Flt. Lt. Redford Henry] Mulock and gave the story verbally to him, who relayed the location to our Royal Navy, who picked up the remains and brought them to St. Pol aerodrome, Dunkirk. The French gave me the *Croix de Guerre*.[3]

Grange's victim was a Sablatnig SF 2 of *Seeflugstaffel* I. The crewmen, *Leutnante zur See* (Ensigns) Otto Soltenborn and Hans Röthig, were killed.

By the end of 1916, Pups were being operated with considerable success by No. 8 Squadron RNAS. Also used by RFC squadrons in 1917, the Pup earned the affection of all who flew it, its performance in dogfights being particularly distinguished by its ability to maintain altitude in a turn. Its only serious weaknesses were its single gun and the slow rate of fire of its synchronizing mechanism.

By the spring of 1917 the German fighters, using more powerful engines and armed with twin machine guns, were rendering the Pup obsolete. By then, however, another fighter had emerged from the Sopwith stable, which was similar to, and at the same time strikingly different from, the Pup. Back in the spring of 1916, Herbert Smith had set to work designing a plane that could climb faster, fly higher, maneuver as well as, if not better than, the present Pup, and, if possible, afford better visibility for the pilot. The result of his efforts, which emerged from the Sopwith hangar on May 30, 1916, was not a biplane but a triplane. Even more surprising than the fighter's configuration was the fact that the added wing did not adversely affect its performance.

Since the overall wing area was being distributed among three wings instead of two, Smith felt that he could make all of them of relatively narrow chord, which provided a correspondingly small change in the center of pressure at various angles of attack. The narrow wings, moreover, interfered less with the pilot's view, and the middle set was mounted in line with the pilot's eyes. In addition to that, the wings' narrow chord gave them a high aspect ratio with an efficient ratio of lift to drag. They were constructed with a shorter span and with ailerons on all three wings, both of which factors endowed the plane with an increased roll rate. The Triplane's maneuverability was further enhanced by Smith's use of a 130-horsepower Clerget engine and his placement of the heaviest weights—engine, pilot, fuel, and armament—near the center of gravity.

Although the Sopwith Triplane prototype was built in only three months, test pilot Harry Hawker was so pleased with it that he looped it just three minutes after

After some time in French use, the Clayton and Shuttleworth-built Sopwith Triplane N5387 was assigned to No. 1 Squadron RNAS at Bailleul and christened *Peggy*. N5425 behind it was flown at various times by Flight Sub-Ltns. Herbert V. Rowley (nine victories) and Stanley W. Rosevear (twenty-five). (*RAF Museum*)

takeoff. Two weeks later the prototype, N500, was dispatched to A Squadron of No. 1 Naval Wing at Furnes, near Dunkerque, for frontline evaluation and flew its first combat mission just fifteen minutes after its arrival.

After some flights over the lines, the prototype Triplane got into its first serious fight on July 1. Its pilot at the time, Lt. Roderic Stanley Dallas, a twenty-four-year-old Australian from Mount Stanley, Queensland, attacked a German two-seater six kilometers off La Panne and sent it down out of control for his fourth victory. On September 30, Dallas took N500 up again and was credited with a German fighter out of control southwest of Saint Marie Capelle. Dallas would later score thirteen more victories in Triplanes with No. 1 Squadron RNAS, as A Squadron was redesignated after being fully equipped with the new fighter in December 1916.

In the early months of 1917, Sopwith Triplanes dazzled their opponents with their performance, and even during the general slaughter of British aircraft known as "Bloody April," they took a disturbing toll of the German *Jagdstaffeln*. As had been the case in 1916 with the Nieuport sesquiplane fighter, the *Inspektion der Fliegertruppen* reacted by requiring German aircraft manufacturers to produce a triplane fighter to counter the Sopwith. The response was perhaps greater than the triplane, with intrinsic limits on its development potential, warranted, but by the end of 1917 virtually all German firms had come up with at least one triplane design. Only two, the Fokker Dr.I and the Pfalz Dr.I, would enter production, and only ten of the latter were ever built.

The most famous triplane in history was created by the man who was the least enthusiastic about building one. In 1917 Anthony Fokker had been investigating the possibilities in concepts of a wooden box-spar wing structure pioneered by Swedish-born engineer Villehad Forssman. This offered the drag-reducing advantages of Dr. Junkers's all-metal cantilever wing structures, and incorporated a lighter weight airframe commensurate with the power that engines of the time could provide. Fokker envisaged a monoplane or a sesquiplane fighter as the ultimate airframe on which to realize the potential of that structure, but in April 1917 he was asked by *Idflieg*—rather to his chagrin—to design a triplane fighter like the Sopwith.

As with the Sopwith Triplane, the fuselage of which was little changed from that of the Pup, Anthony Fokker's triplane was based on an experimental (*Versuchs*) biplane that he had been building for Austria-Hungary, the V4. A simple steel-tube structure covered in fabric, the V4's fuselage offered better stability than the more streamlined fuselages of Fokker's experimental V1, V2, and V3. To that,

Fokker added three wings of torsion-box structure, but in the interests of hastening production and saving weight, he covered them with fabric rather than plywood. Further lift was provided by the undercarriage axle, which was also of aerofoil section. Powered by a 110-horsepower Rhône rotary engine, the altered Fokker V4 displayed such outstanding performance—especially in regard to climb and maneuverability—that Fokker was authorized to proceed with two preproduction aircraft for frontline evaluation, work on which began on July 11. Among the principal modifications on the F.Is (as the preproduction V4s were designated) were increased wing area, more effective, balanced ailerons, and the addition of interplane struts. The F.Is were delivered to *Jagdgeschwader* I on August 28, 1917, and allocated to Germany's two most illustrious aces at the time—JG I's *Geschwaderkommandeur*, *Rittmeister* Manfred *Freiherr* von Richthofen, and *Ltn.* Werner Voss, commander of *Jasta* 10.

Richthofen was not in the best of health or spirits at the time. He had been wounded in the head on July 6, and he was still feeling the effects. He was also depressed because the Albatros D.V, with its unreliable wing structure, remained the only type available to his wing, while the Allied fighters were improving in quality and increasing in quantity. Flying D.V 2059/17 on August 26, he had managed to shoot down a Spad VII of No. 19 Squadron for his fifty-ninth victory, but in a letter to home two days later, he admitted:

> I have made only two combat flights [since returning to the front] and both were successful, but after each flight I was completely exhausted. During the first one I almost got sick in the stomach. My wound is healing frightfully slowly; it

is still as big as a five-Mark piece. Yesterday they removed another splinter of bone; I believe it will be the last.[4]

The arrival of Fokker F.I 102/17 did much to raise the Baron's spirits. He and his "gentlemen" of *Jasta* 11 had already been flying Fokker D.Vs to accustom themselves to the idiosyncrasies of rotary-engine fighters before they arrived in quantity. Voss wasted no time in taking up his Fokker F.I 103/17 and flying it around the vicinity of Marcke to demonstrate its capabilities to all four *Staffeln* of JG I.

British bombers raided German aerodromes west of Courtrai on the night of August 31, but on the following day Richthofen led his pilots into battle in F.I 102/17. He described what followed:

Flying the triplane for the first time [in combat], I and four gentlemen of *Staffel* 11 attacked a very courageously flown British artillery-spotting aircraft. I approached [until] it was 50 meters below me and fired 20 shots, whereupon the adversary went down out of control and crashed on this side [of the lines] near Zonnebeke.

Apparently the adversary had taken me for a British triplane, as the observer stood up in his machine without making a move to attack me with his machine gun.[5]

Richthofen's sixtieth victory had been over R.E.8 B782 of No. 6 Squadron RFC. Its pilot and observer, 2nd Lts. John C. B. Madge and Walter Kember, were both killed.

Voss and Richthofen both took up their triplanes on the morning of September 3, when a JG I patrol pounced on a flight of Sopwith Pups of No. 46 Squadron RFC. One of its members, Lt. Arthur Stanley Gould Lee, described the action in a letter home:

The news is not good. The squadron has take[n] a hammering. . . . The first patrol ran quickly into trouble, five of 'A' Flight . . . met Richthofen's Circus and had a hectic scrap. The Pups were completely outclassed by the D-Vs, and most of their share of the fighting consisted of trying to avoid being riddled. Mac [2nd Lt. K. W. McDonald] and Bird were seen to go down in Hunland. [Lieutenant Richard] Asher might have reached the Lines. Two chaps who got away, badly shot about, said that one of the Huns was flying a triplane, coloured red. It must be a captured naval Tripe, I suppose.[6]

MacDonald was brought down by *Ltn.* Eberhardt Mohnicke of *Jasta* 11 for his sixth victory, while Richthofen got into a long-running battle with Pup B1795 (Z), finally bringing it down south of Bousbecque. Richthofen reported:

I was absolutely convinced that in front of me I had a very skilful pilot, who even at 50 meters did not give up, [but] continued to fire and, even when flattening out [before landing] fired at an infantry column, then deliberately steered his machine into a tree. The Fokker triplane F.I 102/17 is absolutely superior to the British Sopwith.[7]

The uninjured pilot, 2nd Lt. Algernon F. Bird, was taken prisoner. Richthofen, elated at the rare prospect of meeting a live victim, hastened to the scene, as did Anthony Fokker, eager to record the event on his cinema camera. The old Baron was evidently back, smiling confidently. Bird, too, had reason to present the camera with a nervous but self-satisfied smile. Although a vanquished prisoner of war, he had done his duty to the last, including the destruction of his machine, so that the enemy would not get it intact—and, on top of it all, he could

Anthony Fokker sits in the cockpit of F.1 102/17 during a flight demonstration at Marckebeke on August 31, 1917. Observing, from left, are *Generalmajor* Friedrich Karl von Lossberg, *Rittm*. Manfred von Richthofen—who would give it its combat debut the next day—and JG1 adjutant Karl Bodenschatz, commander of *Jasta* 6. (*Greg Van Wyngarden*)

legitimately tell his grandchildren that he had survived a run-in with the "bloody red baron."[8]

Voss also had a successful morning, bringing down a Sopwith Camel of No. 45 Squadron, flown by Lt. Aubrey T. Heywood, north of Houthem at 0925 hours for his thirty-ninth victory. In the afternoon of September 5 Voss, with *Ltn*. Erich Löwenhardt as his wingman, attacked a patrol of Pups from No. 46 Squadron and, though unable to shoot any down, he outmaneuvered them and put bullets into practically everyone before breaking away. Later, at 0350, he shot down another of No. 46 Squadron's Pups over Saint-Julien, although its pilot, 2nd Lt. Charles W. Odell, survived. Fifty minutes after that, Voss sent a Caudron G.6 of French *escadrille* C.53 down in flames over Bixschoote.

On September 6 Voss shot down an F.E.2d, flown by Lt. John O. Pilkington and Air Mechanic 2nd Class Herbert F. Matthews of No. 20 Squadron, over Saint-Julien. On that same day Richthofen went on a four-week leave, designating *Obltn*. Kurt von Döring commander of JG I in his absence. Voss, now the second-ranking German ace after Richthofen, was in an ideal position to catch up with his friendly rival, and he made the most of it on September 10 by downing two Camels flown by 2nd Lts. Arthur J. S. Sisley and Oliver C. Pearson of No. 70 Squadron, followed by a Spad VII twenty minutes later, killing *Sgt*. Jules Tiberghein of Spa.31.

At 1030 hours on September 11, Voss and *Obltn*. Ernst Weigand—who generally dealt with administrative matters for *Jasta* 10 on Voss's behalf—each claimed a "new type Spad" over Langemarck, and Voss downed a Camel over Saint-Julien at 1625 that afternoon, killing Lt. Oscar L. McMaking of No. 45 Squadron.

While Richthofen was away on leave, Fokker F.I 102/17 went to one of his most trusted lieutenants, Kurt Wolff, who at age twenty-two was victor over thirty-three Allied aircraft and commander of Richthofen's old red-nosed squadron, *Jasta* 11. Wolff had been wounded in the left hand during a fight with Sopwith Triplanes of No. 10 Squadron RNAS on July 11. He returned *to Jasta* 11 exactly two months later, on September 11. On the following day, Wolff learned that Kaiser Wilhelm II had promoted him to *Oberleutnant*, but wrote his fiancée:

> I am depressed only because I receive this honor without having shot down one [since July 7]. For until now I have already had the bad luck of scuffling with about twenty Englishmen and have not brought down one of them.[9]

On the afternoon of September 15, Wolff led a patrol in the new triplane when they encountered a flight of eight British biplane scouts. Unknown to Wolff, his opponents were once again from Naval 10, which, significantly, had replaced its Sopwith Triplanes with Camels. At 0315 hours Flt. Lt. Desmond F. Fitzgibbon had taken the flight up, rendezvoused with five de Havilland D.H.4 bombers at 1610 hours and escorted them over the lines to Bryke Wood. At 1630 the British spotted four German fighters and went down to investigate, only to see them dive off. At that point the Camel pilots found themselves under attack by what they described as five Albatros D.Vs and four triplanes (despite the fact that Wolff's was the only German triplane in the area at that time).

A general melee ensued, during which Flt. Lt. Norman M. MacGregor closed to within twenty-five yards of an enemy triplane and sent tracers into it before turning away to avoid colliding with his quar-

ry. When last seen by the British, MacGregor's victim was going down in a nose dive. His fifth credited victory was indeed Kurt Wolff, whose body was found in the wreck of F.I 102/17 near the village of Nachtigal, north of Wervicq. Following a funeral service befitting a knight of the *Orden Pour le Mérite* at Saint Joseph's Church in Courtrai, Wolff's remains were sent to his parents' home in Memel (now Klaipeda, Lithuania).

In the meantime, Voss had done nothing to close the gap on Richthofen's score since September 11. He was still keen to do so, but one of his *Jasta* 10 pilots, *Ltn.* Aloys Heldmann, noticed that the past two weeks' activity was taking its toll: "He was on edge; he had the nervous instability of a cat. I think it would be fair to say that he was flying on his nerves. And such a situation could have but one end."[10]

On Sunday, September 23, Voss took off early, and at 0930 he destroyed a D.H.4, crewed by 2nd Lts. Samuel L. J. Bramley and John M. DeLacey of No. 57 Squadron, south of Roulers for his forty-eighth victory. Upon returning, he joined his brothers Otto and Max for lunch, during which they discussed plans for his upcoming leave. At 1800 that afternoon Voss led the last patrol of the day, accompanied by *Ltn.* Gustav Bellen and *Vzfw.* Friedrich Rüdenberg, while *Obltn.* Weigand followed, leading *Ltne.* Julius Bender and Max Kühn in a second flight. There were cloud layers all over the front as the *Jasta* 10 formation headed toward Zillebeke, but there were plenty of British aircraft in the area, offering Voss the enticing prospect of ending the day with a nice round score of fifty.

On the other side of the lines, eleven S.E.5as, comprising B and C Flights of No. 56 Squadron RFC, had left their aero-

Although Manfred von Richthofen started September 1917 with two victories scored in Fokker F.I 102/17, it was his rival, *Ltn.* Werner Voss of *Jasta* 10, who stole the show that month, culminating in a fatal last dogfight in F.I 103/17 on the twenty-third that made him and the Fokker triplane the talk of the Western Front. (*Greg Van Wyngarden*)

drome at Estrée Blanche at 1700 hours. Leading B Flight toward Houthulst Forest, Capt. James T. B. McCudden spotted a DFW C.V and led his flight to attack it, sending it diving vertically into the ground northeast of Houthem. The flight was about to attack six Albatros scouts when McCudden noticed an S.E.5a over Poelcapelle with an enemy triplane on its tail. "The S.E. certainly looked very unhap[p]y," McCudden remarked, "so we changed our minds about attacking the six V-strutters, and went to the rescue of the unfortunate S.E."[11]

The "unhappy" SE pilot was Capt. Harold Alan Hamersley of No. 60 Squadron, two flights of which were returning from a patrol at 1825 when Hamersley, flying behind Capt. Robert L. Chidlaw-Roberts, noticed what he thought was a Nieuport being dived on by an Albatros. Hamersley reported afterwards:

I swung across to attack the DV and what I had thought was the Nieuport turned towards me and I realized that it was a "Tripe" as we spoke of them. It was a little below me and I put my nose down and opened fire. The 'Tripe' passed under me and as I zoomed and turned the Hun was above me and heading straight at me, firing from about 30 degrees off the bow. There was a puff of smoke from my engine, and holes appeared along the engine cowling in front of me and in the wings. Realizing I could do nothing further in the matter, I threw my machine into a spin. The Hun followed me down, dicing at me while I was spinning and I had to do an inverted dive to get away.[12]

Chidlaw-Roberts tried to aid Hamersley, but only fired a few rounds before the German triplane whipped around behind him and shot up his rudder bar. Thereupon, Chidlaw-Roberts said, "I retired from the fray and that is all I saw of it."[13]

At about that point McCudden's flight intervened, he diving on the triplane from the right and 2nd Lt. Arthur P. F. Rhys-Davids from the left. Voss saw them and reacted with what McCudden described as

> not a climbing nor Immelmann turn, but a sort of flat half-spin. By now the German triplane was in the middle of our formation, and its handling was wonderful to behold. The pilot seemed to be firing at all of us simultaneously, and although I got behind him a second time, I could hardly stay there for a second.

Captain Keith L. Caldwell, No. 60 Squadron's New Zealand-born commander, had led the rest of his patrol to Hamersley's assistance, but at that point he noticed that

> a flight of 56 Squadron's SEs (and a very good one too) had taken Voss off its tail and were busy with him in their midst. It was then really 56's affair and six to one was pretty good odds we felt. We were more or less spectators and in my opinion there was little room to join in.[14]

Nearby, C Flight of "Fighting Fifty-Six," led by Capt. Geoffrey Hilton Bowman, had had several fights of its own when its pilots noticed that McCudden's B Flight was having a surprisingly difficult time with two German fighters—a triplane and an Albatros D.V that had entered the fray, and which had cooperated rather well with Voss for a time. Firing at targets of opportunity, Voss got on 2nd Lt. Verschoyle P. Cronyn's tail, and although the Canadian managed to dive away, his wings were so badly damaged that he spent the night uncontrollably perspiring

and unable to sleep. Lieutenant Keith K. Muspratt was also driven out of the fight and was compelled to land at No. 1 Squadron's aerodrome with a bullet in its radiator. That still left Voss having to deal with such paladins as McCudden, Rhys-Davids, Bowman, Reginald T. C. Hoidge, and Richard A. Maybery. At one point McCudden saw the triplane "in the apex of a cone of tracer bullets from at least five machines simultaneously"; yet, as Maybery expressed it, "he seemed invulnerable."[15]

The Albatros had left the scrap by then, although none of No. 56 Squadron's pilots claimed to have shot it down. Voss, however, was still full of fight and put his plane through a maneuver that several of his opponents remarked about. Bowman, for example, said:

> I, myself, had only one crack at him: he was about to pass broadside on across my bows and slightly lower. I put my nose down to fire him a burst and opened fire, perhaps too soon; to my amazement he kicked on full rudder, without bank, pulled his nose up slightly, gave me a burst while he was skidding sideways, and then kicked on opposite rudder before the results of this amazing stunt appeared to have any effect on the controllability of his machine. Rhys-Davids was then on his tail.

Rhys-Davids, who had already replaced two drums of Lewis ammunition in the course of the fight, fired another drum and an equal quantity of Vickers rounds into the triplane at such close range that the two aircraft almost collided. Rhys-Davids reported:

> I saw him next, with his engine apparently off, gliding west. I dived again and got one shot out of my Vickers; however, I reloaded and kept in the dive. I got in

another good burst and the triplane did a slight right-hand turn, still going down. I had now overshot him (this was at 1,000 feet), zoomed, and never saw him again.[16]

At that juncture Rhys-Davids encountered a red-nosed Albatros to the southeast and attacked that as well, forcing it to crash-land. Its pilot was probably *Ltn.* Karl Menckhoff of *Jasta* 3, who survived. Voss was less fortunate. After having dazzled his opponents for some ten minutes, it is possible that fuel exhaustion—after an hour and a half of combat flying (about the limit of the Fokker triplane's endurance)—had been the ultimate cause of his death before Rhys-Davids's guns. Voss's F.I came down at Plum Farm, about seven hundred meters north of Frezenberg, and because fighting was still going on in the area, his body was hastily buried near the crash site. The wreckage was allocated the captured number G.72 and provided British intelligence with its first, albeit incomplete, look at the new aircraft type, although only a few instruments could be salvaged.

While neither of the Fokker F.Is had long careers, they left an impression on both sides of the lines that has since been sealed into aviation annals. Voss's spectacular last dogfight made him and his triplane the talk of Nos. 56 and 60 Squadrons, and arguably even more legendary among his enemies than among his own countrymen. Richthofen's first successes in the type gave the Fokker triplane a virtual seal of approval for full-scale production and an association with the Red Baron that persists to this day—at the expense of the Halberstadt and Albatros scouts in which he gained the lion's share of his eighty victories.

Fokker Dr.Is made their way into JG I and other units in October, during which time Richthofen had his pilots train in Fokker D.Vs as a transitionary measure to accustom themselves to compensating for the torque of the rotary-engine aircraft. Patrols were being flown over the lines by the end of the month, but on October 29 *Jasta* 11 lost *Vzfw.* Josef Lautenschlager, who was shot down and killed by a German fighter. After personally investigating the matter, Richthofen ascertained that the *Kogenluft-Nachrichtenwesen* (air intelligence and communications service) had failed to inform squadrons in the IV *Armee* area about the new Fokker, and the German pilot had mistaken Lautenschlager's Dr.I for a Sopwith Triplane. Consequently, Richthofen not only saved the pilot from a court-martial but also arranged to have him transferred to JG I, where he allegedly went on to a creditable career—although his name has never been disclosed.

On the following morning, October 30, Richthofen was leading a patrol when his brother Lothar was seen to glide down with his engine off, to make a successful forced landing near Zilverberg. Minutes later, as Manfred tried to land, his triplane broke up and he crashed. Richthofen emerged from the wreckage unharmed and subsequently learned just how lucky he had been. On that same day, *Ltn.* Heinrich Gontermann, thirty-nine-victory ace and *Orden Pour le Mérite* recipient, was test-flying his new Dr.I 115/17 over *Jasta* 15's aerodrome at La Neuville when the ribs broke away from the upper wing spar and he crashed, dying of his injuries that evening.

On October 31, nineteen-year-old *Ltn.d.Res.* Günther Pastor of *Jasta* 11 crashed to his death near Moorsele after the upper wing of his triplane, Dr.I 121/17, suffered structural collapse. On November 2, an alarmed *Idflieg* grounded all Fokker Dr.Is pending an investigation.

The conclusion reached by the crash commission was that the triplane was intrinsically sound, but that condensation collecting inside the wing had compromised the integrity of the glue, which played as important a role as the nails in holding it together. Among other things, Fokker was ordered to strengthen the attachment of the ribs to the box spar by putting greater reliance on transverse nailing rather than on glue, ensuring the integrity of the joints, and also was ordered to varnish the inside of the wing to protect it from water damage. Such measures solved the worst of the Dr.I's problems, and by December deliveries had resumed—not only to JG I but also to other *Jagdstaffeln*.

Although too slow to fully realize its role as an offensive weapon, the Fokker Dr.I became the principal instrument for achieving local air superiority during the German offensive in March 1918. It never fully eclipsed the Albatros or Pfalz scouts that supplemented its meager numbers (a peak of 171 at the front in April 1918), but in the right hands it could be a deadly weapon. A few German pilots, most notably Josef Jacobs, were still flying Dr.Is in preference to faster but less nimble types right up to the final months of the war. Ultimately, the triplane fighter proved to be a developmental dead end; but the exploits of the Sopwith and Fokker pilots gave them a notoriety in their own time, as well as a fame in postwar years all out of proportion to their numbers and actual importance.

Among the Dr.I's many victims was the first pilot to achieve success in a triplane fighter. Following the incorporation of the RNAS into the Royal Air Force in April 1918, Maj. Roderic S. Dallas flew S.E.5as as the commander of No. 40 Squadron. In that capacity he brought his final tally up to at least thirty-two before being killed in action by three Fokker Dr.Is on June 1. Dallas was the sixth of seven victories credited to *Ltn.* Johannes Werner of *Jasta* 14—and the last before Werner got the chance to trade his triplane in for a newly delivered Fokker D.VII biplane on June 9.

Five

Glimpses of Things to Come

FIGHTERS OF 1918

The year 1918 began with months of precarious promise for the Central Powers. Russia, torn asunder by revolution and represented by a Bolshevik regime since November 1917, had sued for peace. The Italian army, disastrously routed at Caporetto on October 26, had retreated to the Piave River and seemed on the verge of ultimate collapse. The United States had declared war on Germany on April 6, 1917, but only the first inexperienced contingents of its small though rapidly expanding army had arrived in the front at the beginning of 1918. The Germans saw the overall situation as an opportunity—perhaps their last—to mobilize their troops on the Western Front for a final all-out push to eliminate the British army, take Paris, and compel both Britain and France to sue for peace before the Americans could bring their resources fully into play.

When Operation Michael was launched on March 21, 1918, German ground forces in the West were the most powerful and murderously efficient that they had ever been since the war began. For the air arm that was expected to clear the skies for the offensive, it was a different matter. Although there were now three

crack *Jagdgeschwader* organized to achieve local air superiority and the number of *Jagdstaffeln* doubled through the so-called *Amerika-Programm*, their personnel and equipment betrayed the realities behind the rapid expansion. Most of the *Amerika-Programm Jastas* formed after June 1917 consisted of a veteran commander and perhaps a small cadre of experienced men, surrounded by hastily trained pilots who would soon have to learn their trade the hard way. The best fighter on hand was the Fokker Dr.I, back in *Jasta* strength since Anthony Fokker had rectified the shoddy production techniques that had lain behind the wing failures they had suffered in October 1917. The Dr.Is, however, soon proved to be less than the world-beaters that were needed to maintain air superiority on the offensive, being slower than all Allied fighters except the Sopwith Camel—and the Camel could match the Fokker in a dogfight. Complementing this dubious vanguard were Albatros D.Vas and Pfalz D.IIIas, which at best were no more than an even match with their Allied counterparts.

Ironically, on the other side of the lines, three new fighters were already making their way to Allied frontline units. Two of them fell disappointingly short of expectations, however, and the third was never given the full utilization it deserved.

Developed in the summer of 1917, the Morane-Saulnier AI was another in a long line of rotary-engine monoplanes built by that company. Generally conventional in construction, the AI was powered by a 150-horsepower Gnome monosoupape 9N engine, with a series of stringers fairing the fuselage smoothly past the rounded metal cowling to a point at the tail. The slightly swept-back upper wing was mounted above the fuselage in a parasol arrangement by means of a truss-like strut arrangement. Static testing of the structure yielded a safety factor of 8.5, and during flight tests at Villacoublay in August 1917, the prototype climbed to 3,000 meters (9,840 feet) in 7 minutes and 45 seconds and reached a speed at that altitude of 134.5 miles per hour. That performance was considered phenomenal for a rotary-engine fighter, and since the Spad XIII was experiencing trouble with its geared Hispano-Suiza engine, the French ordered the Morane-Saulnier AI into production, with anywhere from 1,100 to 1,300 being built. The earliest AI variant, armed with a single synchronized Vickers .303-inch machine gun, was designated the MoS.27.C1, while a later twin-gun version was called the MoS.29.C1.

The Morane-Saulnier's reputation preceded it when *Escadrille* N.156 learned that it would be the first to receive the new fighter on February 4, 1918. Among the first pilots sent to ferry them in was *Caporal* David Endicott Putnam, a Lafayette Flying Corps pilot who was serving in the unit along with fellow American volunteers *Caporaux* Walter John Shaffer and Wallace C. Winter. When Shaffer went to Plessis-Belleville to pick up his plane on February 8, he met Putnam and wrote that: "He was greatly elated over the speed and climbing ability of the Morane and told me how nicely it worked in the air." Shaffer's impression was that "it was tiny, but it is so beautifully proportioned and the lines so racy-looking that a more beautiful 'zang' was never built."[1]

By February 9, N.156 had been fully reequipped and was redesignated MS.156. Two other *escadrilles*, MS.158 and MS.161, were fully equipped with MoS.27s by March 4 and February 21, respectively. One of MS.158's Americans, *Caporal* Rufus R. Rand Jr., was equally impressed after a few frontline flights, stating that with a few alterations, it could be one of the best fighters on the Western Front.

Jubilation gave way to trepidation on February 26, however, when *Lt.* Jean Toutary, MS.156's executive officer, lost his wings while performing aerobatics over the squadron aerodrome near Châlons-sur-Marne. Then, on March 8, five of MS.156's planes encountered two German two-seaters over the lines, and two of them dived to attack. One of the Morane-Saulnier pilots discovered that his gun was jammed and pulled up to clear it. The other pilot, *Caporal* Winter, continued his dive until suddenly his wings folded up, and he plunged like a stone to the ground.

Wally Winter's death saddened the *escadrille* and raised a lot of uneasy questions regarding the MoS.27's capabilities. Postwar German records reveal a curious anomaly—a Morane-Saulnier credited to *Ltn.* Julius Keller of Royal Saxon *Jasta* 21. Had he come to the rescue of the two-seaters and fired on Winter from above and behind unnoticed by Winter's French squadron mate? Whatever the actual cause, the high initial expectations for the Morane-Saulnier AI plummeted as fast as Winter's broken plane.

Most of MS.156's pilots resumed flying missions in Nieuport 27s, but at least one of them was not so readily intimidated by the MoS.27. On March 14, Shaffer noted enviously that: "My roommate, Putman [*sic*], was in a fight today—the first with a Morane and I am anxious to get at the Hun myself."[2] Dave Putnam, who had just been promoted to *sergent*, attacked and claimed an Albatros near Nauroy, although it was only counted as a "probable." On the following day he downed a Rumpler over Beine, which was credited as his third victory.

As combat around Reims intensified, Putnam grew more aggressive, but his tendency to fight deep in enemy territory prevented most of his claims from being homologated. On April 12, he attacked a flight of German fighters over Saint-Hilaire-le-Petit and claimed two of them, but neither was confirmed. He claimed a triple victory on April 23, but again it went unconfirmed. On May 15, he teamed up with *Caporal* David Guy to shoot down a German plane over Somme-Py, after which Guy claimed another over Nogent l'Abbesse. Again, both victories were only recorded as probables. On that same day, *Caporal* Émile Boucheron of MS.158 crashed while trying to land his Morane-Saulnier at Maisonneuve.

The first twin-gun MoS.29s arrived at MS.156 on May 18, but by then their cause was lost. Two days later, Shaffer wrote:

> And now we have changed airplanes again, getting Spads this time. . . . The Moranes, am sorry to say, have been given up, owing to their weak construction which would not stand the strain

"chasse" work entails. I say sorry, because not only was it fast, but so small that as one pilot said, it could be maneuvered around a clothes pin.[3]

Escadrilles 158 and 161 had already given up the last of their MoS.27s on the previous day, May 19, and were redesignated Spa.158 and Spa.161. Subsequent production Morane-Saulnier AIs had their wing structures bolstered with extra wires or struts, but for the rest of the war they were relegated to the advanced training role.

Another elegant French newcomer, the Nieuport 28, represented a break at last from its series of nimble but fragile sesquiplane fighters, which by mid-1917 were reaching the end of their development potential. At that time, Nieuport had tried a two-spar lower wing on its Nieuport 24 in an attempt to make it rugged enough to accommodate a 160-horsepower Gnome rotary engine. More fundamental changes were deemed necessary, however, and on June 14, 1917, a new Nieuport took to the air—this time a true biplane, with a stringered fuselage similar to the 24 and 27 but longer and slimmer, although the tail surfaces were virtually identical to those of the 24 and 27. The leading edges of the wings were covered with three-ply laminated woodsheet and the ribs capped with tapering ply strips. The smaller lower wings held the ailerons.

Test-flown in November 1917, the final version of the Nieuport 28.C1 was one of the best-looking aircraft of World War I; but in spite of a creditable performance, including a maximum speed of 128 miles per hour, the *Aéronautique Militaire* did not regard it as sufficiently improved over the Spad XIII to warrant production. That should have been the end of the story, but like the Hanriot HD.1, which became the

Captain James Norman Hall of the 94th Aero Squadron taxies his Nieuport 28 N6153 onto Gengoult aerodrome for takeoff. On April 29, 1918, the former Lafayette Escadrille pilot scored his fourth victory in concert with future American ace of aces 1st Lt. Edward Rickenbacker. Hall's plane was marked with the 94th's "Hat in the Ring" emblem shortly before he was shot down and taken prisoner on May 7. (*Author*)

principal fighter of the Italian and Belgian air arms, the Nieuport 28 got a reprieve from obscurity by another country that was desperate for a fighter. At that time, the United States had yet to develop a fighter capable of matching its German counterparts over the Western Front, and French Spad production was still insufficient to equip US Army Air Service squadrons. Therefore, Nieuport got an order for a total of 297 fighters for the USAS. The 95th Aero Squadron, based at Villeneuve, received the first complement of the new fighters at the end of February 1918, and the 94th Aero Squadron was fully equipped with them by mid-March.

No machine guns had accompanied the planes to the squadrons, but for a time the Americans flew unarmed patrols in them for the sake of familiarization and pilots' morale. It was soon determined, however, that the 95th's pilots had inadequate gunnery training, and the unit was withdrawn until May 2. Meanwhile, the first machine guns reached the 94th—albeit only enough to mount one, rather than the specified two, on each plane—and the unit flew its first armed patrol on March 28.

On April 9, the 94th was transferred from Villeneuve to Gengoult aerodrome near Toul. Senior officers of the American Expeditionary Force were skeptical about the Nieuport's capabilities and viewed them as a temporary measure until enough Spad XIIIs were available to replace them. Meanwhile, by letting them operate over the relatively quiet Toul sector, they expected the fighter pilots to gain experience and confidence against second-line opposition, before eventually reequipping with Spads and being turned loose against Germany's best.

Besides such forgiving conditions in which to test their mettle, the 94th's pilots had the benefit of an extraordinary cadre of "old hands." The commander, Maj. John Huffer, was a Lafayette Flying Corps volunteer with three aerial victories to his credit. His operations officer, Maj. Gervais Raoul Lufbery, was the leading American ace with sixteen victories, all scored while serving in the famed

Escadrille Spa.124 "Lafayette." Two of the flight leaders, Capts. James Norman Hall and David McKelvey Peterson, were also veterans of the Lafayette Escadrille.

The Nieuport 28's first day of combat was to be a memorable one. It began in the early morning hours of April 14, when Capt. Dave Peterson led 1st Lts. Edward V. Rickenbacker and Reed Chambers over the lines. Thick fog convinced Peterson to abort the mission, but Rickenbacker and Chambers mistook his reason for turning back as having engine trouble, and continued with the patrol. The duo was subjected to antiaircraft fire, and as they turned back toward Gengoult, two fighters were dispatched from Royal Württemberg *Jasta* 64's aerodrome at Mars-la-Tour to intercept them.

Evidence has since surfaced to suggest that one of the German pilots, *Vzfw.* Antoni Wroniecki, had a hidden agenda when he led *Uffz.* Heinrich Simon against the intruders. Of Polish extraction, Wroniecki nursed a secret hatred of the Germans and hoped that an Allied victory might result in the resurrection of his partitioned homeland from the ruins of the German and Austro-Hungarian Empires. Wroniecki later told his captors that he had planned to "lose" Simon in the fog, land at the Allied aerodrome and then join the French air service. Both *Jasta* 64w pilots were still lost themselves, however, when they emerged from the clouds over Gengoult.

Upon seeing an Albatros D.Va and a Pfalz D.IIIa appear over their field, the Americans dispatched two more Nieuports, this time flown by 1st Lt. Douglas Campbell and 2nd Lt. Alan F. Winslow, to engage the Germans. In five minutes it was all over, with both enemy planes brought down, though not without some contradictory details. Winslow was officially credited with bringing down the

Albatros intact, with Simon only suffering a few bruises. Campbell was credited with shooting the Pfalz D.IIIa down in flames, the pilot being badly burned and, according to some accounts, dying of his injuries shortly afterward.

A convincing argument can be made that the Americans, upon learning that Wroniecki wanted to go over to their side, deliberately exchanged the pilots' names and aircraft to give the impression that Wroniecki was dead, in case he were to fall into German hands later. Adopting the pseudonym "Wróblewski," Wroniecki joined Gen. Józef Haller de Hallenburg's "Blue Army," a force of Polish volunteers being organized to serve under the French, and subsequently transferred to an *Escadrille Polonaise* that was being organized at Sillé-le-Guillaume. The war ended before the unit could get into action, and Wroniecki went on to establish an unsuccessful airline in Poland (its failure partly due to his preference for Farman airliners over more efficient Junkers types because of his continuing animosity towards all things German). After that, Wroniecki enjoyed somewhat more success in Germany during the late 1930s—as a Polish spy.

Doug Campbell's victory that morning of April 14, 1918, was the first scored by an American-trained USAS fighter pilot (Winslow was a Lafayette Flying Corps man with previous experience in French *Escadrille* N.152). The Nieuport 28's place in history was assured, and it would later make more, as Campbell, Rickenbacker, and several other pilots went on to become aces in it. The Nieuport 28 would also figure in the dramatic death of Raoul Lufbery as he tried to bring down a German two-seater on May 19, and it would also give two prominent American aces some hair-raising experiences when the wing fabric tore away on the aircraft flown by 1st Lt. James A. Meissner on May 2, and on Rickenbacker's on May 17. A poor grade of glue was found to be the cause of the fabric failure; but a more serious problem was found with the single-valve Gnome monosoupape engine, which required a "blip switch" to slow down, and which worked by cutting the ignition to some of the cylinders. While the ignition was off, fuel would leak from the valves and accumulate under the cowling, often resulting in a fire when ignition was restored to all cylinders. Less well known was a tendency of the engines' rigid and improperly annealed copper-tube fuel lines to crack due to engine vibration, a problem acute enough for Brig. Gen. Benjamin Foulois to ground all Nieuport 28s until more flexible fuel lines could be installed.

In spite of those faults, the Nieuport 28s favorably impressed their pilots with their maneuverability and excellent handling characteristics. The 94th Aero Squadron—and the 95th, which soon rejoined it—established a good record over the Toul sector; and two more squadrons, the 27th and 147th, had been equipped with the type when all four were finally transferred on June 28 to Touqin aerodrome, twenty miles south of Château-Thierry, in anticipation of a new German offensive along the Marne. There, harsh reality set in as the four squadrons, newly organized into the 1st Pursuit Group, suffered heavy losses at the hands of the *Jagdgeschwader* and first-string *Jastas*, equipped with new Fokker D.VIIs. By the end of July, sufficient Spad XIIIs were available to reequip the American units, and in early August the Nieuport 28 disappeared from the USAS frontline roster.

While the Morane-Saulnier AI and Nieuport 28 fell below the high expectations that attended their arrival at the front to various degrees, a British fighter was proving rather better than expected. First flown at Martlesham Heath in June 1917, the Sopwith 5F.1 was an attempt to give the pilot as good an all-around view as possible in a fighter that would employ the two-hundred-horsepower Hispano-Suiza. Structurally conventional like virtually all Sopwith fighters, the 5F.1 was a back-staggered biplane with a relatively deep fuselage and the center section placed directly over the pilot's head. Performance was comparable to that of the S.E.5a, and Sopwith got an initial order for five hundred of the Dolphin, as the 5.F1 was christened.

In order to improve forward visibility further, the frontal radiator of the prototype was replaced with a tapered cowling and radiators in the wing roots, though they were later moved to the fuselage sides. Tail surfaces were modified, and the original armament of two synchronized Vickers guns was doubled by mounting two Lewis guns on the forward spar of the center section. After testing with both four- and two-bladed airscrews in September 1917, the latter was standardized for the Dolphin.

By December 31, 1917, 121 Dolphins had been accepted by the Royal Flying Corps, and the first examples were delivered to No. 19 Squadron in January 1918. In spite of the promise it had shown during trials, many British pilots looked askance on the new fighter's appearance. The negative wing stagger reminded them of the D.H.5, which had proven to be a poor performer in air-to-air combat. The fact that the pilot's head protruded above the upper wing did not sit well with them, either—if the plane turned over on landing, which, given the waywardness of its Hispano engine, was a distinct possibility, the pilot would be exposed to a broken skull or a broken neck. Sopwith addressed the latter problem by adding a protective steel framework over the center section, as well as adding breakout panels on the fuselage side as an emergency exit in the event of the plane turning over on its back.

Things got off to a poor start at 19 Squadron on January 22, 1918, when 2nd Lt. Allan A. Veale was killed in an accident in Dolphin C3826. Nevertheless, the squadron's pilots practiced away at their new machines and soon took a liking to their maneuverability and general flight characteristics, which proved to be superior to those of their old Spads. They also found the Dolphin's cockpit more warm and comfortable than that of the average open-cockpit scout.

The Dolphin's baptism of fire came over Comines at 1005 hours on February 26, when one of 19 Squadron's patrols, comprised of both Dolphins and Spads, engaged an equally mixed German flight that included a Fokker Dr.I and three Pfalz D.IIIs. During the fight, the triplane got on the tail of 2nd Lt. J. L. McLintock's Spad and was about to fire a lethal burst when Canadian lieutenant John D. de Pencier, in Dolphin C3841, got on the German's tail and with a quick burst sent it down out of control. The triplane had already done its damage, however: McLintock was last seen descending in a shallow glide. It was later learned that he had been killed, the third victory for *Ltn.d.Res.* Richard Plange of *Jasta* Boelcke.[4]

The Dolphins' next fight occurred over Gheluvelt on March 8, when a flight came under attack by five red Albatros scouts. Captain Patrick Huskinson turned the tables on one of his adversaries and shot it down. Another Albatros got onto Capt. G. N. Taylor's tail and was about to shoot him down when Capt. Oliver C. Bryson got behind it and with a twelve- to fif-

teen-round burst hit its structure, causing its wings to break and fold up. The German spun down and was credited as Bryson's twelfth and final victory of the war. His victim, who fell at 1010 hours Allied time, was possibly *Offizierstellvertreter* (Officer Deputy) Wilhelm Kampe of *Jasta* 27, who had just downed a D.H.4 of No. 27 Squadron for his eighth victory when he was killed at 1208 hours German time. In contrast to the earlier combat, Bryson had been in time to save Taylor, who returned unhurt. The squadron's only casualty, 2nd Lt. F. J. McConnell, came back wounded.

By the middle of March, the pilots of No. 19 Squadron were "sold" on the Dolphin. In addition to being more nimble than the Spad, it proved to be capable of sustaining an even greater amount of punishment and still return home. A second Dolphin unit, No. 79 Squadron, arrived with its full complement in February, while in April No. 23 Squadron exchanged its Spads for Dolphins and No. 87 Squadron brought its Dolphins to the front. All four units gave outstanding accounts of themselves throughout the war, yet no further Dolphin squadrons were destined to see combat, and the type disappeared from first-line service soon after the armistice. In spite of its unusual appearance, no satisfactory reason has been ascertained for such scant utilization of what may well rate as the most seriously underrated fighter on the Western Front.

So it was that German fighter pilots could hope for no more than quantitative parity with the Allies when March 21 rolled around. There was talk of superior new fighters being developed at home, but none had reached the front when Operation Michael commenced. Over the next month, the *Jagdflieger* did their best with what they had, and spent at least some of their time between missions wondering when those new fighters would finally arrive.

Idflieg had in fact been seeking the answer to their prayers with a fighter competition held at Berlin's Adlershof airfield in January 1918. The thirty-one aircraft entered ranged from improved versions of current mainstays, such as the Albatros D.Va and Pfalz D.IIIa, to original, innovative, and often imaginative designs. Although the majority of the fighters were to have utilized the 160-horsepower Mercedes D.III power plant or its higher compression relative, the D.IIIa, a few entries used rotaries, including the Siemens-Schuckert Werke's SSW D.III.

The Siemens-Schuckert Electrical Corporation had been involved in aviation since 1907, but its first production airplane, built in mid-1916, was essentially a copy of the Nieuport 11. Only 65 SSW D.Is were built before the original 150-plane order was discontinued in June 1917; but a modified variant, the D.Ib, achieved an eye-opening climb rate of 5,000 meters (1,640 feet) in 20.5 minutes by use of an overcompressed 140-horsepower version of the nine-cylinder Siemens-Halske Sh.I engine. This and other Siemens-Halske engines were counterrotaries, in which the propeller and cylinders rotated in an opposite direction to the crankshaft. That resulted in greater propeller and cooling efficiency, better fuel economy, reduced drag, lower weight, and a greatly reduced gyroscopic effect. With Siemens-Halske working on a 160-horsepower, eleven-cylinder counterrotary engine, the Sh.III, *Idflieg* ordered a fighter based on the new power plant as early as November 1916. The engine was not ready for flight testing

until June 1917, but on August 5, twelve-victory-ace and SSW test pilot *Ltn.* Hans Müller flew the D.IIb prototype to an altitude of 7,000 meters in a record-setting 35.5 minutes. Further experiments with various wing configurations and propeller types ultimately led to the D.III, which featured wings of unequal chord and a four-bladed propeller to allow for a shorter undercarriage.

On December 26, 1917, *Idflieg* ordered twenty preproduction SSW D.IIIs, of which four were entered in the first fighter competition in January 1918. Again, Müller put his plane through some impressive paces, displaying exhilarating maneuverability and climbing 6,000 meters in 21.5 minutes on January 21. Pilots who were unused to the SSW fighter's high revolutions per minute and unusual handling characteristics were less favorable in their appraisals, however, finding the D.III a hot handful, especially when it came to landing. SSW modified the D.III further to address their complaints, and in response to Manfred von Richthofen's general lament that all aircraft at the competition were too slow, new wings of shorter span and equal chord (based on the D.III's lower set) were installed on a new version, the D.IV, which sacrificed climb rate for greater level speed. Encouraged by the SSW fighter's overall performance, *Idflieg* ordered thirty D.IIIs on March 1, followed by an order for fifty more on March 23, and fifty of the new D.IVs on April 8.

The first six SSW D.IIIs were delivered to *Jagdgeschwader* III on March 16, but the principal recipient, *Jagdgeschwader* II, got its first nine machines on April 6, and its overall complement was up to thirty-five by May 18. *Jastas* 12 and 19 of JG II were then equipped with Fokker Dr.Is, and consequently their pilots' experience with rotary engines would help in making the transition to their assigned SSW. *Jasta* 19's commander, *Ltn.d.Res.* Walter Göttsch, received D.III 8346/17 on April 6 and had its fuselage painted white, but he never got to fly it in combat. On April 10, he was killed in his Fokker Dr.I while pursuing an R.E.8 over British lines and was subsequently credited to Lts. H. L. Taylor and W. I. E. Lane of No. 52 Squadron, although he may just as likely have been the victim of ground fire. Göttsch was posthumously credited with the R.E.8 for his twentieth victory; but while Lane was wounded in the leg, Taylor had in fact made a hasty landing behind the lines that German observers mistook for his being shot down.

As the German offensive proceeded, the SSWs were kept discretely on their own side of the lines while the pilots familiarized themselves with them in between offensive patrols in Dr.Is or, in the case of *Jastas* 13 and 15, Albatros D.Vas and Pfalz D.IIIas. Two successes were attributed to the SSWs, however, and they were probably achieved on April 20, when *Ltn.d.Res.* Hans Pippart, Göttsch's successor in command of *Jasta* 19, intercepted and destroyed a Breguet 14B.2 of *Escadrille* Br.127 on the German side of the lines west of Chauny; and on the following morning, when *Ltn.d.Res.* Ulrich Neckel of *Jasta* 12 bagged another Br.127 Breguet that had penetrated the lines en route to Saint-Quentin. During a visit by SSW engineer Bruno Steffen to JG II's aerodrome on April 22, the *Geschwaderkommandeur*, *Hptmn.* Rudolf Berthold, praised the D.III for its "brilliant" rate of climb, the "faultless" combination of airframe and Sh.III engine, and how the fighter had gained the trust of its pilots.

Berthold's praise proved to be somewhat premature. After seven to ten flying

hours, the engines began suffering from such ills as overheating, spark plug ejection, faulty magnetos, bearing failure, faulty throttles, piston heads disintegrating, and complete engine seizure. By May 23, JG II's D.IIIs had been withdrawn pending the rectification of their engines' faults. Berthold, however, remained a believer in the type, and urged that "this fighter be made available for Front Line use as soon as possible, for, after rectification of its present faults, it could become one of the most useful fighters."[5]

SSW D.IIIs and D.IVs began returning to frontline service on July 22, their engines replaced by Sh.III(Rh) engines built under license by the Rhenania Motorenfabrik A.G. (Rhemag)—which curiously suffered from none of the problems that had plagued the SSW-built originals—and portions of their cowlings cut away to allow better cooling. Generally, D.IIIs were assigned to *Kampfeinsitzer Staffeln* for home defense because of their superior climb rate, while the faster and more maneuverable D.IVs went to *Jastas* on the Western Front. Once pilots mastered the SSW's handling characteristics, they were almost unanimous in their praise for it as the best fighter at 4,000 meters or higher. Unfortunately for the *Luftstreitkräfte*, only 136 SSWs had reached operational units by November 11, 1918.

While the SSW D.III attracted an unusual share of attention at the fighter competition, the overall star was unquestionably Anthony Fokker's VII biplane. Hastily built for the competition, the VII was based on the fuselage and tail of the Dr.I, but utilized a biplane version of the wooden box-spar construction employed on the triplane. The lower wing was smaller than the upper to improve downward visibility and was built in one piece, with a cutout arranged in the steel-tube fuselage frame to accommodate it. The ailerons, installed only on the upper wing, were also made of steel tube and fabric covered. Streamlined steel tubing also served as interplane and cabane struts, with no bracing wires. The VII also featured a radiator mounted in the nose, rather than on the fuselage sides or the upper wing.

Even after its fuselage was lengthened to compensate for the Mercedes inline engine, Manfred von Richthofen found the VII to be overly sensitive and unstable in a dive. Fokker immediately responded by lengthening the fuselage further and adding a vertical stabilizer. Both the VII and the improved V18 were tested at the competition, and Richthofen and numerous other German fighter pilots were unanimous in their praise for the Fokker biplane's overall performance, including its ability to retain its maneuverability at high altitude and to "hang on its prop."[6] Fokker was immediately given a 25,000-mark contract to build 400 of the new biplane, which was given the military designation of D.VII. Of additional personal satisfaction to Fokker was an order for both the Johannisthal and Schneidemühl factories of his rival, Albatros, to manufacture the D.VII under license, with a 5 percent royalty going to Fokker.

Fokker D.VIIs began to reach the front in April 1918, the first examples, not surprisingly, going to Richthofen's JG I. *Leutnant* Aloys Heldmann of *Jasta* 10 claimed to have used one in combat in the middle of the month, but there is no evidence that the new biplanes scored any official aerial victories until the end of May. Certainly they inspired enthusiasm in every pilot that flew them; yet the Red Baron himself was still flying a Dr.I when he scored his seventy-ninth and eightieth

victories on April 20, and also when he was killed—either by Capt. Arthur R. Brown of No. 209 Squadron or, more likely, by Australian ground fire—while pursuing his erstwhile eighty-first the following day.

A possible reason for the slow transition was that JG I was heavily involved in the German offensive, continually moving from aerodrome to aerodrome. On May 21, the *Geschwader* moved from Cappy to Guise and was given a week's rest before the offensive was renewed, this time against the French along the Marne River around Château-Thierry. The pilots of *Jastas* 4, 6, 10, and 11 made use of the welcome "downtime" to familiarize themselves with the D.VIIs, under orders to limit their orientation flights to rear areas only. As a result, the late Red Baron's "Circus" felt ready when the offensive was launched, and the Fokker D.VIIs unleashed in full force on May 27.

Curiously, JG I recorded only one modest aerial success amid the first day's hectic activities. At 1815 in the afternoon, *Jasta* 4 attacked a formation of Breguet 14B.2 bombers over Pont Arcy, and two were claimed by *Ltn.* Viktor von Pressentin *genannt* von Rautter, of which one was confirmed as his fourteenth victory. *Escadrille* Br.126 lost a plane that day, along with its crew of *Sergents* Louis des Salles and Jean Lingueglia. Amid the confusion of this third phase of the German offensive, a curious double mystery arose. On one hand, the British reportedly acquired a Fokker D.VII, serial number 2184/18, found near Achiet-le-Grand on May 27, and gave it the captured aircraft registration number G/5/12. The only German loss that even begins to match such an acquisition is that of *Ltn.d.Res.* Rudolf Windisch, commander of *Jasta* 66, who was brought down while attacking a French aerodrome, probably by *Sous-Lt.*

François Souleau and *Maréchal-des-Logis* Hypolyte Cavieux of Spa.76.

Typifying officers placed in command of *Amerika-Programm* units, Windisch was a twenty-two-victory ace. He was also photographed in the cockpit of a Fokker D.VII, bearing a personal marking of a leaping stag on the fuselage side, suggesting that as a seasoned *Kanone*, he may have had the rare privilege for that time of receiving one of the new fighters. Perhaps it was his plane, or the remains thereof, that were recovered by the British in the sector. Adding to the day's anomalies were conflicting reports that Windisch had been taken prisoner and that he had been killed—shot while trying to escape, perhaps? In any case, the general configuration of the Fokker D.VII did not remain a German-kept secret for long.

That was a moot point, however, because after more than a month of cautious introduction to the front, the last few days of May saw the Fokker D.VII burst into almost instant prominence. At 1730 on the afternoon of May 28, *Ltn.* Fritz Friedrichs of *Jasta* 10 was credited with destroying a balloon south of Chevigny for his sixth victory, although Allied records identified no corresponding loss. *Jasta* 10 lost one of its pilots on May 29, when *Ltn.d.Res.* Rudolf Rademacher came down in Allied lines and was taken prisoner; his plane, however, was one of the *Staffel*'s few remaining Albatros D.Vas. On the following day, *Ltn.d.Res.* Johann Janzen of *Jasta* 6 downed a Spad over Beauvardes for his ninth victory.

The afternoon of May 31 saw another engagement with the Breguets southwest of Soissons, with *Leutnant* von Rautter downing another of the bombers at 1255, killing *Sous-Lt.* Maurice Béranger and his gunner, *Sgt.* Edouard Wolf of Br.29. Moments after scoring his fifteenth victo-

An early production Fokker D.VII awaits assignment to a frontline unit. Although it dominated the fighter competition in January 1918, it did not get the chance to realize its full combat potential until late May. (*National Archives*)

ry, however, Rautter himself was shot down and killed, either by return fire from the Breguets or by a Spad flown by *Adj.* Gustave Daladier of Spa.93, who claimed a Fokker Dr.I, but who may have confused its silhouette with that of the new, unfamiliar D.VII. *Jasta* 4's commander, *Ltn.d.Res.* Ernst Udet, downed another Breguet 14 five minutes later for his twenty-fourth victory, his victims being *Sgt.* Hippolyte Martin and *Soldat* Jean Galbrun of Br.29. At 1435 hours it was *Jasta* 6's turn, as *Ltn.d.Res.* Hans Kirschstein of *Jasta* 6 claimed a Breguet over Grand Rozoy, his victims possibly being *Caporal* René Lecomte and *Maréchal-des-Logis* Garciete of Br.129. At 1940 that evening, *Ltn.* Martin Skowronski of *Jasta* 6 scored his first success over a Breguet near Marizy-Saint-Mard, and *Hptmn.* Wilhelm Reinhard, Richthofen's chosen successor as *Geschwaderkommandeur*, scored his fourteenth victory over a Spad near Bonneuil five minutes later. *Oberleutnant* Erich Rüdiger von Wedel of *Jasta* 11 completed the day's activities by downing another Spad for his third victory over the Bois de Barbillon at 2040 hours.

June saw a proliferation of Fokker D.VIIs, accompanied by glowing reports from their pilots and awed reactions from their Allied opponents. "We were astonished by what Fokker, the *Geschwader's* favorite builder, had once again squeezed out of the long-antiquated Mercedes [engine]," recalled *Ltn.d.Res.* Richard Wenzl of *Jasta* 6 after receiving his first D.VII. "Although the new biplane was not as maneuverable as the triplane, it was still a bit faster; it climbed slower at lower altitude, better at higher altitude; above all, as we had to work at higher altitudes, it was fitted with a super-compressed engine."[7]

There seemed to be only three things wrong with the D.VII. The first was a tendency for its incendiary ammunition to overheat and explode, which was alleviated by cutting ventilation holes in the cowling and was ultimately remedied with improved ammunition. Many pilots thought the Fokker would benefit from a better engine, and later that summer it got one—the 185-horsepower BMW IIIa, which gave the Fokker D.VIIf that it powered markedly better performance, especially at altitudes of 18,000 feet or higher, and which truly made it the terror of the Western Front.

The third great complaint German airmen expressed about the Fokker D.VIIs was that there were not enough of them to satisfy the demand, and the fighters produced to supplement them almost invariably suffered in comparison when appraised by the pilots who had to fly them. In mid-May the Roland D.VI, a product of the Luftfahrzeug Gesellschaft GmbH, featuring a fuselage built up of wooden clinkers like a boat, was assigned to *Jastas* 23b, 32b, 33, and 35b. Its performance was not better than that of the Albatros D.Va, and in late May *Jasta* 23b's commander, *Ltn.d.Res.* Otto Kissenberth, flew a captured Sopwith Camel in preference to his unit's Rolands. August saw the

appearance of the Pfalz D.XII at the front, although its frontal radiator and N-shaped interplane struts caused Allied pilots to frequently mistake it for the Fokker D.VII. The Pfalz, however, featured the same semimonocoque plywood fuselage as the earlier D.III; and wings with a drag-reducing thin airfoil section inspired by the Spad, conventionally wire-braced with two bays of interplane struts. In his postwar memoir, *Die Jagdstaffel unsere Heimat: Ein Fliegertagebuch aus dem letzten Kriegsjahr* (*The Jagdstaffel of Our Homeland: An Airman's Diary of the Last Year of the War*), *Ltn.* Rudolf Stark mentioned that his command, *Jasta* 35b, accepted Pfalz D.XIIs on September 1, after "a long telephone conversation," and that every pilot "climbed into the new machines with a prejudice against them and immediately tried to find as many faults as possible." His mechanics were already so "spoiled" by the Fokker D.VII's cantilever wings that they complained of the renewed labor required to keep the Pfalz's guy wires adjusted between missions. Later, Stark admitted that the Pfalz D.XII turned out to be a fairly good plane which "flew quite decently and could always keep pace with the Fokkers; in fact they dived even faster. But they were heavy for turns and fighting purposes, in which respect they were not to be compared with the Fokkers."[8]

Two noteworthy exceptions among the Fokker D.VII's maligned stablemates were the SSW D.III and D.IV, once their engine problems were rectified. Another was another Fokker product—a monoplane, which, like the Morane-Saulnier AI, elicited great expectations, only to fall disappointingly short of them.

When *Idflieg* held a second fighter competition at Adlershof between May 27 and

June 21, 1918, Anthony Fokker and his staff entered the V28, a parasol monoplane powered by a 110-horsepower Oberursel Ur.II rotary engine. Fokker's experiments with cantilevered wing structures bore fruit with a lightweight plywood-covered wing that gave the plane outstanding climb and maneuverability, while its placement at eye level by means of streamlined steel-tube struts gave the pilot a superb view in all directions. Judged the best rotary-engine fighter of the competition, the V28 was accepted for production as the E.V on July 3, but Fokker had already taken the liberty of starting production two weeks prior to receiving the official order. On August 5, the first E.Vs arrived at JG I.

One of the Fokker E.Vs assigned to *Jasta* 6 in August 1918 displays that unit's black and white markings. Its combat career with the *Jagdstaffel* was destined to be disappointingly brief. (*Greg Van Wyngarden*)

By that time, the last German offensive had been stopped, and it was the Allies who were on the advance. JG I fell back to the aerodrome at Berne on August 11. Once the unit had settled in there, *Jasta* 6 chalked up its first success in the new monoplane on August 16, when *Ltn.d.Res.* Emil Rolff shot down a Camel for his third victory, its pilot, Sgt. P. M. Fletcher of No. 203 Squadron, being taken prisoner. Later that same day, however, another E.V crashed during a test flight and its pilot, *Vizefeldwebel* Lechner, was injured.

On August 18, the RAF intelligence staff began alerting frontline units that "Fokker monoplanes have been issued to at least two pursuit flights now working in the Somme area." Then suddenly things went frightfully wrong for the E.V. During a fight with No. 54 Squadron on August 19, *Leutnant* Matzdorf, apparently flying a D.VII, was credited with a Camel, and *Leutnant* Wenzl was lined up to down a second one when his E.V's rotary engine overheated and seized, forcing him to disengage. Worse, on the same day, Rolff was putting one of the E.Vs through its paces over Berne when his wing suddenly broke up at three hundred meters altitude, and he crashed to his death.[9]

Sensing a reprise of the misfortunes that had befallen the Dr.I, *Idflieg* immediately grounded all E.Vs. A special crash commission concluded that condensation moisture which had entered through wing breathing holes had begun to rot the flying surfaces from within. Improper spar dimensions and faulty construction at Fokker's Perzina factory were also blamed for the flaws in an intrinsically sound design.

On August 24, E.V production was suspended until the shoddy production techniques were rectified. Fokker responded by building a new, more carefully assembled wing with strengthened spars and by varnishing its interior. It was successfully load tested on September 7, and *Idflieg* permitted production of the monoplane, now redesignated the D.VIII, to resume on September 24. The first D.VIII was accepted on October 8, and additional wings were built to be retrofitted to the 139 grounded E.Vs. Too much time had been lost, however, and Rolff's fleeting success remains the only positively recorded aerial victory attributed to the much-touted cantilever Fokker monoplane before Germany surrendered on November 11, 1918. Likewise, there is no evidence that an even more advanced design, that is, the Junkers D.I all-metal low-wing monoplane, even saw combat before the Germans agreed to the armistice.

While the Germans—in a curious prelude to the closing months of a later war—experimented with innovative fighter designs that came too little and too late, Britain was introducing one more new type which could not have been less radical in concept. Begun in mid-1917 as a Camel successor, the Sopwith 7F.1 featured wings of equal dihedral, with the upper wing center section lowered to the pilot's eye level in order to improve his forward and upward vision, which had been a long-standing shortcoming of the Camel. The plane's prototypes evolved over the months that followed: adapting the 230-horsepower Bentley B.R.2 rotary engine, increasing the wingspan with a two-bay rather than the original single-bay interplane strut layout, and altering the tail surfaces. After testing at Martlesham in February 1918 and operational frontline evaluation at Saint-Omer in March, the 7.F1 "Snipe" was ordered into production, with seven companies being contracted to build 1,700 of the new fighter. Although its fundamental structural makeup was little changed from the Pup of 1916, the Snipe chanced upon a historic distinction after the RFC merged with the RNAS on April 1, 1918—as the first new fighter to enter service with the Royal Air Force.

The first unit to employ the new Sopwith in battle, No. 43 Squadron, received its first Snipe on August 12 and reached its full strength of twenty-four in mid-September. For several weeks, the squadron's pilots alternated between orientation flights behind the lines in the new planes and combat patrols in their old Camels until the latter were fully phased out. Pilots found the B.R.2 engine to be commendably reliable and the Snipe, with a maximum speed of 121 miles per hour at 10,000 feet, a better overall performer, though a little more docile than the Camel. Full activity was resumed with a frontline patrol on September 23.

On the morning of the twenty-seventh, No. 43 Squadron's Snipes and the S.E.5as of No. 1 Squadron were escorting No. 107 Squadron on a bombing raid against the railroad station at Bohain when the de Havilland D.H.9s were attacked by Fokker D.VIIs at 0915 hours. Number 1 Squadron intervened first, Lt. B. H. Mooney claiming a D.VII in flames over Bertry and Lt. C. W. Anning sending one down out of control near Bévillers. Ten minutes later, it was 43 Squadron's turn as Capt. Cecil Frederick King and Lt. Charles C. Banks drove a Fokker D.VII down out of control near Cambrai. Five minutes later another Snipe pilot, Lt. R. S. Johnston, sent a second D.VII down out of control southeast of Cambrai. Two days later, Capt. Augustus Henry Orlebar sent another D.VII down out of control over Renaucourt, bringing his wartime total up to seven. In all, No. 43 Squadron claimed ten enemy planes in the last six days of the month without loss.

Number 4 Squadron, Australian Flying Corps, got into action with its Snipes in October, and No. 208 Squadron RAF was just commencing operations when the armistice was signed on November 11, 1918. The most famous Snipe of all, however, was E8102, a personal machine assigned to Maj. William George Barker while attached to the still Camel-equipped No. 201 Squadron. Flying it, the Canadian ace fought an epic duel with at least fifteen Fokker D.VIIs on October 27, and survived to receive the Victoria Cross.

An alternative Snipe variant, completed in April 1918, was powered by the 320-horsepower ABC Dragonfly radial engine, which gave it a startling maximum speed of 147.8 miles per hour at 10,000 feet. This type, too, was ordered into production, eventually being renamed the Sopwith Dragon. Unfortunately for the British, the Dragonfly engine proved to be as unreliable as it was powerful, handicapping the careers of the Dragon and several other promising fighters that had been designed around it. The Dragon never achieved squadron service, whereas the Snipe would remain a postwar RAF mainstay until 1927.

When World War I began, the fighter plane did not exist. The war's last year showed how far aviation in general had come, and the sleek, swift fighting machines of 1918 had showcased many of its most noteworthy structural refinements. Nevertheless, a good many of the innovations managed to be too far ahead of their time. While the immediate future was claimed by conventional wood-and-wire biplanes like the Bristol F.2B and the Sopwith Snipe, airplanes featuring such refinements as radial engines, monoplane wings, cantilever wing structures, and all-metal construction generally proved to be less practical as fighting machines, but which would serve more as passing glimpses of things yet to come. It would not be long, however, before the state of the art caught up with the concepts they had pioneered.

Six

A New Generation

Fighters of the Spanish Civil War, 1936–39

The airplane had proven itself to be a useful addition to armies' arsenals from the earliest months of World War I, and had developed rapidly in the crucible of that conflict by the time it officially ended on November 11, 1918. Less than eighteen years later, the airplane would play an essential role in starting another war.

When conservative political, religious, and military factions, led by *General* Francisco Franco y Bahamonde, rebelled against Spain's five-year-old Republican government on July 18, 1936, it was the prompt dispatch of German transport planes to Franco's aid that made it possible for him to airlift 8,000 of his troops from Morocco between July 20 and the end of August, thereby establishing a viable base for his Nationalist forces in southern Spain. The civil war that ensued became a test laboratory for aircraft designs and tactics that would figure in the second world conflict, which would begin just months after the Spanish war ended.

Three fighters of the Spanish Civil War stand out as classic embodiments of the changing state of the art during the 1930s. The Fiat CR.32 was typical of its time, yet was essentially no more than a sleeker, higher powered, structurally more advanced variation on the standard World War I formula—a biplane with fixed landing gear and an open cockpit. Nevertheless it performed its duties very well—perhaps too well, in retrospect, for the good of the Italians who built it. In contrast, the Soviet Polikarpov I-16 represented a radical advance, being the first low-wing monoplane fighter with retractable landing gear and an enclosed canopy to enter production. The Messerschmitt Bf 109, though essentially no more than a structural and aerodynamic refinement of I-16's basic formula, was nonetheless sufficiently advanced to be the prototype for most of the aircraft that dueled for the sky in World War II—and improved versions of the Messerschmitt fighter would remain among them right to the end.

Although the League of Nations leveled an embargo against the shipment of arms to the warring Republican and Nationalist forces in Spain, great quantities of French, Czechoslovakian, American, and British weaponry, including aircraft, found their way to the two sides by various subterfuges. Three dictatorial powers that perceived an ideological stake in the conflict were more open in their support. With a strong communist constituency ensconced in the elected Spanish government, Soviet leader Josef Stalin shipped tanks, aircraft, and airmen to aid the Republican cause. His archenemies, Adolf Hitler of Nazi Germany and Benito Mussolini of Fascist Italy, saw a kindred spirit in Spain's *General* Franco and began sending thousands of fighting men, as well as equipment, to bolster the Nationalist army.

It was the Germans who made the first aerial intervention—in the form of Junkers Ju 52/3m transports that, between July 29 and October 11, 1936, airlifted almost fourteen thousand Nationalist soldiers and some five hundred tons of matériel from Spanish Morocco to Seville. This, the first airlift of its size in history, was so critical to the Nationalists' prospects of success that in 1942 Hitler remarked that "Franco ought to erect a monument to the glory of the Junkers 52."[1] As the Nationalists moved on their next objective, Madrid, the Germans also sent in the first contingent of Heinkel He 51B fighters and a cadre of pilots to fly them and train Nationalists on them. Developed by Siegfried and Walter Günter from their He 49 "trainer/sport plane," which had first flown in November 1932, the He 51 was a single-seat, metal-framed, fabric-covered biplane with fixed landing gear, powered by a 750-horse-power BMW V12 liquid-cooled engine. First flight-tested in the summer of 1933, the He 51A soon dropped its sport plane pretense when two synchronized 7.9mm MG 17 machine guns were installed over the engine.

On August 6, 1936, the cargo ship SS *Usaramo* arrived at Cádiz, and the next day it off-loaded the first contingent of six He 51B-1s, which arrived by rail at Tablada airfield, outside Seville, on the ninth. Upon their assembly the next day, the Germans began training the first five of eighteen Nationalist volunteers: *Capitanes* (Captains) Luis Rambaud and Joaquin García Morato; and *Tenientes* (Lieutenants) Manuel García Pardo, Ramiro Pascual, and Julio Salvador Díaz-Benjumea.

Capitán García Morato was already a seasoned veteran, having flown a Nieuport-Hispano NiH.52 (a French Nieuport-Delage NiD.52, built in Spain under license by Hispano-Suiza) that had fallen into rebel hands, to shoot down a Vickers Vildebeest over Antequera on August 3. During a solo patrol in an He 51 on August 18, he shot down a Potez 540 twin-engine bomber and an escorting NiH.52, and dispatched another Nieuport on September 2.

Up to that point, the Germans had been barred from engaging in combat but soon obtained permission from Nationalist air force general Alfredo Kindelán y Duany to participate. Even then, the Spaniards, who had formed a loose-knit unit they called the *Escuadrilla Rambaud*, were loath to give up "their" Heinkels until the end of September, when Kindelán disbanded the *escuadrilla* and transferred its personnel to train on Italian-supplied Fiat CR.32 fighters.

The Germans finally flew a combat mission in their own planes on August 25, when *Obltn.* Kraft Eberhardt led *Obltne.* Johannes Trautloft and Herwig Knüppel on a patrol over Guaderrama under summer heat in which Knüppel recalled, "I sat in my aircraft in shorts and a T-shirt—my tennis clothes!" Suddenly, Eberhardt signaled his wingmen to attack, having spotted three Republican Breguet 19 biplane reconnaissance bombers, license-built by the Spanish Construcciones Aeronáuticas

SA (CASA), five hundred meters below them outside of Madrid. As the Germans dived on them, Trautloft recalled:

> As I approach I see the gunner aiming his gun at me and then the muzzle lights up as he opens fire. It all looks rather harmless. With my first burst, the gunner disappears—his machine gun points vertically toward the sky. The 'Red' now pushes over into a steep dive. My second burst is brief, but on target, because all of a sudden the Breguet rears up, rolls over, roars towards the earth in a steep, uncontrolled dive and smashes into the ground north of the village of Comenar [sic; Colmenar].[2]

Upon returning to base at Escalona del Prado, all three Germans claimed a victory, but Knüppel had to disengage with jammed guns and his claim was disallowed. Trautloft was credited with the first German fighter victory since World War I—and his first of an eventual fifty-eight—while Eberhardt was credited with the second. Eberhardt and Knüppel downed two more Breguets the next day, but the latter then found a Dewoitine D.372 parasol monoplane fighter on his tail, and got a further shock. "Just as I was about to turn onto him," Knüppel reported, "he shot upwards. I was unfortunately unable to catch him, as his aircraft climbed better and was faster."[3]

Knüppel did better on August 27 when he downed an NiH.52, an older design that was outclassed by the He 51. The Germans found the Republicans' French-built Potez 54 twin-engine bombers a somewhat more difficult prospect; but they soldiered on, adding to their scores and serving as the vanguard to the growing Luftwaffe contingent that became the Legion Condor.

Meanwhile, Italian aircraft were also finding their way into Spain, including Fiat CR.32s. Their pedigree dated to 1923, with the introduction of the CR (*Caccia Rosatelli*), first of a line of fighters designed by Celestino Rosatelli, featuring two wings of unequal span, braced by a girder-like set of Warren truss struts. The CR.20 of 1926 introduced an all-metal airframe, and the CR.30, introduced in 1932, featured a 600-horsepower Fiat A.30 liquid-cooled engine, streamlined spats for the main and tail wheels, and an armament of either two 7.7 or two 12.7mm Breda-SAFAT machine guns synchronized to fire above the engine. The CR.32, which first flew on April 28, 1933, was basically a scaled-down, more compact and refined CR.30. Its 592-horsepower Fiat A.30 RA engine gave it a maximum speed of 237 miles per hour at 10,000 feet. Ailerons, balanced by "park-bench" tabs, were installed in the upper wing only, and the rudder and ailerons were statically and dynamically balanced. Stable and yet phenomenally maneuverable as well as extremely sensitive on the controls, the Fiat also had an exceptional diving speed, while its robust wing structure lent confidence to the pilot.

The CR.32 was the most numerically important fighter in the *Regia Aeronautica* by the time hostilities broke out in Spain. Wasting little time in providing support to Franco's cause, Mussolini dispatched twelve Savoia-Marchetti S.81 trimotor bombers to Spanish Morocco on July 31—three of which were lost en route—followed by twelve CR.32s, which arrived at Melilla aboard the freighter SS *Nereid* on the night of August 12. Secretly flown to Tablada, near Seville, the Fiats formed the 1*a Escuadrilla de Caza de la Aviación del Tercio* (1st Fighter Squadron of the Foreign Legion Air Arm) under *Capt.* Vincenzo Dequal on August 21.

Initially flown by Italians only, the Fiats—nicknamed *Chirris* by the

An aluminum finish CR.32 of the 101*a Squadriglia*, 10*a Gruppo di Caccia Baleari*, flown by *Sottotenente* Aurelio Vevodi over Palma de Mallorca. On May 31, 1937, Palma was bombed by a Potez 540 piloted by Czechoslovakian volunteer Jan Ferak, and it was in turn attacked by *Capitano* Guiseppe d'Agostini, *Sottotenente* Ippolito la Latta, and Vedovi. Badly shot up, the Potez crash-landed near Andratx. (*Paolo Varriale*)

Spaniards after the Italian pronunciation of the letters "CR"—were soon asserting themselves over the mixed bag of aircraft then available to the *Fuerzas Aéreas de la República Española*. Their first victim was an NiH.52 downed near Cordobá on August 21 by *Sottotenente* Vittor Ugo Ceccherelli. *Sergente* Giovanbattista Magistrini scored the next victory for the Fiats on August 27, when he shot down an NiH.52 that was escorting CASA-Breguet 19 bombers over Guadix aerodrome, killing its Spanish pilot, *Capitán* Antonio de Haro López.

The Italians' first setback occurred on August 31, when three CR.32s encountered two NiH.52s and a Hawker Fury of the Republican *Escuadrilla Mixta* near Talavera de la Reina, resulting in two Fiats shot down and the third returning damaged. Disabled by the Nieuports flown by *Cabos* (Corporals) Roberto Alonso Santamaria and Rafael Peña Dugo, *Sgt.*

Bruno Castellani force landed near Villanueva de la Serena and managed to make his way back to Nationalist territory on foot. The other Italian, *Tenente* (Lieutenant) Ernesto Monico, was less fortunate. Shot down in flames by *Sargento* (Sergeant) Andrés García Lacalle in the Fury, he bailed out and was captured by Republican militia, who claimed to have shot him as he tried to escape. The Nationalists issued a different report, saying that when Monico asked to see the Italian ambassador in Madrid, the militiamen shot him out of hand.

Meanwhile, more CR.32s were arriving in Spain, and by mid-September a second squadron had been formed, under the command of *Capt.* Dante Olivera. By then, the pilots of the 1*a Escuadrilla* were starting to refer to their unit as *La Cucaracha* (The Cockroach) after the popular song of the time. Three Spanish pilots—*Capitán* Joaquin García Morato,

Capitán Ángel Salas Larrazábal, and *Teniente* Julio Díaz-Benjumea—had joined the unit at that point. García Morato first flew the CR.32 on September 6.

September 11 saw the Fiats in full stride, claiming two Breguet 19s and five fighters. Two of the Republican aircraft were credited to Magistrini, and García Morato was credited with an NiH.52 over Talavera for his ace-making fifth victory. By the end of the month, García Morato's score stood at eight, making him the war's leading ace—a status that he never relinquished. The Nationalists, with the generous assistance of the German and Italian allies, had achieved air superiority, but that situation was about to undergo an unexpected change.

The two fighters that the Soviet Union sent to Spain in October 1936 represented a transition in aircraft configuration. One, the radial-engine *Istrebitel* (Fighter) I-15 developed by Nikolai N. Polikarpov in 1933, had brought indigenous Soviet fighter design on a par with that of Western nations for the first time, but was nevertheless a biplane with fixed landing gear and an open cockpit, its principal departure from convention being the gulled upper wing that earned it the Russian nickname of *Chaika* (Gull). In dramatic contrast the other type, the I-16, was the world's first low-wing monoplane fighter with an enclosed cockpit and retractable landing gear.

Development of the I-16 began as a result of conceptual studies begun in 1932 by Professor Andrei N. Tupolev, head of the TsKB (*Tsentralny Konstruktorskoye Byuro*), leading to the development of the ANT-31, or I-14, in May 1933. Polikarpov, who had assisted in the I-14's development, was subsequently transferred to the TsKB at Factory 36 to

Polikarpov I-15 CA-118 of Republican *1a Escuadrila de Chatos, Grupo de Caza num. 26*, based at Murcia in July 1937, was flown by Yugoslav volunteer Bozidar "Bosko" Petrovic. (*Branko Kmic via Haakan Gustavsson*)

develop the I-15 biplane. Polikarpov, however, was convinced that he could create a fighter incorporating all of the I-14's innovations in a lighter airframe. Joined by A. G. Trostyansky, he set about developing a prototype, the TsKB-12, which was first flown by test pilot Valery Pavlovich Chkalov on December 30, 1933. Powered by a 480-horsepower Shvetsov M-22 nine-cylinder radial engine (a license-built Gnome-Rhône Jupiter), the TsKb-12 reached a maximum speed of 190 miles per hour. A second prototype, designated the TsKB-12bis, used a more powerful 710-horsepower M-25 (license-built Wright Cyclone SGR-1820-F3), which yielded an impressive performance when it was test-flown by Chkalov in January 1934, including a speed of 218 miles per hour. When difficulties were encountered in the I-14, the Soviet government authorized production of the TsKB-12 under the military designation I-16.

The first thirty I-16 Type 4s, powered by M-22 engines, were primarily meant as trainers to help biplane pilots make the transition to the new monoplane. The first major production model, the Type 5, used the M-25 engine and began to reach Soviet squadrons in the summer of 1935.

To appreciate the enormity of the Polikarpov design team's achievement, it's worth noting that the first Messerschmitt Bf 109 would not take to the air until September 1935, the first Hawker Hurricane not until November, and the first Supermarine Spitfire not until March 1936.

Advanced though it was in overall configuration, the structure of the I-16 Type 5 was not quite state of the art. At a time when all-metal flush-riveted stressed-skin construction was becoming common in western Europe and the United States, its monocoque fuselage was comprised of four longerons and eleven half frames of pine, over which were glued strips of birch to form a varnished surface called *shpon*. Its wings were of chrome-molybdenum steel alloy, with duralumin wings and control surfaces. Although the center section and leading edges of the wings were aluminum skinned, the rest of the wings, ailerons, and control surfaces were fabric covered.

The landing gear had to be laboriously hand-cranked up or down by means of cables. The forward-sliding canopy was extremely narrow and confining for the pilot, and exposure to propeller-driven sand and dust tended to give it all the clarity of bottle glass. Pilots had such difficulty gaining confidence in the enclosed canopy that it was eventually dispensed with on later I-16 production models. The Type 5 was armed with two wing-mounted 7.62mm ShKAS machine guns.

Nationalist pilots who first encountered the I-16 over Spain quickly dubbed it the Boeing, implying that it was essentially a copy of the American Boeing P-26. The sobriquet was as inaccurate as it was disparaging, because for all its novelty in its brief time (1933–35), the P-26 had nothing in common with the I-16 other than being a radial-engine low-wing

The earliest Polikarpov I-16 Type 5 to fall into Nationalist hands intact in 1936 was evaluated by them and was still flying at Cuatro Vientos airfield outside of Madrid in 1939. (*National Archives*)

monoplane. For one thing, the I-16 Type 5's speed of 283 miles per hour was almost 50 miles per hour faster than the P-26A's maximum of 234. In spite of its superior speed and climb compared to those of its biplane contemporaries, the I-16 was surprisingly nimble as well, with feather-light ailerons and an outstanding roll rate. Unlike the I-15 and other biplanes, however, it was difficult to fly, being unforgivingly sensitive on the controls and stalling easily if not given constant attention by its pilot from takeoff to landing.

The Russians made no secret of their new fighter, unveiling it in the May Day flypast of 1935 and at the *Salone Internazionale Aeronautica* in Milan, Italy, in October. Western observers, however, dismissed the stubby little fighter as a hollow piece of Soviet propaganda. When the Spanish Civil War broke out, the I-16's creators found an opportunity to put their design to a real test—and a venue where it would have to be taken seriously.

The Soviet government established diplomatic relations with Spain in August 1936, and voted to send military aid to the Republic later that month. Stalin neither expected nor wished to make significant communist inroads in Madrid, but in the wake of Hitler's Anti-Comintern Pact with Mussolini, he had no desire to see Franco's forces, aided and influenced by the Germans and Italians, come to power. Stalin's other motive was based more on capitalistic than socialist principles—he drove a hard bargain for his military aid package, for which the Republic paid in Spanish gold, for example, the equivalent of $35,000 for each I-15. The first twenty-five I-15s and thirty-one I-16 Type 5s were shipped to Spain in mid-October 1936, along with 141 pilots and almost 2,000 technicians and mechanics, who would operate the aircraft in combat even while they were training Spanish Republican pilots in their use. In February 1937, the Republic obtained a license to build I-15s. Spain also began producing its own I-16 Type 10s, with 730-horsepower M-25A engines and the armament doubled by the addition of two synchronized 7.62mm machine guns, in the summer of 1938.

Officially dubbed the *Yastrebok* (Falcon), the I-16 was more popularly, if unofficially, known as the *Mushka* (Fly) among its Russian pilots, and the Republicans promptly referred to it as *Mosca* as well, while the I-15 was referred to as the *Chato* (Snub-Nose). After initially calling it the "Boeing," the Nationalists came to refer to the I-16 as the *Rata* (Rat), a term that became universal among the Germans, Italians, and their allies in the years to come.

Once the aircraft were assembled, the first contingent of Soviet fighters was deployed in twelve-plane squadrons around Madrid, which was about to face a major Nationalist offensive. The I-15-equipped 1*a Escuadrilla de Chatos*, under *Starshy Leitenant* (Senior Lieutenant) Pavel Rychagov (under the thin Spanish alias of Pablo Palancar), was set up at Campo XX at El Soto estate near Algete, sixteen miles northwest of Madrid, while the 2*a Escuadrilla de Chatos*, under *Mayor* (Major) Sergey Tarkhov (a.k.a. Antonio), was based at Alcalá de Henares, east of Madrid. The I-16s equipped the 1*a Escuadrilla de Moscas*, under *Mayor* Aleksandr I. Tarasov at Campo XX; the 2*a Escuadrilla de Moscas*, under *Mayor* Sergey P. Denisov at Alcalá de Henares; and the 3*a Escuadrilla de Moscas*, under *Mayor* Konstantin Kolesnikov, still working up at Albacete, where the Soviet chief aviation advisor, *Polkovnik* (Colonel) Yakov V. Smushkevich (a.k.a. General Douglas), set up his air headquarters. In original theory, the two fighter types were expected to complement each other, with the I-16s primarily going after enemy bombers, while the more tractable I-15s dealt with any fighter escort. They were also to escort bombers of their own, in the form of Tupolev SB-2s, known to the Russians as *Katiuskas*, or "Sophias" or "Martin bombers" to the Spaniards. Until sufficient numbers arrived to form separate air groups, the fighters and bombers were combined within a temporary organization called *Grupo* 12.

The bombers, led by Swiss-born *Major* Ernst Schacht, struck *Grupo* 12's first blow against the Nationalists when they attacked Tablada airfield near Seville on October 28. Bad weather and poor serviceability of their aircraft limited the activities of the Soviet fighter pilots to discrete flights on their side of the lines, familiarizing themselves with the terrain. Then, on November 4, Rychagov led ten of the 2*a Escuadrilla*'s I-15s on a mission to escort a squadron of SB-2s. As the *Chatos* descended through a cloud, they ran into a formation of six German-flown and six Spanish-flown Junkers Ju 52/3ms of *Hptmn.* Rudolf *Freiherr* von Moreau's *Kampfstaffel* Moreau, which were climbing to cross the Sierra de Guadarrama en route to bomb Republican targets north of Madrid. The Russians immediately attacked and shot down two of the German Junkers. They also damaged one of the Spanish-flown bombers, which was forced to land at Esquivias, and *Starshy Leitenant* Georgy N. Zakharov was credited with an He 51, returning in spite of his own plane being badly damaged in the encounter. The Germans did not actually lose a fighter, but the demonstrated superiority of the new Republican fighters was sobering enough for General Kindelán to order the He 51s not to engage them unless they had clear numerical superiority.

Resuming their flight, the I-15s next encountered a Nationalist Romeo Ro.37bis light bomber. The Soviets damaged the plane and wounded its observer before they themselves were attacked by two CR.32s from the reorganized Italian XVI *Grupo de Caccia "La Cucaracha,"* which had just come upon the scene. The *Chatos* soon overwhelmed and shot down both CR.32s, whose pilots were none other than early squadron leader *Capitano* Dequal, who, in spite of suffering injuries, parachuted to earth and eventually regained Nationalist territory, and early scorer *Sergente* Magistrini, who force landed but subsequently died of his wounds. In a later patrol, Rychagov also claimed an He 51. The I-15s were credited with a total of seven victories without any combat losses in that eventful first day, but two Soviet pilots became disoriented, and when their fuel ran out, made emergency landings near Segovia, in Nationalist territory. In spite of the capture of the two wayward I-15 men, a report from the crew of the damaged Ro.37 of being attacked by American Curtiss fighters resulted in the *Chatos* being referred to as "Curtisses" by the Nationalists for a long time to come.

The sudden appearance of advanced Soviet aircraft over Madrid gave the Nationalists, Italians, and Germans an unpleasant shock, but it did not last long. On the very next day, nine Italian and Nationalist CR.32s were escorting Ro.37s to Casa del Campo when they were attacked over Leganés by sixteen I-15s that had been escorting three Potez 54 bombers, precipitating the largest dogfight yet seen over Spain. The Italians lost a squadron commander, *Capt.* Carlo Alberto Maccagno, who was captured with a wounded right leg that had to be amputated. Of the seven enemy fighters claimed by the Fiat pilots, García Morato scored one for his thirteenth victory, also damaging a Potez bomber's engine, while Salas was credited with a fighter in flames before diving away from another assailant. In actuality, only two I-15s had been shot down, *Leitenant* (Lieutenant) Piotr A. Mitrofanov bailing out of his burning plane but nevertheless falling to his death to become the first Soviet fatality in Spain. The other Russian survived his crash landing unhurt. Several other *Chatos* returned to Campo Soto with various degrees of damage.

Aerial encounters between the *Chirris* and *Chatos* continued on an almost daily basis. On Friday, November 13, a formation of Ju 52s and Ro.37s escorted by fifteen CR.32s were jumped by thirteen I-15s over Madrid. The claims of both sides—six fighters and an SB-2 bomber by the Nationalists and five fighters by the Republicans—were inflated, but two Russians, *Leitenanti* Karp I. Kovtun and Piotr A. Purtov, were killed. Among the Italians, *Capt.* Goliardo Mosca claimed a fighter in flames and saw the pilot bail out before he was badly wounded in the right thigh. He limped back to Talvera before crash-landing, while another Italian force landed his Fiat near Getafe airfield. Spanish *capitán* García Morato claimed a "Curtiss fighter" for his fifteenth victory, as did Salvador for his fifth, while Salas also achieved acedom with credit for an SB-2.

Friday the thirteenth also saw the first major clash between the Russian fighters and the German contingent which was formally designated the Legion Condor that month of November. Nine He 51s had left their base at Ávila to escort five Ju 52s of *Kampfstaffel* Moreau and three Heinkel He 46s in an attack on Republican forces on the west bank of the Manzanares River when they encountered sixteen I-15s, led by Rychagov, and twelve newly arrived I-16s, led by Tarkhov. In the confused dogfight that followed, *Uffz.* Ernst Mratzek claimed to have downed an I-16 and *Obltn.* Oskar Henrici claimed another. One disabled I-16 did force land in Nationalist territory intact; the next day, the Nationalists dropped the chopped-up pieces of its pilot, *Lt.* Vladimir Bocharov, into Republican lines, in a box.

Attacked by I-15s, Knüppel led his *Kette* (flight) down through the clouds and, upon reemerging, attacked the I-15s from below. *Oberleutnant* Knüppel, *Ltn.* Hennig Stümpel, and *Obltn.* Dietrich von Bothmer each claimed an I-15 in that pass, as did *Uffz.* Erwin Sawallisch, although he came back with his tail riddled.

Yet another Soviet fighter fell victim to the German *Staffelführer* (Squadron Leader), *Oberleutnant* Eberhardt, for his fifth confirmed victory, but moments later he collided with his adversary and was killed. The Soviet pilot, *Mayor* Tarkhov, took to his parachute over Madrid, only to be hit six times by Republican ground fire, then was mistaken for a German by angry members of the

city's populace and so brutally treated that he died in hospital several days later. The incident led to the issuance by the Republican army of an order against the shooting of downed pilots, on the grounds that they might provide useful information—to say nothing of their possibly turning out not to be the enemy.

In a final tragic coda to what the Germans considered a "Black Friday," *Oberleutnant* Henrici, though shot through the back in a lung by future I-16 ace *Lt.* Sergey Chernykh, managed to land his He 51 in friendly territory, emerged from the cockpit, took a few steps from his plane and then collapsed and died. Oskar Henrici and Kraft Eberhardt were posthumously awarded the *Cruz Laureada de San Fernando* by the Nationalists. In Moscow, Sergey Tarkhov and Vladimir Bocharov were posthumously awarded the Gold Star of the Hero of the Soviet Union on December 31.

Later, the Germans learned that the Republicans had claimed to have shot down four of their planes for the loss of two I-16s. The He 51s were clearly outclassed, as one of the shaken survivors of that first major encounter, *Ltn.* Otto-Heinrich *Freiherr* von Houwald, admitted:

> We downed five of them, but what were these victories when compared with the loss of our *Staffelführer*? This only served to show that our good old He 51s were too slow compared with the new Ratas—they could play with us as they wanted. Furthermore, the Soviet 'Martin Bombers,' which were arriving daily, were 50 km/h faster than us, and the people were scared of them. Feverishly, we waited for the Bf 109s to arrive from Germany.[4]

The I-16 pilots had their first encounter with the CR.32s at 1530 on the afternoon of November 15, when four of them, led by *Leitenant* Chernykh, attacked a trio of Fiats escorting bombers over Villa del Navalcarnero. With exceptional teamwork the Italians—*Sottotenenti* Bernardino Serafini and Giuseppe Cenni and *Sergente* Berretta—managed to fend off the monoplanes and were collectively credited with shooting one down; its pilot, *Lt.* Vladimir N. Vzorov, crash-landed in an olive grove, and he destroyed it before making his way back to base. Chernykh and *Leitenant* Sergey Denisov were each credited with an enemy fighter, even though all the Italians returned unharmed.

The innovative Soviet fighter met its redundant Italian counterpart again the next day, when four I-16s flying "top cover" for nine I-15s of Rychagov's 1*a Escuadrilla de Chatos* saw the biplanes come under attack by CR.32s. The monoplanes, in turn, dived on the Fiats and claimed two of them, while three others were also claimed by the counterattacking I-15 pilots, though the only Italian casualty was a wounded pilot who managed to make it home. Rychagov, whose I-15 had already been hit by return fire from Ju 52 gunners, was credited with two CR.32s before *Sottotenente* Brunodi Montegnacco shot him down. Bailing out 150 feet over Madrid's Paseo de la Castellana, the wounded Rychagov had to be hospitalized, and Piotr Pumpur took his place in command of the I-15 squadron. *Sergente Maggiore* (Sergeant Major) Vittorino Daffara also downed an SB-2 that day for his first of an eventual six individual and four shared victories.

With six personal victories and fourteen shared to his credit when he rotated back to the Soviet Union, Rychagov was promoted to major and awarded the Gold Star, subsequently flying against the Japanese in China and swiftly rising to brigadier general. After the German invasion of the Soviet Union, however, he was one of many Spanish Civil War veterans to fall out with Stalin. Arrested on June 24, 1941, he was tortured and finally shot in Kuibyshev prison on October 28, 1941. Among others sharing his fate was fellow Spanish Civil War veteran *General-Leitenant* (Lieutenant General) Yakov Smushkevich, a two-time Gold Star recipient who rose to deputy chief of the general staff before his arrest on June 8. He was executed on personal orders from NKVD chief Lavrenty Beria on the same day Rychagov died.

The I-16 pilots got one more reminder of the folly of underestimating the Fiat on November 17, when two fighters of the 2*a Escuadrilla de Moscas* attacked a squadron of Ju 52 bombers over the Madrid suburb of Fuencarral, only to be pounced on by seven CR.32s. Montegnacco was credited with one, while *Sergente Maggiore* Eugenio Salvi and *Sergenti* Guido Presel and Baccara shared credit for the other. For the first time, the Republicans publicly admitted to the loss of one I-16 and its pilot, *Leitenant* Pavlov. Montegnacco now had six victories, and he would survive the civil war as its leading Italian ace with fifteen.

The last major air battle of 1936 occurred on November 19, when eighteen Ju 52s, four S.81s, and eight Ro.37s dropped forty tons of bombs on Madrid. The bombers' escort of nine Nationalist He 51s and sixteen CR.32s engaged about an equal number of I-15s and I-16s, resulting in the downing of a Ju 52, whose Spanish crew survived, against Italian claims of seven Soviet fighters. Actual losses on the Republican side were *Kapitan* Dmitry I. Zhedanov, killed in his I-16 by *Tenente* Corrado Ricci, and Spanish I-15 pilot *Sgt.* Fernando Roig Vilalta.

Bad weather curtailed aerial operations for the rest of the year, and both sides took the time to reappraise the situation and develop tactics to compensate for their respective deficiencies. The Italians, intimidated by both the I-15 and I-16, made it a policy to seldom go outside of their lines and not to engage the Republican fighters unless they flew in formations of at least fifteen planes. The Spanish Nationalists, in contrast, continued to fly their *Chirris* aggressively against *Chatos* and *Ratas* alike. García Morato took command of his own Spanish unit in May 1937; officially designated *Grupo 2-G-3*, it would become legendary as the *Patrulla Azul* (Blue Patrol).

One of the CR.32 pilots' principal methods of dealing with the I-16s was to attack from above, using the Fiat's sturdy structure and high diving speed to advantage. Another popular tactic among the Italians was to engage the Republican fighters head on, using the greater range and striking power of their 12.7mm Breda-SAFAT machine guns. Many Spanish pilots carried a combination of one 7.7mm and one 12.7mm gun on their CR.32s, although García Morato preferred to keep both of the lighter 7.7s on his plane.

The Soviets were equally chastened by their own losses, which left only fifteen I-15s operational by November 20. The I-16 pilots resolved to avoid being drawn into a turning dogfight with the Fiats, in which the CR.32s were more nimble and the I-16 had proven to be a less stable gun platform. Whenever possible, the I-16 pilots reverted to the formula that had brought them their first success on November 15—flying their *Moscas* about 1,000 feet above the I-15s, making diving attacks on whatever enemy aircraft engaged the *Chatos*, and then using the

momentum of the dive to climb back up for another pass.

As for the German fighter pilots, the action of November 13 had painfully but decisively rammed home what they already knew—that their He 51B fighters were barely enough to handle an NiH.52 and no match for an I-15, let alone for the newer, faster I-16. Fortunately for them, a new fighter was on the way, of a design that represented an advance beyond the I-16—the Messerschmitt Bf 109.

The "Great White Hope" of the Legion Condor in 1936 had begun as a "dark horse" just a few years earlier. On May 1, 1933, thirty-five-year-old Wilhelm Emil Messerschmitt was co-manager of the Bayerische Flugzeugwerke Allgemeine Gesellschaft (BFW), a firm that had been revived after being driven to bankruptcy through the cancellation of a contract with Deutsche Lufthansa two years earlier. BFW's future prospects still seemed bleak, however, because the National Socialist Party's undersecretary of aviation, Erhard Milch, had been Lufthansa's managing director—and one of Willy Messerschmitt's chief detractors—back in 1931.

Messerschmitt was working on a light civil transport in the spring of 1934 when he learned that the German state technical office, or *C-Amt*, was holding a competition of aircraft designs. With little to lose, he rushed his project, the BFW M.37, to completion and entered it in the fourth *Challenge de Tourisme Internationale* as the Bf 108 *Taifun* (Typhoon). Although the Bf 108 did not win any of the events, its overall configuration and performance was impressive enough to land a production contract.

Even before the Bf 108 had made its first flight in the spring of 1934, Messerschmitt learned that the RLM (*Reichsluftfahrtministerium*, or Air Trans-

Messerschmitts assigned to fighter squadron J/88 of the Legion Condor in Spain in 1938 include Bf 109C-1 6-49, bearing the name *Jaguar* on the cowling and victory markings on the vertical stabilizer. (*Bundesarchiv*)

port Ministry) was about to issue a specification for a fighter, to be powered by a Junkers Jumo 210 engine and to be capable of at least 280 miles per hour. Although most German airplane manufacturers were invited to submit designs in February 1934, only the established firms, like Arado, Heinkel, and Focke-Wulf, could expect serious consideration. Given BFW's lack of experience in fighters and Milch's long-standing antagonism toward Messerschmitt, it seemed like a waste of time and effort for Messerschmitt to enter such a competition. Milch had not even informed Messerschmitt of the competition, but his superior, Aviation Minister Hermann Göring, did—in a confidential message cryptically requesting Messerschmitt to develop "a lightning-fast courier plane which needs only to be a single-seater."[5]

Deducing just what Göring was hinting at, Messerschmitt and his design team at BFW's Augsburg factory—most notably Robert Lüsser, and Richard and Hubert Bauer—set about incorporating the Bf 108's features into a low-wing monoplane fighter with a monocoque fuselage of as narrow a cross section as possible, and including retractable landing gear—which in contrast to the I-16's hand-cranked undercarriage, operated hydraulically—an enclosed cockpit, leading-edge slots, and trailing-edge flaps in the wings. While work proceeded on the *Versuchs* (prototype) Bf 109 fighter, Germany officially established the Luftwaffe on March 1, 1935, and on the sixteenth Hitler publicly renounced the Treaty of Versailles's restrictions on German rearmament.

The prototype Bf 109V1 was completed in August 1935, and evaluation flights began at the RLM's test center at Rechlin, initially using a 675-horsepower Rolls-Royce Kestrel engine in place of the Jumo. The Bf 109V2, completed in October, introduced the 610-horsepower Jumo 210A as well as a strengthened undercarriage; and the Bf 109V3, delivered in June 1936, was the first to be armed with an engine-mounted 7.92mm MG 17 machine gun.

In spite of its high wing loading, which handicapped its maneuverability at low speeds, the Bf 109 yielded such outstanding performance that the RLM quickly eliminated the Arado Ar-80 and Focke-Wulf Fw 159 from consideration, leaving only the Heinkel He 112 in contention. Ten preproduction Bf 109B-0s were ordered, at which point two events occurred that would alter the Bf 109's fate.

Coincident with the Bf 109V3's appearance, June 1936 also saw the issuance by Britain's Royal Air Force of production contracts for 600 Hawker Hurricane fighters and 310 Supermarine Spitfires, the latter of which, first flown on March 5, had a similar performance to that of the Bf 109V1. The potential threat posed by those British fighters added urgency to the German fighter's development, and armament in the Bf 109V4, introduced in November, was increased from two to three MG 17s.

The other event of significance was, of course, the outbreak of the Spanish Civil War in July. Here was an opportunity to test the Bf 109 under combat conditions, developing not only the aircraft but also practical tactics for using it to the best effect—an idea lent an element of urgency by the appearance of the I-15 and I-16 over Madrid in November. In consequence, the Germans rushed the Bf 109V4 to Spain in December, to be followed by Bf 109B-1s, the first of which left the production line in February 1937.

The first operational Bf 109 squadron in Spain, the 2nd *Staffel* of *Jagdgruppe* 88 (2.J/88), received its first three planes on March 12. On March 19, the *Staffel* was placed under the command of *Obltn.* Günther Lützow and, with its strength up to seven, was stationed at the northeastern corner of Vitoria airfield, operating with the 1st and 3rd *Staffeln*, equipped with ten He 51s each.

Early engine troubles were resolved on March 19; but even then the high torque produced by the Jumo engine, combined with the narrow track of the new fighter's undercarriage, gave it an alarming tendency to drop its left wing during takeoff and landing, causing a rash of accidents at first. Eventually, however, its pilots learned to compensate through liberal application of rudder to the right. Once they had overcome the Bf 109B's idiosyncrasies, they gained confidence in it and were ready to begin operations over a new front being opened by the Nationalists in the industrial and mining regions along the Basque coast between Gijon and Bilbao, commencing on March 31, 1937.

The Messerschmitt drew first blood in the air at 1715 in the afternoon of April 6, when Lützow, leading a three-man *Kette*, shot down one of four I-15s encountered northwest of Ochandiano. The pilot, who parachuted down in Nationalist territory, was described by Lützow as "an 18-year-old raw beginner on an overland ferry flight." Rain impeded air operations until April 22, when *Ltn.* Günther Radisch and *Fw.* Franz Heilmeyer claimed an I-15 each over Bilbao.[6]

Republican documents identified only one victim in its account of the action: "At 1600 hrs a flight patrolling near the airfield sighted nine twin-engine bombers and several fighters heading for Bilbao. The government aircraft forced them into a combat lasting 20 minutes and shot down two enemy fighters. One was a Heinkel and the other was unknown. In this combat one aircraft was lost."[7]

One of the I-15 pilots, *Teniente* Julián Barbero López, reported having Bf 109s on his tail but evaded them by feigning terminal damage and then force landing at Lamiaco. Barbero's flight leader, *Capitán* Felipe del Río Crespo, was less fortunate, crashing to his death. Officially credited with seven victories, he was slated for promotion to *mayor* that very day.

The Messerschmitts, joined by a single He 112 that proved to be cleaner but more complex and heavier than the Bf 109, continued operations in support of the offensive, adding more I-15s to 2.J/88's tally by the time Bilbao fell to the Nationalists on June 19. On the Madrid front, however, the Republicans launched a counteroffensive on July 6 that took the city of Brunete, precipitating a bloody battle of epic proportions. Among the Republicans' assets were 150 of their most up-to-date available aircraft.

On the same day Brunete fell, Franco telegrammed the Legion Condor commander, *Generalmajor* (Brigadier General) Hugo Sperrle, at Vitoria: "It is urgent that the Bf 109 fighter is put into action tomorrow at the front from Ávila. I request that you order the transfer from the said airfield."[8] Sperrle rushed 2.J/88 from Burgos to Ávila. After three months of relatively easy pickings, the Bf 109s would get their first real test against their nearest counterparts, the I-16s, in air battles involving as many as two hundred planes at a time.

The first clash came on July 8, when *Uffz.* Guido Höness and *Ltn.* Rolf Pingel were each credited with an SB 2, although the Republicans attributed only one of their two losses that day to a Bf 109, the other having fallen victim to a CR.32. The Republicans later claimed the first Bf 109 shot down in aerial combat, crediting it jointly to Soviet I-16 pilot Evgeny Ptukhin and to Božidar Boško Petrovic, a Serbian-born Yugoslav communist flying I-15s in the 1*a Escuadrilla de Chatos*, *Grupo de Caza Num.* 26, although the Germans recorded no such loss for July 8.

Boško Petrovic, for whom this was his fifth victory, exemplified how the ideological nature of the Spanish Civil War

was drawing an international hodgepodge of volunteers to both sides. In May 1937, Soviet personnel began rotating home from *Kapitan* Ivan T. Yeryomenko's 1*a Escuadrilla de Chatos*, and before the Soviet replacements arrived, their places were largely being taken by three Spanish Republican pilots, two Austrian volunteers, two from the United States, and Petrovic.

July 12 started out with Petrovic, Soviet *leitenant* Leonid Rybkin, and Spanish pilot Luis Sandino sharing in the destruction of an He 111B. That afternoon saw a sprawling dogfight near El Escorial with twenty-nine I-16s and eight I-15s tackling about forty He 51s and CR.32s, the last mentioned including García Morato's *Grupo* 2-G-3. The Nationalists claimed four *Ratas* and five *Chatos*, while the Republicans claimed eight CR.32s and one He 51. In actuality, only one plane was lost on either side. Petrovic spotted a CR.32 attacking his commander, Yeryomenko, and shot it off his tail. This plane, credited as the seventh victory for history's only Yugoslav ace, was apparently García Morato, who nevertheless survived unhurt. Moments later, Petrovic went down out of control and crashed to his death, probably shot down by *Capitán* Miguel García Pardo of 2-G-3, who would survive the war with thirteen victories.

The Germans were also active on the twelfth, Höness, in Bf 109B-1 No. 6-4, downing two Polikarpov R-Z observation biplanes, surviving members of which, having never seen a Bf 109 before, described their attackers only as "high-speed monoplanes." Escorting I-16s then engaged the Messerschmitts, and when Höness made the mistake of trying to out-turn rather than outrun an attacking *Rata*, the Soviet flight leader, *Lt.* Piotr Burtym, sent him plunging 2,500 meters into the ground near Villaverde—the first of thousands of Messerschmitt pilots to die in combat over the next nine years. In another action, Pingel shot down an SB-2, and then he and his *Kette* engaged the escorts, resulting in an I-16 each credited to Pingel, *Fw.* Peter Boddem, and *Fw.* Adolf Buhl.

Foreign volunteers in the Republican ranks were again in evidence during a sprawling dogfight on July 13, which lasted nearly an hour. In addition to two Ro.37s, one Ju 52, and nine fighters claimed by the Republicans, a Bf 109 was credited to *Teniente* Frank G. Tinker, an American attached to *Kapitan* Ivan A. Lakeyev's 1*a Escuadrilla de Moscas, Grupo de Caza Num.* 21.

The Germans lost no Bf 109s on July 13, although a subsequent confusion of dates may have linked Tinker's claim to Höness's death the day before. On the other side, *Fw.* Heinz Braunschweiger, assigned a Bf 109B marked 6-13 because nobody else in his *Staffel* wanted to fly a plane with such an unlucky number, shot down an I-16 in the July 13 melee and made a point of it to keep the same plane thereafter.

Another success for the Bf 109s on the thirteenth was an I-15 shot down by *Feldwebel* Boddem, whose pilot turned out to be another American, *Teniente* Harold Evans Dahl of Yeryomenko's 1*a Escuadrilla de Chatos*. Condemned to death by the Nationalists, "Whitey" Dahl was the subject of a protracted appeal—including one by his show-girl wife, Edith Rogers—before Franco finally relented and rescinded the death sentence. Dahl was finally released from Spanish captivity in 1940.

The Bf 109B and its principal rival, the I-16, were at first closely matched. The Bf 109B was faster in level flight and in a dive, while the I-16 had a better climb rate and superior maneuverability. Republican ace and fighter leader Andres Garcia Lacalle commented in his memoirs that the I-16 was superior to the Messerschmitt up to 3,000 meters (9,840 feet); but from that altitude upward, the Bf 109B's performance achieved complete mastery.

During the second Ebro campaign, between July and October 1938, *Obltn.* Werner Mölders of 3.J/88 developed a fighter tactic of far-reaching significance. By combining two *Rotten*, the basic two-man elements within a *Staffel*, into a loose but mutually supportive team, he created an infinitely flexible offensive and defensive unit which he called the *Vierfingerschwarm* (four-finger formation). That fundamental concept, allegedly originated by the Finns but used operationally for the first time by Mölders, would become the basis for numerous variations to the present time. Mölders himself emerged as the leading ace of the Legion Condor with fourteen victories and later became the first fighter pilot to pass the hundred mark on July 15, 1941. When he died in a transport plane crash on November 22, 1941, his score stood at 115.

Guided by lessons learned in Spain, Messerschmitt incorporated a rapid succession of improvements into his new fighter. Following the Bf 109B, widely nicknamed "Bertha," came "Cesar," the Bf 109C-1, with a fuel-injected Jumo 210Ga engine and four machine guns, which arrived in Spain in the Spring of 1938, followed by the Bf 109C-2, which mounted a fifth machine gun in the engine. The Bf 109D "Dora," five of which joined 3.J/88 in August, combined the Bf 109C-1's four-gun armament with the Bf 109B-1's carburetor-equipped Jumo 210Da engine.

Meanwhile, Messerschmitt experimented with the fuel-injected Daimler Benz DB 600 and DB 601 engines, whose initial cooling problems were ultimately solved by installing two radiators under the wings, leaving only an oil cooler under the fuselage. In addition, the DB 601A-powered Bf 109V14 increased the armament to two MG17 machine guns in the nose and two 20mm MG FF cannon in the wings, along with installing a three-bladed controllable pitch VDM airscrew. The result was put into production in early 1939 as the Me 109E-1 "Emil."

The fighter's revised designation, which has caused confusion and controversy among aviation historians for decades, reflected the complete acquisition of BFW stock by Willy Messerschmitt in late 1938. According to the Luftwaffe's own historical records, the old "Bf" reference was retained for the Bf 108, the Bf 109B through D, and for the Bf 110A and B *Zerstörer* twin-engine fighters. All subsequent Messerschmitt products, starting with the Me 109E and Me 110C, officially used the "Me" prefix, although the issue would continue to be confused in the years to come by the appearance of the "Bf" prefix on stamped plates on various Me 109 components as late as 1945.

The Germans wasted no time in sending three Me 109E-1s to La Cenia late in December 1938. Twenty of an eventual grand total of forty-four Me 109Es—including the latest Me 109E-3s—were operational with the Legion Condor's fighter arm in March 1939. By then, however, Nationalist air superiority was virtually absolute, and the Luftwaffe's newest fighter saw relatively little action prior to March 28, 1939, when the last of the Republican forces capitulated. They would, however, have plenty to do on September 1, 1939, and in almost six years to follow.

Maggiore Mario Bonzano's Fiat G.50 of the Gruppo Sperimentale de Caccia stands ready for action at Escalona in January 1939. Although it made its operational debut during the Spanish Civil War, the G.50 saw no aerial combat until January 1940—with the Finnish air force. (*Author*)

The later careers of the three transitional fighters of the Spanish Civil War give insights into the conclusions drawn by the countries that built them. Having often overcome the performance handicaps of the CR.32s by skill and adaptive tactics, the Italian pilots remained hard to convince that the days of the biplane fighter, let alone that of the lone dogfighter, were on the wane. Consequently the next important Fiat fighter, the CR.42, would be another biplane with fixed landing gear and an open cockpit—even while that same company was producing a more advanced design with retractable landing gear and an enclosed canopy, the G.50.

The Fiat G.50 was built in response to specifications issued by the *Ministerio dell' Aeronautica* in 1936, calling for a lightly armed interceptor, a long-range escort fighter, and a fighter-bomber. While other manufacturers submitted four designs to address each of those requirements, Fiat's Giuseppe Gabrielli designed his G.50 to satisfy all three. Built around the new 840-horsepower Fiat A.74 twin-row fourteen-cylinder radial engine, the

G.50, which first flew on February 26, 1937, became the first Italian monoplane fighter to enter production, with an initial order for forty-five machines. After the second prototype crashed in September, the G.50's competitor, the Macchi C.200, was judged the better fighter, but G.50 production continued as insurance against any problems in getting the C.200 into operation.

During a visit to Italy, García Morato test-flew a G.50 at Guidonia in October 1937. He then returned to the *Catrulla Azul* and his trusy CR.32, in which he brought his tally to forty on January 19, 1939. On April 4, days after the Nationalist victory, Spain's ace of aces performed aerial stunts for the newsreel cameras at Griñón. While flying inverted at low altitude, his Fiat's engine suddenly cut out and Morato crashed to his death.

The first G.50s entered *Regia Aeronautica* service at the end of 1938, and ten were promptly shipped to Spain, where they were formed into a *Gruppo Sperimentale de Caccia* (Experimental Fighter Group) under the command of *Maggiore* (Major) Mario Bonzano. The unit was based at Escalona alongside the CR.32s of Bonzano's old unit, the XXIII *Gruppo*, and consequently some of the G.50s were marked with that group's "*Asso di Bastoni*" (Ace of Spades) emblem. Flying as escort to the CR.32s at an altitude of 8,000 meters, the G.50s saw some service in the last fortnight of the war but encountered no aerial opposition. The principal operational evaluation consisted of pilot complaints about inadequate visibility from the enclosed cockpit, which resulted in the adoption of a traditional open cockpit for all subsequent production batches of the G.50.

Even after World War II broke out, the *Regia Aeronautica* was remarkably reticent about committing its Fiat G.50s to com-

bat. During the Battle of Britain, for example, the 20o *Gruppo*'s G.50s only flew discrete patrols over the English Channel, while the 18o *Gruppo*'s CR.42 biplanes escorted Fiat BR.20 bombers over Britain in November 1940—with predictably disastrous results when they encountered Hurricanes and Spitfires. By that time, however, another country had been less shy about blooding the G.50 in combat.

After the Soviet Union attacked Finland on November 30, 1939, Italy—which unlike Germany had not signed a nonaggression pact with Stalin—shipped some of its G.50s to the beleaguered Finns. The first Fiats were organized into a *Koelentue* (Test Flight) under *Kapteeni* (Captain) Erkki Olavi Ehrnrooth, and were soon "tested" in battle. On January 13, 1940, Ehrnrooth, appropriately flying a Fiat bearing the serial number FA-1, shot down an SB-2 bomber over Sisä-Suomi, followed by an Ilyushin DB-3 on January 29.

By February, G.50s were actively serving in a regular squadron, *Lentolaivue 26*, which added a number of additional Soviet aircraft to the butcher bill before Finland finally capitulated on March 13, 1940. On February 26, a Spanish Civil War match—which had not, up till that time, occurred—was finally achieved when *Luutnantti* (Lieutenant) Risto Olli Petter Puhakka, in Fiat FA-4, took on an I-16 and shot it down over Etelä-Suomi for his fifth victory (he had scored four earlier in Fokker D.XXIs). On February 29, *Vääpeli* (Sergeant Major) Lasse Erik Aaltonen, also flying FA-4, downed a DB-3, followed by an I-153 on March 2. Two days before the armistice Puhakka, in FA-21, destroyed a DB-3. The Finnish Fiats would serve on in the Continuation War as well, with considerably more distinction than they achieved in *Regia*

Aeronautica service. Puhakka would score another eleven victories in G.50s before going on to Me 109Gs and finishing out the war as Finland's sixth-ranking ace with a final tally of forty-two, while Aaltonen's tally would total at least a dozen.

The Red Army Air Force (*Voyenno-Vozdushny Sili*) came away from the Spanish Civil War with mixed feelings about the monoplane as the fighter of the future. To hedge its bets, Polikarpov was ordered to manufacture a cleaner version of the I-15 with retractable landing gear. Built after the intrinsically more advanced I-16, the I-153 biplane was to be the second most numerous fighter in the V-VS at the time of the German invasion of the Soviet Union on June 22, 1941. However, neither the I-153 nor late-model variants of the I-16 would prove to be a match for the Me 109E "Emils," Me 109F "Friedrichs," and DB 605-powered Me-109G "Gustavs" that served as the vanguard of the rampaging Luftwaffe.

While the CR.32's and I-16's relatively short careers were enough to earn them a place in aviation history, the Me 109 achieved an uncommon measure of longevity as well as immortality. After being adopted as the Luftwaffe's principal fighter, it underwent a number of engine and airframe changes, culminating in the Me 109K "Karl," which kept it an adversary to be reckoned with right up to the fall of the Third Reich on May 8, 1945. By then, 35,675 Me 109s of all models had been built, making them among the most heavily produced airplanes in history. Moreover, the ubiquitous Me 109 was credited with shooting down more enemy aircraft and producing more aces than any single fighter in the annals of aerial warfare—including the leading ace of all time, Erich Hartmann, with 352 victories. Even after the war, a Czech-built Jumo-engine version, the Avia Š-199, had the ironic distinction of being the first fighter to serve in the Israel Defense Force/Air Force during that new nation's fight for survival in 1948–49.

The Me 109's operational career ended where it began—in Spain. In 1945, Hispano Aviación installed 1,300-horsepower Hispano-Suiza 12-Z-89 engines in the airframes of the Me 109Gs it had imported from Germany, and subsequently manufactured its own version. The final Spanish variant, the HA-1112-MIL *Buchon*, powered by a 1,400-horsepower Rolls-Royce Merlin 500/45 engine driving a Rotol four-bladed propeller, was built until 1956 and soldiered on into the 1960s. As if using the same engine as its most famous enemy was not final irony enough, in 1969 the Merlin-engine Me 109 represented its German forebear in a film re-creation of its earlier duels with the Spitfire: *The Battle of Britain*.

Seven

First Shots of a Second World War

CZECHOSLOVAKIAN, POLISH, GERMAN, FRENCH, DUTCH, AND YUGOSLAV FIGHTERS, 1939–41

Although World War II officially began with the German invasion of Poland on September 1, 1939, there had been conflict aplenty in Europe earlier that year. The spring of 1939 saw both the Spanish Civil War winding down and the flare-up of a brief territorial struggle between Hungary and the newly formed republic of Slovakia. The latter conflict saw the less than auspicious combat debut of an Eastern European fighter that had earned a place alongside the classic biplanes of aviation's golden era: Czechoslovakia's Avia B.534.

The Czechoslovakian republic, formed after the disintegration of the Austro-Hungarian Empire on October 28, 1918, wasted no time in building up an army—along with an air force, the *Ceskoslovenské Vojenské Letectvo*, and its own aviation industry, sustained by talented designers and engineers who had gained previous experience in Austro-Hungarian service. One of its manufacturers, Avia, was established in 1920 and was bought by the powerful Skoda armaments firm in 1926. In 1930, Avia designer František Novotny produced the B.33 fighter, followed by the B.34 in 1931. First trials of the B.34 were

disappointing; but Novotny subjected it to a series of modifications and engine changes, ultimately leading to the use of a supercharged 775-horsepower Hispano-Suiza 12 Ybrs engine in the B.534/1. All the pilots who flew the B.534/1 in August and September 1933 were impressed not only by the aesthetically pleasing, well-balanced contours of the plane, but also by its speed, climb, and diving characteristics. On July 17, 1934, the Ministry of Defense decided to make the B.534 the standard fighter of the *Vojenské Letectvo*, and in the following month placed an order for 34 planes, which was later increased by another 147.

Avia B.534s came in several models, with open cockpits or Plexiglas canopies, and armament ranging from two to four 7.62mm Model 28 or 30 synchronized machine guns. There was also the Bk.534, with a Swiss-built Oerlikon FFS 20 20mm cannon firing through the reduction gear shaft of its Hispano-Suiza 12 Ycrs engine, as well as two machine guns in the fuselage and racks for six 44-pound bombs under the wings.

The principal variant, the B.534 Series IV, had a maximum speed of 252.12 miles

Still decked in Czechoslovakian national markings, an Avia B.534-IV of the new Slovakian air force's *Stíhací Letka* (Fighter Flight) 45 sits at Spišká Nová Ves airfield on March 16, 1939, shortly before seeing its first combat against the Hungarians. (*Súdny via Csaba B. Stenge*)

per hour empty, but with its full armament of four machine guns and 220 pounds of bombs, that speed dropped to 236 miles per hour. During the Fourth International Flying Meet held at Zürich-Dübendorf, Switzerland, between July 23 and August 1, 1938, the B.534 distinguished itself among the biplanes in almost every event, but almost invariably came off second best against Germany's new Messerschmitt Bf 109. Novotny responded by initiating work on a new monoplane fighter of his own, the B.35, but time was running out—both for Avia and for Czechoslovakia.

In spite of the buildup and modernization of the German armed forces, Czechoslovakians were prepared to fight when Hitler demanded the Sudetenland, with its large German population, to be incorporated into his Reich in 1938. As a result of the Munich Agreement signed on September 30, however, Czechoslovakia had to cede the Sudetenland and its western borders. On March 15, 1939, the Germans occupied the entire country. All of the fighters in the occupied "Protectorate of Bohemia and Moravia" were confiscated by the Germans, but some were given to the breakaway fascist state of Slovakia.

On March 14, another country with which Hitler was trying to ally, Hungary, invaded Subcarpathia, securing it on the eighteenth. The Hungarians then laid plans to advance fifteen to twenty kilometers further west to protect the Ungvár-Uzsok railway line, a vital link to Hungary's ally, Poland. Although Hungary had recognized the new Slovakian state, this incursion would entail the seizure of land that, although dominated by a Hungarian-speaking population, was now regarded by Slovakia as part of its territory.

As both nations mobilized for war, the Slovaks formed a new air arm, the *Slovenské Vzdušné Zbrane* or SVZ, mainly from the former Czechoslovakian 3rd *Letecky Pluk* (Air Regiment), which had been stationed at Piešťany since 1938. Comprised of 164 fighter, bomber, and reconnaissance aircraft, the regiment was reorganized into specialized flights, including Fighter Flights (*Stíhací Letkou*)

37, 38, 39, 45, and 49, equipped with a total of sixty B.534s and one obsolete B.33. The Slovaks then hurriedly transferred their newly acquired warplanes to airfields at Spišská Nová Ves and Nižny Šebeš (Šebastová), near Prešov.

At 0500 on the morning of March 23, 1939, Hungarian troops and armor crossed the Uh River into Slovakia. Three Letov Š.328 two-seat biplanes flew a reconnaissance mission at 0800 hours and then, at 1220, three more Š.328s, escorted by three B.534s, took off to strike at the advancing Hungarian ground forces. At 1500 hours, three B.534s of 45 *Letka*, flown by *Poručík* (Lieutenant) Ján Svetlík, *Rotmajster* (Sergeant) Ján Hergott, and *Desiatnik* (Leading Aircraftman) Martin Danihel, attacked a tank and armored car column of the Hungarian 9. Independent Brigade, which responded with intense antiaircraft fire. Svetlík was shot down and fatally crashed near Ulič. At 1630 a trio of Avias from 49 *Letka*, flown by its acting commander, *Poručík* Ján Prháček, as well as *Desiatnici* Štefan Devan and Cyril Martiš, attacked the same area, and Devan was hit. Although he managed to force land his stricken plane near the railroad station at Stakčin, he died from loss of blood soon after, and Slovak troops destroyed his plane in the face of the advancing enemy. The Hungarian air force had been conspicuous by its absence throughout the day, but in response to the Slovak air attacks, the 1/1 *Vadászrepülö-Század* (1st Squadron of the 1st Fighter Group) at Ungvár (Uzhorod) airfield was placed on alert.

The first air-to-air combat of the conflict began at 0630 the next morning, when three B.534s of 45 *Letka* left Spišská Nová Ves and attacked the Hungarian 2. Motorized Brigade near Sobrance. Telephoned by ground observers of the Avias' presence at 0700, the 2nd Section

of 1/1 *Vadászrepülö-Század*, comprised of three Fiat CR.32s flown by *Föhadnagy* (Senior Lieutenant) Aladár Negró and *Örmesterok* (Sergeants) Árpád Kertész and Sándor Szoyák, took off. Meeting at 0740, the opposing fighters engaged in a swirling dogfight—and the Slovaks took the worst of it. Quickly outmaneuvering one Avia, Negró got on *Poručík* Prháček's tail, and his two 8mm Gebauer machine guns set the B.534 on fire. Badly wounded, Prháček tried to force land near Sobrance, but was killed when one of the bombs that was still toggled under his wings exploded. Szoják scored eight hits on *Desiatnik* Martiš's engine and oil cooler, but after jettisoning his remaining bombs he made an emergency landing near Lúčky, where his plane overturned in the muddy ground. The remaining Avia, piloted by *Slobodnik* (Aircraftman First Class) Michal Karas, used his higher speed to escape Kertész.

At 1000 hours another three B.534s of 45 *Letka*, flown by *Poručík* Jozef Páleníček, *Rotmajster* Ján Hergott, and *Slobodnik* Jozef Zachar, took off to attack the Hungarian troops, but a machine gun from the ground struck Zachar's plane, compelling him to land in a field where he was taken prisoner. The other B.534s had also suffered damage when they returned to their air base.

Air activity reached its crescendo that afternoon. At 1545 three Š.328s of 12 *Letka* left Spišská Nová Ves to observe and attack Hungarian forces along the Ungvár-Michalovce road, accompanied by three B.534s of 45 *Letka*. They were engaged in attacking various targets in the area when nine CR.32s of *Század* 1/1 led by *Föhadnagy* Béla Czekme arrived on the scene. The Hungarian formation was made up of three V-shaped elements, the first of which flew on into a cloud without noticing the Slovaks. In the Third

Section led by *Föhadnagy* László Palkó, one of his wingmen, *Hadnagy* (Lieutenant) Mátyás Pirity, spotted the six Slovakian aircraft to their left and attacked. Just then, the Slovak fighters spotted Negro's Second Section and left the Avias to attack those Fiats—unwittingly leaving the Letovs vulnerable to Palkó's section.

Again, the fight went badly for the Slovaks. Palkó accounted for a Letov, the pilot of which, *Slobodnik* Gustáv Pazicky, was killed. The observer, *Poručík* Ferdinand Švento, bailed out but was shot by members of the Hungarian 1. Hussar Regiment, jumpy after having recently repulsed a Slovak assault, who thought he was reaching into his jacket for a pistol—only to discover afterward that he had only been reaching for his identification papers. Attacking from behind and three meters below the leftmost Letov, crewed by *Slobodnik* Ján Drlička and *Poručík* Ladislav Šronk, Pirity last saw his opponent streaming white smoke, indicating burning oil. It subsequently force landed near Stražske.

Meanwhile, the Avia pilots were fighting for their lives, while *Föhadnagy* Negró and *Örmester* Szoyák each claimed his second victim of the day. *Rotmajster* Hergott, probably attacked by Negró, force landed southeast of Bánovce nad Ondavou with his engine burning and light wounds to his right arm and left leg. Szoyák's victim, *Čatnik* (Corporal) František Hanovec, took engine hits and force landed southeast of Senné. *Föhadnagy* Ántál Békássy was credited with a third B.534, probably flown by *Desiatnik* Danihel, who force landed unhurt and whose plane was subsequently recovered. *Föhadnagy* Czekme and *Örmester* Kertész also claimed Avias, but there were no further Slovak losses. For their part, Hanovec later claimed that both he and Danihel shot down Fiats in

the action, but none of the Hungarian fighters were even damaged that day.

Meanwhile, during an interrogation of the captured *Desiatnik* Zachar, the Hungarians learned of the concentration of Slovak aircraft at Spišská Nová Ves. At 1600 hours nine Hungarian Junkers Ju 86K-1s bombed the airfield, damaging hangars and twelve aircraft. One Š.328 pilot tried to take off but was wounded by machine gun fire, and his observer was killed. *Čatnik* František Cyprich managed to take off in a B.534, but did not take part in the fighting.

The Slovaks assembled twenty-eight aircraft for a retaliation raid on Miskolc the next day; but before it could be launched, the Germans persuaded both sides to agree to an armistice, which went into effect on March 26. Negotiations concluded on November 2, 1939, with Hungary allowed to keep 4,200 square kilometers of the Slovakian territory it had occupied, along with the million people living therein. The Hungarians repaired and test-flew Zachar's captured B.534 against the CR.32, and found the Avia to be faster and more maneuverable, but more vulnerable to gunfire than the Fiat.

The Avia B.534 went on to a curious career throughout World War II. When the Germans invaded Poland on September 1, 1939, 35,000 Slovakian troops moved into the regions of Orava and Spiš—Slovakian territories that had been taken by Poland in the wake of the 1938 Munich Agreement—and twenty B.534s participated in the invasion. Two were lost, but on September 26 *Čatnik* Viliam Grún scored the SVZ's first confirmed victory—and its only one of the campaign—when he brought down an RWD.8 liaison plane of the 13th *Eskadra Szkolna* near Prešov. The Polish crew were taken prisoner but eventually escaped to

Romania. The B.534 was still the SVZ's principal fighter when Slovakia committed its *Rychlá Divise* (Express Division) to support German Army Group South in the invasion of the Soviet Union on June 22, 1941. By June 1942, however, the B.534 was acknowledged as obsolete and was mainly used against Soviet Ukrainian partisans behind the lines, while Slovak fighter pilots switched to the Me 109E.

When elements of the Slovakian armed forces rose up against the Germans on August 29, 1944, the insurgents operated a handful of aircraft, including two B.534-IVs and a Bk.534, from the airfield at Tri Duby (Three Oaks). The Germans crushed the uprising by the end of October, but not before one of Tri Duby's defenders had added a curious footnote to history. On August 31, *Štábny Rotmajster* (Staff Sergeant) František Cyprich took off in response to reports of an enemy plane south of the field, and attacked what turned out to be a Hungarian Junkers Ju 52/3m transport flying from Budapest to Krakow with a load of mail and half a dozen military passengers. Cyprich set the plane's right engine on fire and killed two passengers on his second firing pass, after which the Ju 52's pilot, *Föhadnagy* György Gách, force landed in a meadow near Radvan. As Gách and his surviving passengers emerged to be taken prisoner by the villagers, Cyprich flew over and waved before turning back for Tri Duby. He may as well have been waving a belated farewell to an era that most people thought had ended years earlier.

Avia B.534s were also used by the Royal Bulgarian Air Force, but by 1943 they were patently useless as fighters. After Bulgaria switched its allegiance from the Axis to the Allies in September 1944, about a dozen B.534s served as close-support planes in operations against the Germans until January 1945. For a biplane that had been obsolete before the war began, the B.534 had had a remarkably long, if less than outstanding, fighting career.

While Czechoslovakia's Avia B.534 was destined to serve on the peripheries of World War II's decisive campaigns, a Polish fighter, the PZL P.11c, met the first one head on.

Built by the *Pánstwowe Zaklady Lotnicze* (National Aviation Establishment) and first flying in August 1931, the PZL P.11 was the descendant of a series of clean monoplanes designed by Zygmunt Pulawski, incorporating a unique gull wing that was thickest near the point where four faired steel struts buttressed it from the fuselage sides. When the first PZL P.1 flew on September 26, 1929, it thrust Poland to the forefront of progressive fighter design. In 1933 Poland's air force, the *Lotnictwo Wojskowe*, became the first in the world to be fully equipped with all-metal monoplane fighters as the improved P.6 and P.7 equipped its *eskadry*. When the production P.11c, powered by a 645-horsepower Škoda-built Bristol Mercury VI S2 nine-cylinder radial engine, entered service in early 1935, it still rated as a modern fighter, with a maximum speed of 242 miles per hour at 18,045 feet and a potent armament of four 7.7mm KM Wz 33 machine guns, although its open cockpit and fixed landing gear were soon to become outdated. By 1939 the P.11c was clearly obsolete, and efforts were already under way to develop a successor to replace it within the year. Poland did not have a year, however—on September 1, time ran out as German forces surged over her borders.

A morning fog over northern Poland thwarted the first German air operation, as *Obltn.* Bruno Dilley led three Junkers Ju 87B-1 Stukas of 3rd *Staffel, Sturzkampfgeschwader* 1 (3./StG 1) into the air at 0426, flew over the border from East Prussia and at 0434—eleven minutes before Germany formally declared war—attacked selected detonation points in an attempt to prevent the destruction of two railroad bridges on the Vistula River. The German attack failed to achieve its goal and the Poles blew up the bridges, denying German forces in East Prussia an easy entry into Tszew (Dirschau). The "fog of war" also handicapped a follow-up attack on Tszew by Dornier Do 17Z bombers of III *Gruppe, Kampfgeschwader* 3 (III./KG 3).

Weather conditions were better to the west, allowing *Luftflotte* 4 to dispatch sixty Heinkel He 111s of KG 4, Ju 87Bs of I./StG 2, and Do 17Es of KG 77 on a series of more effective strikes against Polish air bases near Kraków at about 0530, Rakowice field being the hardest hit. Assigned to escort the Heinkels was a squadron equipped with a new fighter of which Luftwaffe *Reichsmarschall* Hermann Göring expected great things: the Messerschmitt Me 110C-1 strategic fighter, or *Zerstörer*.

The Me 110 had evolved from a concept that had been explored during World War I but which was only put into successful practice by the French with their Caudron 11.A3, a twin-engine, three-seat reconnaissance plane employed as an escort fighter in 1918. The strategic fighter idea was revived in 1934 with the development of the Polish PZL P.38 *Wilk* (Wolf), which inspired a variety of similar twin-engine fighter designs in France, Germany, Britain, the Netherlands, Japan, and the United States.

Göring was particularly enthralled by what he dubbed the *Kampfzerstörer* (battle

Although obsolete by the time it fired its guns in anger on September 1, 1939, the PZL P-11c scored the first Allied victory of World War II. After claiming a total of 129 German aircraft, scores of its pilots went on to become aces in French, British, and American service. (*RAF Museum*)

destroyer), and in 1934 he issued a specification for a heavily armed twin-engine multipurpose fighter capable of escorting bombers, establishing air superiority deep in enemy territory, carrying out ground-attack missions, and intercepting enemy bombers. BFW, Focke-Wulf, and Henschel submitted design proposals; but it was Willy Messerschmitt's sleek BFW Bf 110, which ignored the bombing requirement to concentrate on speed and cannon armament, that won out over the Fw 57 and the Hs 124. Powered by two Daimler Benz DB 600A engines, the Bf 110V1 was first flown by Rudolf Opitz on May 12, 1936, and attained a speed of 314 miles per hour, but the unreliability of its engines required a change to 680-horsepower Junkers Jumo 210Da engines when the preproduction Bf 110A-0 was completed in August 1937.

Although more sluggish than single-seat fighters, the Bf 110A-0 was fast for a twin-engine plane, and its armament of four nose-mounted 7.9mm MG 17 machine guns and one flexible 7.9mm MG 15 gun aft was considered impressive. Prospective *Zerstörer* pilots were convinced that tactics could be devised to maximize its strengths and minimize its shortcomings, just as the British had done with the Bristol fighter in 1917. The Bf 110B-1, which entered production in March 1938, was even more promising, with a more aerodynamically refined nose section housing a pair of 20mm MG FF cannon. Later, in 1938, the 1,100-horsepower DB 601A-1 engine was finally certified for installation, and in January 1939 the first Messerschmitt Me 110C-1s, powered by the DB 601A-1s and bearing a new prefix to mark Willy Messerschmitt's acquisition of BFW, entered service. By September 1, a total of eighty-two Me 110s were operating with I *Gruppe* (*Zerstörer*) of *Lehrgeschwader* (Operational Training Wing) 1 (I(Z)./LG 1) commanded by *Maj.* Walter Grabmann, and I *Gruppe, Zerstörergeschwader* 1 (I./ZG 1) under *Maj.* Joachim-Friedrich Huth, both assigned to *Luftflotte* 1; and with I./ZG 76 led by *Hptmn.* Günther Reinecke, attached to *Luftflotte* 4 along the Polish-Czechoslovakian border.

Intensely trained for their multiple tasks, the *Zerstörer* pilots, like those flying the Stuka, had been indoctrinated to think of themselves as an elite force. Therefore, the Me 110C-1 crewmen of the 2nd *Staffel* of ZG 76 were as eager as Göring himself to see their mettle tested as they took off at 0600 hours to escort KG 4's He 111s. To the Germans' surprise and disappointment, they encountered no opposition over Kraków.

During the return flight, 2./ZG 76's *Staffelführer, Obltn.* Wolfgang Falck, spotted a lone Heinkel He 46 army reconnaissance plane and flew down to offer it protection, only to be fired at by its nervous gunner. Minutes later Falck encountered another plane, which he identified as a PZL P.23 light bomber. "As I tried to gain some height he curved into the sun and as he did I caught a glimpse of red on his wing," Falck recalled. "As I turned into

him I opened fire, but fortunately, my marksmanship was no better than the reconnaissance gunner's had been, [for] as he banked to get away I saw it was a Stuka. I then realized that what I had thought was a red Polish insignia was actually a red E. I reported this immediately after landing and before long the colored letters on wings of our aircraft were overpainted in black."[1]

As the Stukas of I./StG 2 were returning from their strike, they passed over Balice airfield just as PZL fighters of the III/2 *Dywizjon* (121st and 122nd *Eskadry*), attached to the Army of Kraków, were taking off. By sheer chance one of the Stuka pilots, *Ltn.* Frank Neubert, found himself in position to get a burst from his wing guns into the leading P.11c's cockpit, after which he reported that it "suddenly explode[d] in mid-air, bursting apart like a huge fireball—the fragments literally flew around our ears." Neubert's Stuka had scored the first air-to-air victory of World War II—and killed the commander of the III/2 *Dyon, Kapitan* Mieczyslaw Medwecki.[2]

Medwecki's wingman, *Porucznik* (Lieutenant) Wladyslaw Gnys of the 121st *Eskadra*, was more fortunate, managing to evade the bombs and bullets of the oncoming trio of Stukas and get clear of his beleaguered airfield. Minutes later, he encountered two returning Do 17Es of KG 77 over Olkusz and attacked. One went down in the village of Zurada, south of Olkusz, and Gnys was subsequently credited with the first Allied aerial victory of World War II. Shortly afterward, the wreckage of the other Do 17E was also found at Zurada and confirmed as Gnys's second victory. None of the German bomber crewmen survived.

In spite of the adverse weather that had spoiled its first missions, *Luftflotte* 1 launched more bombing raids from East

Prussia, including a probing attack on Okacie airfield outside Warsaw by sixty He 111Ps of *Lehrgeschwader* 1, escorted by Me 110Cs of the wing's *Zerstörergruppe,* I(Z)./LG 1. As the Heinkels neared their target, the Polish *Brygada Poscigowa* (Pursuit Brigade), on alert since dawn, was warned of the Germans' approach by its observation posts, and at 0650 it ordered thirty PZL P.11s and P.7s of the 111th, 112th, 113th, and 114th *Eskadry* up from their airfields at Zielonka and Poniatów to intercept. Minutes later, the Poles encountered scattered German formations and waded in, with *Kapral* (Corporal) Andrzej Niewiara and *Porucznik* Aleksander Gabszewicz sharing in the destruction of the first He 111. Over the next hour, the air battle took the form of numerous individual duels, during which *Kapitan* Adam Kowalczyk, commander of the IV/I *Dyon*, downed a Heinkel, and *Porucznik* Hieronim Dudwal of the 113th *Eskadra* destroyed another.

The Me 110s pounced on the PZLs, but the *Zerstörer* pilots found their nimble quarry to be most elusive targets. *Podporucznik* (Sub-Lieutenant) Jerzy Palusinski of the 111th *Eskadra* turned the tables on one of the *Zerstörer* and sent it out of the fight in a damaged state. Its wounded pilot was *Maj.* Walter Grabmann, a Spanish Civil War veteran of the Legion Condor and now commander of I(Z)./LG 1.

In all, the Poles claimed six He 111s, while the German bombers were credited with four PZLs; their gunners had in fact brought down three. Once again, Göring's vaunted *Zerstörer* crews returned to base empty handed. When the Germans sent reconnaissance planes over the area to assess the bombing results at about noon, *Porucznik* Stefan Okrzeja of the 112th *Eskadra* caught one of the Do 17s and shot it down over the Warsaw suburbs.

Messerschmitt Me-110C M8-DK of the 2nd *Staffel,* *Zerstörergeschwader* 76, lurks in a wooded glade shortly before being unleashed against Poland on September 1, 1939. (*Bundesarchive*)

As the weather improved, *Luftflotte* 1 struck again in even greater force, as two hundred bombers attacked Okecie, Mokotow, Goclaw, and bridges across the Vistula. They were met by thirty P.11s and P.7s of the *Brygada Poscigowa*, which claimed two He 111Ps of KG 27, a Do 17, and a Ju 87 before the escorting Me 110Cs of I(Z)./LG 1 descended on them. This time the *Zerstörer* finally drew blood, claiming five PZLs without loss, and indeed the Poles lost five of their elderly PZL P.7s. One Me 110 victim, *Porucznik* Feliks Szyszka, reported that the Germans attacked him as he parachuted to earth, putting seventeen bullets in his leg. The Me 110s also damaged the P.11c of Hieronim Dudwal, who landed with the fuselage just aft of the cockpit badly shot-up; two bare metal plates were crudely fixed in place over the damaged area, but the plane was still not fully airworthy when the Germans overran his airfield.

For most of September 1, the Me 109s were confined to a defensive posture, save for a few strafing sorties. For the second bombing mission in the Warsaw area, however, I. *Gruppe* of *Jagdgeschwader* 21 was ordered to take off from its forward field at Arys-Rostken and escort KG 27's He 111s. The Me 109s rendezvoused with the bombers, only to be fired upon by their gunners. When the *Gruppenkommandeur* (Group Commander), *Hptmn.* Martin Mettig, tried to fire a recognition flare, it malfunctioned, filling his cockpit with red and white fragments. Mettig, blinded and wounded in the hand and thigh, jettisoned his canopy—which broke off his radio mast—and turned back. Most of Mettig's pilots saw him head for base, and being unable to communicate with him by radio, they followed him. Only upon landing did they learn what had happened.

Not all of the *Gruppe* had seen Mettig, however, and those pilots who continued the mission were rewarded by encountering a group of PZL fighters. In the wild dogfight that followed, the Germans claimed four of the P.11cs, including the first victory of an eventual ninety-eight by *Ltn.* Gustav Rödel. The Poles claimed five Me 109s, including one each credited to *Podporuczniki* Jerzy Radomski and Jan Borowski of the 113th *Eskadra*, and one to *Kapitan* Gustaw Sidorowicz of the 111th. *Podpolkovnik* (Lieutenant Colonel) Leopold Pamula, already credited with an He 111P and a Ju 87B earlier that day, rammed one of the German fighters and then bailed out safely. *Porucznik* Gabszewicz was shot down by an Me 109 and, like Szyszka, subsequently claimed that the Germans had fired at him while he parachuted down.

In addition to challenging the waves of German bombers and escorts that would ultimately overwhelm them, PZL pilots took a toll on the army cooperation aircraft which were performing reconnaissance missions for the advancing panzer divisions. *Podporucznik* Waclaw S. Król of the 121st *Eskadra* downed a Henschel Hs 126, while *Kapral* Jan Kremski shared in the destruction of another. After taking off on their second mission of the day to

intercept a reported Do 17 formation at 1521 hours, *Porucznik* Marian Pisarek and *Kapral* Benedykt Mielczynski of the 141st *Eskadra* spotted an Hs 126 of 3.(H)/21 (3 *Staffel* (*Heeres*), *Aufklärungsgruppe* 21, or 3rd Squadron Army of Reconnaissance Group 21), attacked it and sent it crashing to earth near Torun. The pilot, *Obltn.* Friedrich Wimmer, and his observer, *Obltn.* Siegfried von Heymann, were both wounded. Shortly afterward, two more P.11cs from their sister unit, the 142nd *Eskadra*, flew over the downed Henschel, and one of the Poles, *Porucznik* Stanislaw Skalski, later described what occurred when he landed nearby to recover maps and other information from the cockpit:

> The pilot, Friedrich Wimmer, was slightly wounded in the leg; his navigator, whose name was von Heymann, had nine bullets in his back and shoulder. I did what I could for them and stayed with them until an ambulance came. The prisoners were transferred to Warsaw. After the Soviet Union invaded Poland on 17 September, they became prisoners of the Russians, but were released at the end of October. When they were interrogated by the highest Luftwaffe authorities, Wimmer told them of my generosity. The Germans, who later learned that I had gone to Britain to fight on, said if I should become their prisoner, I would be honored very highly.
>
> The observer, von Heymann, died in 1988. . . . I tried to get in touch with the pilot for three years. The British air attaché and Luftwaffe archives helped me to contact Colonel Wimmer. I went to Bonn to meet him in March 1990, and the German ace Adolf Galland also came over at that time. In 1993, Polish television went with me, to make a film with Wimmer. Reporters asked why I did it—why I landed and helped the enemy, exposing my fighter and myself to enemy

air attack. I was young, stupid and lucky. That is always my answer!

I came back late in the afternoon and I had to land on the road close to a forest—Torun aerodrome had been bombed already. I then gave [*General Dywizji* Wladyslaw] Bortnowski, commander of the *Armia Pomorze*, the maps that I had captured from the Hs 126, which gave all the dispositions and attack plans of German divisions in Pomerania. He kissed me and said this was all the information his army needed.[3]

On the following day, Skalski came head on at what he described as a "cannon-armed" Do 17 in a circling formation of nine and shot it down, then claimed a second bomber minutes later. Dorniers were not armed with cannon; but Me 110s were, and Skalski subsequently recalled that the Poles were completely unfamiliar with the *Zerstörer*—nobody had seen them in action until September 1. Moreover, I/ZG 1 lost a Bf 110B-1, its pilot, *Hptmn.* Adolf Gebhard Egon Claus-Wendelin, *Freiherr* von Müllenheim-Rechberg, commander of the 3rd *Staffel*, being killed, while his radioman, *Gefreiter* Hans Weng, bailed out and was taken prisoner of war (POW). Skalski's "double" was the first of four and one shared victories with which he would be officially credited during the Polish campaign. Later, flying with the Royal Air Force, he would bring his total up to 18 11/12, making him the highest-scoring Polish ace of the war.

Although Poland was overrun in three weeks, its air force occasionally put up a magnificent fight, though its efforts were rendered inconsistent by poor communications and coordination. Polish fighters were credited with 129 aerial victories for the loss of 114 planes, and many of the pilots who scored them would fight on in

the French *Armée de l'Air* and the Royal Air Force.

The fall of Poland terminated the career of the PZL P.11c, but only marked the beginning for the Me 110, which, after a further run of success, finally met its nemesis in the form of the Hurricane and Spitfire. Relegated to fighter-bomber and photoreconnaissance duties after the Battle of Britain, the *Zerstörer* would undergo a remarkably productive revival as a night fighter.

Although France and Britain had let Hitler get away with his takeovers of Austria and Czechoslovakia, he overplayed his hand by invading Poland, and by September 3 Germany found itself at war with both powers. Apart from an abortive French offensive into the Saar region, however, the armies on both sides of the Franco-German border spent months in a state of stalemate that was called the *Drôle de Guerre* by the French, *Sitzkrieg* by the Germans, and the Phony War by the British. The opposing air arms were less shy about trading shots, however, and at 0730 hours on September 8 six Curtiss H.75A Hawks of *Groupe de Chasse* II/4, led by *Adjudant-Chef* (Chief Warrant Officer) Robert Cruchant, were on a mission to escort a reconnaissance plane over the Saar region when they came under heavy antiaircraft fire and then were bounced by a four-plane *Schwarm* of Me 109E-1s. Using the superior maneuverability of their H.75As to advantage, the French managed to turn the tables on their attackers and claimed two of the Messerschmitts—one jointly credited to Cruchant and *Adj.* Pierre Villey and the other to Cruchant and *Sergent-Chef* (Master Sergeant) Jean Casenobe.

The German protagonists in this first encounter over the Western Front were from the 1st *Staffel* of JG 53, and in fact their only casualty occurred when the leader of the four-aircraft *Schwarm* was hit in the engine and, as he tried to force land in a field near Wölfersweiler, his wheels sank in the soft ground and his plane somersaulted over onto its back. It took the combined effort of three burly farm laborers and members of a nearby flak unit to free the pilot, *Obltn.* Werner Mölders, who had to spend the next few days in bed with a strained back.

On the same day, two Bf 109Ds of *Jagdgruppe* 152 were strafed on the German airfield at Saarbrücken by a Morane-Saulnier MS.406 of GC I/3. Three more MS.406s of the *groupe*'s 1*ème Escadrille* attacked the same field on the ninth, destroying several German planes on the ground, for which *Lt.* Gaston Lacombe, the squadron's commander who had led the strike, was reprimanded by his superiors for having carried out an unauthorized mission.

That marked the combat debut of France's principal indigenous fighter. Its precursor, the MS.405, was conceived in secret in 1934 and marked a departure from Morane-Saulnier's long line of parasol monoplanes when it first flew on August 8, 1935. An improved version, the MS.406, entered production on February 8, 1938. Powered by an 860-horsepower Hispano-Suiza 12Y-31 V-12 liquid-cooled engine with a retractable ventral radiator, the MS.406 had a maximum speed of 302 miles per hour and was armed with a 20mm Hispano-Suiza HS 9 or HS 404 cannon firing through the propeller hub and two wing-mounted 7.5mm MAC 1934 machine guns. Most of the airplane was skinned with Plymax, a light alloy bonded to plywood, but the rear of the fuselage was fabric covered. A total of 1,081 MS.406s were built, and 1,074 had been delivered to ten *groupes de chasse* by

May 10, 1940, making them the most numerous fighters in the *Armée de l'Air*.

When it entered service on March 1, 1939, the MS.406's performance was roughly a match for that of the Messerschmitt Bf 109D, but by September of that year, it was seriously outclassed by the Me 109E and several other contemporaries—including the Curtiss H.75A, which a desperate France had ordered as a stopgap measure until a better home-built successor, the Dewoitine D.520, became available.

After recovering from his minor back injury, Mölders eventually got his revenge—and his first victory of World War II—by downing one of two H.75As claimed by the Germans on September 20. The French involved were from GC II/5—*Sergent* Péchaud force landed near Saint-Mihiel with twenty-two bullet holes in his plane, while *Sgt.* Roger Quéquiner went down in flames, bailing out and surviving with severe burns on his face and hands. *Sergent* André Legrand was credited with an Me 109 in the fight, compelling *Uffz.* Martin Winkler of 3./JG 53 to make an emergency landing in German lines.

Also on September 20, *Oberfeldwebel* (Sergeant First Class) Willibald Hien of 3./JG 53 was credited with a "Morane," although the French did not record any MS.406 losses. They would on the following day, however, when the Me 109Es of 1./JG 53 had their first major air-to-air encounter with the MS.406, and quickly established the disparity between the two fighters. The lower wing loading and smooth responsiveness of the MS.406 made it the more nimble plane; but the Me 109E's performance was superior in all other respects, and the fight ended in the Germans' favor, with *Sous-Lt.* Marius Baizé of 1ère *Escadrille*, GC I/3 falling near Bliesbrück, the war's first French

The Morane-Saulnier MS.406 of *Adj*. Jacques Vinchon of the 2*ème Escadrille*, *Groupe de Chasse* I/3 awaits maintenance at Vélaine-en-Haye airfield in September 1939. On the 22nd Vinchon shared in bringing down two Messerschmitt Me 109Es of *Jagdgeschwader* 76, and subsequently scored a third victory flying a Dewoitine D.520. (*Collection Bernard Philippe via Christophe Cony*)

fighter pilot fatality. His was one of three Moranes claimed by 1./JG 53's *Staffelführer*, *Hptmn*. Erich Mix, though only one seems to have been confirmed. Mix, a forty-one-year-old World War I veteran, had scored three victories flying Fokker D.VIIs with Royal Saxon *Jasta* 54, and would raise his overall tally to at least eleven before being retired from his second war.

The MS.406s were more successful the next day, September 22, when aircraft of GC III/2 attacked a Do 17P of 3.(F)/22 (3rd *Staffel* [*Fernaufklärung*], *Aufklärungs-gruppe* 22, or 3rd Squadron, Long-Distance Reconnaissance Group 22), escorted by four Me 109s of 6./JG 53. *Adjudant* Maurice Romey claimed the Dornier but lacked enough witnesses for a confirmation—which in fact is somewhat ironic since the plane crashed at Landweiler-Reden with two crewmen dead and one wounded. Also unwitnessed was the loss of one of 6./JG 53's Messerschmitts, which force landed and burned. The French lost their first MS.406 when *Sergent* Duclos accidentally landed at Ensheim and found himself to be in German territory. He was taken prisoner before he could destroy his plane, but French artillery was in range, and it did the job for him.

Another thirteen MS.406s would be lost before the Germans launched their offensive in the West on May 10, 1940, though the Messerschmitts did not have things entirely their way. On September 24, MS.406s of the 1*ère Escadrille* of GC I/3 were escorting three Mureaux 115s on a reconnaissance mission when they were attacked by six Bf 109D-1s of *Jagdgruppe* 152, a unit slated for reequipment with new Me 110Cs and redesignation as I./ZG 52, but for the time being flying border patrols from Illesheim. In the ensuing fight, *Sgt*. Antonin Combette had a fierce dogfight with *Ltn*. Kurt Rosenkranz of 3./JGr 152, who put twenty-eight holes in his Morane-Saulnier before Combette drove him down to land at Rimling, where Rosenkranz was taken prisoner. *Capitaine* Roger Gérard was also credited with a Bf 109 fourteen kilometers east of Sarreguemines, whose pilot, *Gefreiter* Hesselbach of 1./JGr 152, was also taken prisoner. *Sergent* Garnier claimed a third victory that could not be confirmed. The Germans were a bit more prone to overclaiming on this occasion, with Rosenkranz and his *Staffelführer*,

Obltn. Werner Schnoor, being credited with Moranes, and two others, *Leutnant* Hagen and *Obfw.* Johannes Oretel, making claims that were not confirmed.

The cannier Messerschmitt pilots soon learned not to let themselves be enticed into traditional turning engagements with MS.406s. Although the Morane-Saulnier pilots would claim a grand total of 191 confirmed victories and 93 "probables"—figures not borne out by German loss records—about 100 MS.406s were lost in aerial combat and another 200 destroyed on the ground or written off due to other causes by the time France capitulated on June 18, 1940.

When the Germans finally launched their offensive in the West on May 10, 1940, two powerful new Allied fighters were ready to stand in their way, though there was not enough of either to do so for long.

Designed by a team headed by Dr. Erich Schatzki, the prototype of the Fokker G-1 twin-engine fighter was first unveiled at the 1936 Paris Salon, where it caused a sensation. A twin-boomed heavy fighter with a central nacelle that could be modified to fulfill a variety of tasks, the G-1 made its first flight on March 16, 1937, and entered service with the Royal Netherlands Air Force in May 1938. Officially referred to as a *Jachtkruiser*, it came to be nicknamed *Le Faucheur* (The Reaper) by its crews.

The original G-1, powered by two 830-horsepower Bristol Mercury VIII nine-cylinder radial engines, had no less than eight 7.9mm FN-Browning M36 machine guns in the nose of the nacelle, as well as a ninth gun in a rotating tail cone, in addition to which it could carry an internal bombload of 880 pounds. Another G-1 variant, twelve of which

had been intended for use by the Spanish Republican forces before the Dutch government placed an embargo on their export, was powered by two 750-horsepower Pratt & Whitney R-1595-SB4-G Twin Wasp Junior fourteen-cylinder radials and had a nose armament of two 23mm Madsen cannon and two 7.9mm FN-Brownings.

The primary duty of Dutch aircraft during the first months of the war was to guard the country's neutrality, and it was in that pursuit that the Fokker G.1 first fired its guns in anger. At 2305 Greenwich time on the night of March 27, 1940, Armstrong Whitworth Whitley Mark V N1357, call sign KN-H of No. 77 Squadron, departed Driffield to drop propaganda leaflets, but strayed into Dutch air space on the homeward leg and came under attack at 0630 by a Mercury-powered Fokker G.1 piloted by 1e *Luitenant-Vlieger* (First Lieutenant) Piet Noomen of the 3e *Jachtvliegtuig Afdeling*. Set on fire, the Whitley came down on the Vondelingenweg at Pernis. The bomber's observer, Sgt. J. E. Miller, was killed and is believed to have fallen from the plane seconds before it crashed. The rest of the crew—Flying Officers T. J. Geach and W. P. Copinger, Leading Airman S. E. E. Caplin, and Airman 2nd Class R. B. Barrie—were interned but soon released and returned to Driffield by the Dutch, who may have little suspected that the unwelcome intruders Noomen had intercepted would be his allies just seven weeks later.

A total of twenty-three serviceable Fokker G.1s were available to the Dutch—eleven with the 3e *JaVA* at Rotterdam/Waalhaven and twelve with the 4e *JaVA* at Bergen—when the Germans invaded. Expecting trouble, pilots of the 3e *JaVA* had started their engines at 0300 hours Amsterdam time on May 10, just to warm them up before shutting down again. That would be to their benefit at 0350 (0530 Berlin time), when twenty-eight He 111Ps of II./KG 4, in a succession of three-plane Vs led by the *Geschwaderkommandeur, Oberst* Martin Fiebig, skirted the Dutch coast, then turned shoreward over the Maas estuary and came from the southwest to bomb the Koolhoven aircraft factory near Waalhaven.

The first of 3*e JaVA*'s G.1s to get off the runway was apparently No. 312 crewed by *Luitenant* Noomen and *Korporaal* (Corporal) H. de Vries, who attacked the three lead Heinkels, damaging one and wounding three crewmen before attacking Fiebig's He 111P 5J+DA. Noomen was credited with both bombers before return fire compelled him to land with one damaged engine and two punctured fuel tanks after ten minutes in the air. Retiring to the southeast, Fiebig belly-landed south of Zwartedijk fifteen minutes later with his rear gunner mortally wounded; he and the rest of his crew were taken prisoner.

Hard on Noomen's heels, at 0351 1*e Luitenant* Jan Pieter Kuipers scrambled up in G.1 309 and engaged a second wave of bombers from KG 4's 5th *Staffel*. He had to abort his first attack when his rear gunner, *Sgt.* Jan Reinder Venema, reported three German aircraft approaching from behind and to the left. Kuipers made a climbing turn to the left, found himself behind three He 111s and opened fire at two hundred meters distance. He subsequently reported:

> The enemy gunners immediately responded. The combat offered a fascinating spectacle: all the bullets that the antagonists served up were tracers. My first reaction had to be to try to put the machine-gunners out of action. For that

Fokker G.1 312 was the first of its type to scramble up and engage German bombers on May 10, 1940, two being credited to the crew of 1*e Luitenant-Vlieger* Piet Noomen and *Korporaal* H. de Vries in ten minutes before they were forced to land with a damaged engine and two punctured fuel tanks. Seven weeks earlier Noomen had scored the first Fokker G.1 victory when he brought down a British bomber that had violated Dutch air space. (*Netherlands Institute of Military History*)

effect, I fired successively on all three airplanes. During that action, the distance between the squadron and I finally came down to 25 to 50 meters.

As we flew over Rotterdam (Charlois quarter), the squadron turned south, all maintaining a tight formation. Southwest of Waalhaven aeroport, the first bomber was finally forced to land on its belly east of Pernis, another Heinkel went into a pronounced turn and fell into a dive. I would not observe the result further because I had already thrown myself into the pursuit of the third machine. However, my Fokker had not left the fight without harm and at a given moment my left motor's power diminished and then it stopped completely. Forced to make a half turn, it was only with great effort that I succeeded in landing at Waalhaven airport.[4]

It was 0410 as Kuipers and Venema scrambled out of their disabled plane and joined the ground forces defending the airfield. Kuipers assisted an antiaircraft section at the northeast part of the field until Ju 88As of KG 30 attacked them so vigorously that he was forced to take cover in a crater, miraculously emerging unhurt. Most of the antiaircraft gunners were less fortunate. Only after the war did Kuipers learn that Venema was among those killed in the attack.

Meanwhile, at 0352, 2e Lt. Gerben Sonderman, an experienced G.1 test pilot, took off in No. 311, with Sgt. H. Holwerda as his gunner, and headed west to engage a plane he'd spotted milling around the area. This was apparently a Do 17M of the *Fernaufklärerstaffel* (Long-Range Reconnaissance Squadron) of *General* (Lieutenant General) Kurt Student's 7. *Fliegerdivision*, who was there to observe the airborne assault about to occur. Sonderman drove the Dornier off in a damaged state and claimed to have shot down an "Me 110."

While the G.1s that scrambled up were engaging the bombers, at 0450 a horde of Ju 52/3m transports of 9th *Staffel*, *Kampfgeschwader zur besonderen Verwendung* (Special Purpose Battle Wing) 1 arrived over Waalhaven. G.1 315, crewed by 1e Lt. G. A. van Oorschot and Sgt. W. P. Wesly, attacked an He 111P of 1./KG 4 that had just bombed Ypenberg, chased it as far as Arnhem, damaging it and wounding the rear gunner, and then returned to the Rotterdam area in time to shoot down a Ju 52. Van Oorschot then landed at De Kooy, only to damage 315 on an obstruction on the airfield. Less fortunate was 2e Lt. Johannis van der Jagt in G.1 319, who was last seen attacking a Ju 52 when he was shot down, probably by an escorting Bf 109D flown by *Obfw.* Hermann Förster of the 12th *Staffel* (*Nacht*) of *Jagdgeschwader* 2.

Sergeantmajoor-Vlieger (Sergeant Major) Jan J. Buwalda was preparing to take off in G.1 No. 330 at 0400 when he saw three unidentified single-engine planes approaching, while Dutch antiaircraft gunners, equally uncertain of their

nationality, held their fire. Then the trio—which turned out to be Me 109Es—began strafing the field. Gunning his engine and slaloming between barrages of cannon shells and machine gun bullets, Buwalda managed to get his plane airborne. While he fought for altitude his gunner, Sgt. J. Wagner, noted German bombers coming, while two kilometers to the left, flights of Ju 52s were dropping paratroopers over the airfield. Buwalda recalled:

> The Germans destroyed Waalhaven aerodrome to assure their airborne operation. It was 4:00 in the morning, and as the bombardment reached its paroxysm, I succeeded in taking off and found myself in the middle of a packet of bombers flying at 150 meters. In my first attack, I downed a Heinkel . . . then I saw another and I got on his tail, my eight machine guns spitting three short volleys at a distance of 100 meters.[5]

Buwalda was credited with the second plane, a Dornier Do 215 of 2nd *Staffel* (*Fernaufklärung*) of the *Oberbefehlshaber* (Supreme Commander) *der Luftwaffe*, but then he came under attack by twelve Me 109Es. He dived with the enemy in pursuit, his gunner firing at each in turn, allegedly causing one to explode in the air and shooting down a second. "Then they fired at us from above," Buwalda said, "hitting both motors and forcing me to the ground. I was unhurt, but Sergeant Wagner was wounded."[6]

Buwalda's was probably the Fokker G.1 credited as downed over Zevenbergen at 0450 hours to *Obltn.* Richard Leppla of 3./JG 51. The other three Fokkers fared somewhat better. Sonderman claimed a Ju 52 and two fighters before being damaged by fighters. One of his assailants, *Fw.* Peter Keller of 10(N)./JG 2, crash-landed his Bf 109D near Rotterdam, was subsequently transferred from a Dutch compound to England, and spent the rest of the war in a Canadian POW camp. *Sergeant* H. F. Souffrée in G.1 328 brought down He 111P 5J+DN of 5./KG 4 and claimed an Me 109E, while *Lt.* K. W. Woudenberg in 329 claimed two Junkers. Unable to return to Waalhaven, the three planes landed on the beach west of Oostvoorne, where they were hurriedly camouflaged. Amid the confusion of the German offensive, however, it was not until the morning of May 14 that Souffrée and his gunner, *Sgt.* J. C. de Man, managed to return to the beach with fuel, oil, ammunition, and ground personnel for the three G.1s—only to discover that they had been strafed and set afire by Me 109Es just half an hour earlier.

At Bergen, the Luftwaffe found all twelve Fokker G.1s of the 4e *JaVA* parked wingtip to wingtip when they attacked at 0359. One Fokker was destroyed and ten damaged, leaving only aircraft No. 321 to take off, with *Lt.* J. W. Thijsse at the controls, to intercept the next wave of bombers. As he did, however, he came under fire not only from enemy fighters but from Dutch antiaircraft gunners, who were already assuming anything in the air to be German. Thijsse therefore gave up the idea of fighting and sought a safe haven at Schipol airfield—only to find it in flames. He then headed for the beach at Katwijk, where he found three newly landed Ju 52s, which he strafed and set afire. After reconnoitering Ypenberg and Schipol airfields, he opted to return to Bergen, which was having a momentary respite from German attack, and landed at 0620.

Considering the circumstances, the Fokker G.1s—of the 3e *JaVA*, at least—gave an extraordinarily good account of themselves, shooting down at least a dozen German aircraft in their first chaotic two hours of combat. Heavily armed and easy to fly, though too slow to compete with single-engine fighters, the G.1 had lived up to its nickname of *Le Faucheur*, but it would only have four more days in which to fight before the Netherlands was overrun. After that, most surviving G.1s became part of a growing trove of war booty, to serve the Germans as trainers for their twin-engine fighter pilots.

Although numerically better prepared than Belgium or the Netherlands, by 1940 France found most of its fighters to be outclassed by the Luftwaffe's Me 109Es. In addition to the outdated MS.406, the French had just introduced the Bloch MB.152, packing a formidable pair of 20mm cannon in the wings as well as two 7.5mm machine guns. Although the Bloch was robust and handled well, it proved to be underpowered and was consequently outperformed by the Me 109s it encountered.

It was a sad commentary on the state of France's military aircraft development that the best fighter available to the French at the start of 1940 was the Curtiss H.75A—ironically, imported from the United States, which in the previous war had had to purchase its fighters from France—and that too was outperformed by the Messerschmitt in all respects save maneuverability. The great exception was the Dewoitine D.520, an outstanding little fighter for its time, but unfortunately for France, a case of almost too late and decidedly too little.

Emile Dewoitine had built an innovative line of monoplane fighters with fixed undercarriage during the early 1930s. But in June 1936 he set up a new design bureau, headed by Robert Castello, with the intention of creating a new fighter with retractable landing gear and using the 900-horsepower Hispano-Suiza 12Y-21 engine to achieve a speed of 500 kilometers per hour (311 miles per hour). When the *Armée de l'Air* rejected his design due to an upgraded speed specification of 520 kilometers per hour (323 miles per hour), Castello and staff redesigned the plane with a reduced wingspan and an airframe to accommodate a newer 1,200-horsepower engine expected from Hispano-Suiza. Designated D.520 (in reference to the speed requirement), the new fighter was rejected in favor of the already chosen MS.406; but Dewoitine persevered, building two prototypes at his own expense, and finally obtained a government contract on April 3, 1938. The D.520.01 prototype first flew on October 2, 1938, with Marcel Doret at the controls, and after being modified with a single central radiator in place of the original twin underwing radiators, a new HS 12Y 29 engine, and a three-bladed metal propeller in place of the original two-bladed wooden one, it satisfied the speed requirement. After further alterations the French authorities—who were at last coming to the realization that the quality of their fighter force was falling dangerously behind that of Britain and Germany—placed an order for two hundred D.520s in April 1939.

The first production D.520, powered by an 850-horsepower HS 12Y-45 engine for a maximum speed of 332 miles per hour and armed with one 20mm HS 404 cannon firing through the propeller hub and four wing-mounted 7.5mm MAC 1934 machine guns, did not fly until November 1939. The tempo of production rose slowly but steadily from there, and by May 10, 1940, thirty-six D.520s were on the strength of *Groupe de Chasse* I/3 at Cannes-Mandélieu, while GC II/3,

Sergent Paul Bellefin of the *2ème Escadrille*, *Groupe de Chasse* I/3, poses beside his new Dewoitine D.520 at Cannes in March 1940. Flying the MS.406, Bellefin had shared in downing two Me 109Es on September 22, 1939, but he was shot down and killed in a D.520 on May 15, 1940. (*Collection Bernard Philippe via Christophe Cony*)

GC II/7, and GC III/3 were engaged in conversion training in the new type. With the start of the German offensive, GC I/3 was hastily moved to Wez-Thuisy in the *2ème Armée* sector.

The D.520's first day of battle, May 13, went encouragingly well. *Sous-Lieutenant* Georges Blanck was leading *Adj.* Jacques Vinchon and *Sgt.* Lucien Rigalleau on a mission to escort some Lioré-et-Olivier LéO 45 bombers when they encountered an Hs 126 of 1.(H)/14 and drove it down in German lines near Bras-de-la-Semois. GC I/3's pilots claimed another two Hs 126s and an He 111 without loss that day. On May 14 the *groupe* fought over Sedan, downing four Me 110s, two Me 109s, two Do 17s, and two He 111s for the loss of *Sous-Lt.* Lucien Potier and *Adj.* André Carrier, who were killed, the latter probably being an "MS.406" credited to *Obfw.* Johann Schmid of 1./JG 2, his first of forty-five victories. Among the victorious pilots that day were *Sous-Lt.* Michel Madon, who downed an Me 110 of ZG 26 over Brinon, and *Sgt.* Marcel Albert, who teamed up with an MS.406 pilot to bring down a Do 17 north of Suippes. Both Madon and Albert would become aces, Albert eventually flying Yakovlev fighters with the Normandie Regiment on the Eastern Front to become the leading French fighter pilot of the war with a total of twenty-three victories.

Blanck scored again on May 15, sharing in the destruction of an Me 109 near Hastière-sur-Meuse and then downing an He 111 northeast of Reims, while *Sous-Lt.* Émile Thierry accounted for a Do 17 north of Reims. The Dewoitines were credited with a total of six Germans that day, but three were shot down—including that of *Adj.* Antonin Combette, whose score stood at four when he was brought down wounded and taken prisoner—and three others were lost in crash landings. On May 17 the Luftwaffe struck back, attacking Wez-Thuisy in force and destroying seven D.520s on the ground, after which GC I/3 was withdrawn to Meaux-Esbly.

GC II/3, based at Bouillancy, entered the fighting on May 21, joined by GC II/7 and GC III/3 in June. At Le Luc airfield in southern France, GC III/6 received its first D.520 on June 10—the same day on which Italy declared war and invaded southern France. The unit put up a brief but creditable resistance, *Adj.* Pierre Le Gloan of GC III/6's 5ème *Escadrille* adding four Fiat CR.42s and two Fiat BR.20s (including five victories on June 15) to the four German aircraft he had previously downed while flying MS.406s.

A total of 437 D.520s had rolled off Dewoitine's assembly line at Toulouse by June 25—demonstrating how much inertia can be overcome by a national emergency—but by then France had capitulated. The D.520s went on to curious careers in the wake of the armistice, actively serving in such Axis air arms as Italy's, Romania's, Bulgaria's, and that of Vichy France itself, engaging the Royal Air Force over the Levant in 1941, and

the US Navy and Army Air Forces over North Africa in 1942. The fortunes of the surviving D.520s would come around full circle in the summer of 1944, as airmen of the *Forces Françaises de l'Interieur* formed its first fighter unit, commanded by former Dewoitine test pilot Marcel Doret, to participate in the liberation of their country.

One other fighter was produced by a country that had to face German invasion—Yugoslavia, which after young officers overthrew its pro-Axis government on March 27, 1941, and refusing to allow German troops through its territory to assist the Italians in their stalled invasion of Greece, became the target of *Unternehmen Strafgericht* (Operation Punishment) at the orders of an enraged Hitler, on April 6. Among the aircraft thrown up by the Royal Yugoslav Air Force (*Jugoslovensko Kraljevsko Ratno Vazduhoplovstvo*, or JKRV) during the brief but spirited struggle was the indigenously built IK-3.

The IK-3 was the brainchild of three engineers: Ljubomir Ilic, Kosta Sivčev, and Slobodan Zrnic. Ilic was attending the national high school of aeronautics in Paris when he conceived of a new fighter to replace the Czechoslovakian-built Avia BH.33E, then the mainstay of the JKRV. At the same time, he met Sivčev, a pilot and engineer who was then working with the Breguet and Hispano-Suiza firms. The two friends returned to Yugoslavia in 1931, and after overcoming skepticism at air force headquarters, they got permission to build the IK-01, a high-wing monoplane with fixed, spatted undercarriage and a fully enclosed cockpit, powered by an 860-horsepower Hispano-Suiza 12Y-Crs engine and armed with one 20mm cannon firing through the propeller hub and two 7.9mm machine guns in the front upper fuselage. After assembly at the Ikarus factory in Zemun, the prototype flew for the first time on April 24, 1934, but crashed on its third flight. A new prototype, the IK-02, with wings covered with duralumin instead of fabric, was completed in June 1936 and displayed very good performance with a maximum speed of 435 kilometers per hour—15 kilometers per hour higher than originally planned. During comparative tests held in June 1937, the IK-02 proved to be not only faster than the Hawker Fury Mark I but was more agile and had a better rate of climb. Twelve IK-2s were produced and seven were still operational in April 1941.

While they were working on the IK-01 in 1933, Ilic and Sivčev started on a new project. This time it would be a streamlined, low-winged aircraft with retractable landing gear, and enough speed and power to intercept the latest generation of bombers. That project was also started as a private venture when Slobodan Zrnic-Zrle, chief of the construction bureau of the airplane factory at Kraljevo, joined the team. Ilic worked on aerodynamic calculation, Sivčev on equipment, tail surfaces, and ailerons, while Zrnic worked on the airframe and engine installation.

The new aircraft was designated the IK-3, again according to the first letters of the constructors' names—"3" standing in for the "Z" in Cyrillic writing but later being transformed to the number "3" as a serial mark. Powered by a 980-horsepower Hispano-Suiza 12Y-29 engine, the prototype was completed in the Rogožarski factory in Belgrade, and after test-flying it in May 1938, *Kapetan* (Captain) Milan Bjelanovic stated that the plane was very maneuverable and the control efficiency was very good in all

respects, while maximum speed was 527 kilometers per hour. It could climb to 5,000 meters in seven minutes. Armament was comprised of one Oerlikon 20mm cannon firing through the propeller hub and two 7.9mm Browning/FN machine guns in the upper front fuselage. Eighteen production IK-3s had been delivered by the end of July 1940. It was planned to produce forty-eight IK-3s in 1941 and 1942, but twenty-five aircraft were still under construction when the Germans struck.

Air-raid sirens awoke Belgrade on the early morning of Sunday, April 6, 1941, and at 0650 the first German bombers flew over the capital. Outside the city at Zemun airport, less than ten minutes after the alert was sounded, sixteen fighters of the 51st *Grupa*, 6th *Lovački Puk* (Fighter Wing) took off, comprised of six IK-3s of the 161st and 162nd *Eskadrili* and leading ten German-built Messerschmitt Me 109E-3s of the 102nd *Eskadrila*.

After taking off, the formation scattered in pairs and intercepted the Germans, just as the latter arrived over Belgrade. *Potporučnik* (Second Lieutenant) Dušan Borčic of the 161st *Eskadrila* and his wingman, *Potporučnik* Eduard Bamfic, separated over Senhak, one plane heading toward Romania, while the second continued toward the old fortress at Kalemegdan. As Borčic reached an altitude of about 4,000 meters, a formation of eighteen Do 17Zs appeared to the north. He turned his aircraft and closed on them from the rear. When he opened fire with his cannon and two machine guns, the nearest German bomber shook, fell out of formation and plunged down to the banks of the Danube River, where it exploded.

Having scored the first aerial victory for the JKRV, Borčic continued flying north, where more German bombers were coming from Hungary. His IK-3 plunged head on into the enemy formation but was soon caught by escorting Messerschmitt fighters in what the Yugoslavs called the "Devil's Circle." Borčic's short-lived moment of glory ended as his crippled IK-3 crashed on Sarajevska Street. Meanwhile his wingman, Bamfic, had been wounded and his plane damaged in a fight above the village of Batajnica. Bamfic broke off the action and tried to land but was attacked by two Me 109s, and although he evaded them for awhile by a series of steep banking turns, during which his wingtips almost touched the ground, he finally crash-landed on the airfield. His IK-3 was completely demolished, but Bamfic survived.

Yugoslavia's most advanced fighter, the IK-3 put up a spirited but short-lived resistance to the Luftwaffe during *Unternehmen Strafgericht* in April 1941. (*Museum of Aviation, Belgrade*)

Kapetan Sava Poljanec, commander of the 161st *Eskadrila*, lost his wingman, *Podnarednik* (Sergeant) Dušan Vujičic, when the latter suffered engine problems. Climbing alone to 4,000 meters, Poljanec dived at a German formation he spotted over the village of Krnjaca, set the left engine of a German bomber afire and saw it go down into Pančevački Rit. Its crew survived to be taken prisoner. At that moment, tracers shot past Poljanec's cockpit as about twenty yellow-nosed Me 109Es dived toward him. Poljanec, a flight instructor and one of the three best aerobatic pilots in Yugoslavia at that time, rolled and spun his aircraft but soon was caught in a Devil's Circle. One Messerschmitt made the mistake of passing in front of the IK-3, and Poljanec discharged his machine guns into it. Glycol vapor began trailing from the Me 109's fuselage as it fell into an erratic dive, suggesting that the pilot was dead. Poljanec tried to land at Zemun, but his IK-3 was hit again, this time by an Me 110 that had been strafing the airfield. Upon landing, Poljanec jumped from the cockpit of his smoking plane. The Me 110 was chased away by *Podnarednik* Sava Vujovic of the 101st *Eskadrila*.

In spite of the prominent yellow markings employed to distinguish them from their Yugoslav counterparts, German Me 109s were understandably reticent about engaging Yugoslav Messerschmitts; on the other hand, the IK-3s tended to attract their undivided attention. After another air battle with a hundred German attackers at 1000 hours, *Maj.* Adum Romeo was relieved of command of the 162nd *Eskadrila* by his brigade commander for insufficient air activity in the face of the enemy, and replaced by *Kapetan* Todor Gogic. As another wave of thirty-seven Ju 87Bs and thirty Me 109s attackers came on at 1400 hours, Gogic and *Potporučnik* Veljko Vujičic attacked a Stuka formation, shot down one of them, then used up their fuel and ammunition in a fight with the escorting Messerschmitts before returning to base. A third IK-3, flown by *Podnarednik* Milislav Semiz, also downed a Ju 87 that day—the first of four victories credited to him in seventeen sorties before the brief campaign was over.

German bombers mounted a total of 484 sorties in four waves over Belgrade alone in the course of April 6, and small groups or single bombers continued to make nuisance raids in the evening.

According to the 6th *Puk*'s commander, *Podpukovnik* (Lieutenant Colonel) Božidar Kostic, a total of ten German aircraft were downed by pilots of the 51st *Grupa* in that first day. The Germans themselves recorded the loss of two Do 17s from 8./KG 3, four Ju 87s of II./StG 77, one Me 109 of *Stab.*/JG 54, and five Me 110s of I./ZG 26 in combat, as well as a sixth Me 110 that crashed on landing. Determining German claims over IK-3s is handicapped by their unfamiliarity with the fighter, which bore a superficial resemblance to the Hurricanes with which the Yugoslavs were also equipped, but a claim over a "Dewoitine" by *Obfw.* Erwin Riehl of III./JG 77 is one likely candidate.

German raids continued on April 7, but with significantly smaller bombing formations. The Yugoslav fighters attacked in larger groups, breaking up formations and downing a substantial number of German aircraft, but their own losses were also heavy due to increased enemy fighter activity over Belgrade.

The old IK-2s also became involved in the fighting at 1400 hours on April 9, when twenty-seven Me 109Es of III./JG 54 struck at the 4th *Puk*'s airfield at Nova Topola. *Podnarednik* Branko Jovanovic was landing when he came under attack by nine Me 109s but evaded them by pulling his IK-2 into a steep turn. Five of the 107th *Eskadrila*'s IK-2s joined six Hurricanes taking off to defend the airfield, during which *Podnarednik* Zivorad Tomic claimed an Me 109, while *Kapetan* Dragisa Miljevic, commander of the Hurricane-equipped 106th *Eskadrila*, claimed a "probable" before being killed. One IK-2 and two Hurricanes were shot down, while a second IK-2, its pilot wounded, crash-landed. *Leutnant* Erwin Leykauf was credited with a Hurricane in the fight, while *Obltne.* Hans-Ekkehard

Bob and Gerhard Koall of 9./JG 54 were credited with the IK-2s. Tomic's victim, *Gefreiter* F. Fabian, was listed as missing by 9./JG 54, but he had survived and rejoined the unit a few days later.

After several days of attrition, the remaining fighters of the 51st and 32nd Fighter Groups (three IK-3s and eleven Me 109Es) flew to the auxiliary airfield at Veliki Radinci near Ruma, where they were burned on the night of April 11 to prevent their falling into German hands. Some of the remaining IK-2s were destroyed by their crews prior to the retreat, while three planes were captured by the Germans, who later turned them over to the air arm of the *Nezavisna Država Hrvatska* (Independent State of Croatia).

The German occupation of Yugoslavia did not bring an end to the story of the best prewar Yugoslav aircraft. Blueprints for the IK-3 spent the war carefully hidden under the floor of Slobodan Zrnic's house in Belgrade. After the war, relations between Yugoslavia and the Soviet Union gradually deteriorated because of Marshal Yosip Broz Tito's constant refusal to be under Josef Stalin's "protection." In 1948, the dispute culminated in Yugoslavia being under an economic blockade by all Socialist countries, and Ilic, Sivčev, and Zrnic found themselves in business again. The design team built Model 211 and 212 trainers, as well as the S-49A fighter, a direct modification of the IK-3. Forty-five of those fighters were produced, followed by 130 of the more powerful S-49C fighter-bomber, which would serve the Yugoslav Air Force until 1965.

Eight

The Immortals

HAWKER HURRICANE AND SUPERMARINE SPITFIRE,

1939–41

Great Britain was still in the process of modernizing its Royal Air Force when Germany invaded Poland on September 1, 1939, and nowhere was this transitional state more evident than in its fighter squadrons. Some still had the Gloster Gladiator, a biplane with fixed landing gear, on which the only concessions to modernity were an enclosed cockpit and a second pair of .303-inch machine guns that looked like they had been stuck under the wings as an afterthought. A few squadrons, in marked contrast, had equipped with the new Vickers Supermarine Spitfire Mark I, a sleek, shapely thoroughbred with landing gear that hydraulically retracted into remarkably thin wings for a cantilever monoplane. In between those extremes of old and new, filling out the majority of RAF squadrons, was the Hawker Hurricane Mark I.

The Hurricane's designer, Sydney Camm, had been a believer in monoplanes since he joined Harry G. Hawker's engineering company as a draftsman in 1923. The RAF did not agree, however—ever since the bad experiences that No. 60 Squadron, Royal Flying Corps, had

had with the Morane-Saulnier N in 1916, the senior officers of Britain's air arms had been dead set against monoplanes as a matter of principal. Consequently, after becoming Hawker's chief aircraft designer in 1925, Camm dutifully produced a series of high-quality single- and two-seat biplane fighters, with names like Hart, Fury, Hornet, Audax, and Hotspur.

By 1933, Camm was so convinced that the biplane fighter had reached the logical zenith of its performance that he began conceiving a "Fury monoplane." At about that same time, at Vickers Aviation's Supermarine subsidiary, Reginald J. Mitchell was coming to the same conclusion and began working on a fighter based on his experience with a highly successful series of seaplane racers. On January 30 of that fateful year, Adolf Hitler was elected chancellor of Germany. Following the consolidation of his National Socialist Workers' Party in power, Hitler started repudiating terms of the Treaty of Versailles and formally withdrew from the Disarmament Conference and the League of Nations on October 14.

If the prospect of German rearmament was not enough to stir the RAF's air marshals into reconsidering their prejudices against monoplanes, the fact that Hawker's Hart light bomber could outpace most fighters should have. Even so, Camm's and Mitchell's efforts over the next few years remained private ventures. Mitchell entered a monoplane with fixed "trousered" landing gear and powered by a 660-horsepower Rolls-Royce Goshawk engine, the Type 224, in a fighter competition to satisfy the RAF's F7/30 specification; but it failed to find favor, the production contract ultimately going to the Gladiator biplane.

Camm forged ahead with his "Fury monoplane," later called simply the "Monoplane Interceptor." Rolls-Royce's new 900-horsepower PV12 engine was substituted for the Goshawk, its extra weight necessitating the underwing radiator to be moved back to compensate. In so doing, Camm found room under the front of the thick cantilever wing for hydraulically operated inward-retracting landing gear. Expecting higher speeds from his design, he also added a glazed sliding canopy for the pilot. By then, only the shape of the nose and tail surfaces remained to hint at the fighter's Hart and Fury ancestry.

Meanwhile, during discussions on armament at the Air Ministry, Squadron Leader Ralph Sorley, a member of its Operational Requirements Branch, suggested that modern fast bombers would be capable of speeds that would allow an intercepting fighter pilot only two seconds to keep one in his sights, for which reason an armament of eight guns would be necessary to bring the enemy down. Although Air Chief Marshal Sir Robert Brooke-Popham, commander of Britain's air defenses, thought that "eight guns was going a bit far," early in 1935 the Air Ministry at last issued a specification, F5/34, for a monoplane fighter capable

of a speed of 275 miles per hour at 15,000 feet, and of mounting eight .303-inch machine guns.

As far as Camm and Mitchell were concerned, the RAF had finally caught up with them. In fact, they were both so confident of exceeding the specification's expectations that the Air Ministry accommodated them by issuing a revised specification, F36/34, raising the speed requirement from 275 to 330 miles per hour.

Even before Camm submitted his plans to the Air Ministry on September 4, 1934, and displayed a wooden mock-up at Kingston on January 10, 1935, he had been optimistically forging ahead with his "high-speed monoplane." Consequently, he was ready when the Air Ministry finally gave him an official contract on February 12. On November 6, Flt. Lt. Paul Ward Spencer Bulman took the prototype, K5083, for its first flight from Brooklands aerodrome. After further flight testing at Martlesham Heath, the RAF ordered six hundred of the new fighter on June 3, 1936, officially christening it the Hurricane on June 27. The first production Hurricane, L1547, flew on October 12, and fifteen months later No. 111 Squadron became the first RAF unit to be exclusively equipped with a monoplane fighter since 1916.

While the Hurricane and Spitfire were undergoing acceptance trials at Martlesham Heath, a visiting RAF group captain described them in memorable fashion. Passing a Spitfire, he remarked, "There's a racehorse for you." Then, as he approached a Hurricane, he declared, "That's more like an aeroplane."[1]

The more sophisticated Spitfire may have been the better performer of the two, but the Hurricane's simpler construction allowed it to be mass-produced in greater quantities at a time when Britain had a lot of catching up to do. The

Supermarine Spitfire Mark Is of No. 602 Squadron, Auxiliary Air Force, at Royal Air Force base Abbotsinch on September 3, 1939, with K9970 LO-D in the foreground. Flight Lieutenant George Pinkerton flew L1019 LO-S, which he named "Grumpy," to score the first Spitfire victory over an enemy plane. (*Dugald Cameron*)

thin wing that Mitchell had designed for the Spitfire was an aerodynamic marvel which proved capable of handling speeds approaching the sound barrier; but its distinctive elliptical shape—described by more than one pilot as the most graceful thing in the world (with the possible exception of a nice pair of feminine legs)—only served to complicate the manufacturing process. Consequently, by the time Britain declared war on Germany on September 3, 1939, it had a total of seventeen fully equipped Hurricane squadrons, as compared to nine equipped with the Spitfire.

Few military aircraft can be said to have had a more embarrassing combat debut than the Hurricane and Spitfire. Given the precedent set during World War I, the British were not unjustified in expecting German bombers to strike at their cities within hours of the declaration of the new war. Squadrons were put on alert, and false alarms led to numerous abortive interception flights.

Then, on September 6, two flights of Spitfires from No. 74 (Trinidad) Squadron—A Flight, led by Flt. Lt.

Adolph Gysbert Malan, and B Flight, under Flt. Lt. Wilfred Patrick Francis Treacy—took off from the unit's base at Hornchurch, Essex, to intercept enemy aircraft reported to be moving up the Thames Estuary. Radar had detected the intruders, and soon afterward antiaircraft guns at Clacton reported that they were firing at enemy twin-engine bombers, one of which had been shot down.

With Malan in the lead, the two Spitfire flights sighted an oncoming fighter formation and rapidly closed to engage. Suddenly, Malan realized that the approaching aircraft were not German and immediately turned away. Not all of his pilots followed his lead, however—Flying Officers Bryan Vincent Byrne and John Connell Freeborn opened fire and saw their two targets go down. After landing, the entire squadron was horrified to find its suspicions confirmed. The Spitfires' "opponents" had been Hurricanes of No. 56 (Punjab) Squadron, which had taken off from North Weald to intercept the same imaginary enemy force that had brought 74 Squadron on the scene.

It took almost an hour to sort everything out, but the final sobering realization was that no German aircraft had come near the Thames Estuary that day. Six Heinkel He 111Hs of I *Gruppe*, *Kampfgeschwader* 26 (I./KG 26) had set out to reconnoiter that very location, but adverse weather conditions had caused them to abort the mission and return home with full bomb loads.

The origin of the "Battle of Barking Creek," as it came to be called, lay with the Chain Home radar station at Canewdon near Southend, which had received echoes from aircraft to the west and, failing to filter them out, came away with the impression that aircraft were approaching from the east. As more and more British aircraft scrambled up to meet the reported threat, their echoes added to the confusion, making the supposed enemy force seem to grow progressively larger. As word of the attack spread and became fixed in the minds of the defenders, a case of mistaken identity, followed by the first shot from a trigger-happy antiaircraft gunner that would set off a chain reaction of shooting from the ground and air, became inevitable. The whole affair would have been comical had it not been for the loss of a Bristol Blenheim of No. 64 Squadron to the antiaircraft gunners, and the destruction of two of No. 56 Squadron's Hurricanes and the death of one of their pilots, Pilot Officer Montagu Leslie Hulton-Harrop, at the hands of the Spitfires.

Byrne and Freeborn were subsequently court-martialed, but both were acquitted. A court of inquiry ultimately determined that the solution to such tragedies in future would be the development of IFF (identification, friend or foe) radar equipment for the aircraft, the production and installment of which was given highest priority.

Johnny Freeborn would later redeem himself during the Battle of Britain, shooting down eleven German aircraft and sharing in the destruction of two more. "Paddy" Byrne's misfortunes were not over, however. Leading a section of Malan's flight during a sortie over Dunkirk on May 21, 1940, he was downed by antiaircraft fire, wounded in the leg and taken prisoner by the Germans. Byrne made several escape attempts—including the "Great Escape" from Stalag Luft III on March 24, 1944—but was repeatedly recaptured and in the end was lucky just to have survived the war.

Both of RAF Fighter Command's principal weapons had gotten off to an inauspicious start, and were sorely in need of redemption. The opportunity would not be long in coming.

Aside from being simpler to build, the stolid Hurricane's airframe proved to be more robust than the Spitfire's, and its wider-track undercarriage more stable, making it better suited for dealing with the rough airstrips placed at the RAF's disposal by the French when World War II began. Those virtues resulted in the Hurricane being the first to be committed to France, while the Spitfire stayed in Britain. Numbers 1, 73, 85, and 87 Squadrons were the first to go, with Nos. 1 and 73 moving up to become the fighter element of the Advanced Air Striking Force, providing escort for twelve Bristol Blenheim and Fairey Battle bomber squadrons.

In spite of the Hurricanes being closer to potential action, however, it was the stay-at-home Spitfires that would draw first blood from the Luftwaffe. On October 16, nine Junkers Ju 88As of I./KG 30, led by *Hptmn.* Helmut C. W. Pohle, left their base at Westerland on the Isle of Sylt with orders to attack British

During an interview in 1989, George Pinkerton uses models to demonstrate how he brought down *Hptmn*. Helmut Pohle's Junkers Ju 88A on October 16, 1940. (*Dugald Cameron*)

warships in the Firth of Forth. Their principal victim was to be the battle cruiser *Hood*, which was reported to have been there, but only if she was in open waters; in that early stage of the war, the Luftwaffe High Command wanted to avoid inflicting civilian casualties.

The Germans did not expect serious opposition, since their intelligence had stated that only obsolescent Gladiators had been committed to Scotland's defense. Unknown to them, a flight of Spitfires of No. 602 (City of Glasgow) Squadron, Auxiliary Air Force, had arrived at Drem on October 13. Nor could the Germans have known that a section from No. 603 (City of Edinburgh) Squadron was refueling at Turnhouse after unsuccessfully chasing a Heinkel He 111P of KG 26 over the Lothians.

Serving in No. 602 Squadron at that time was Flt. Lt. George Pinkerton, a Renfrewshire farmer who typified the unit's local membership, as he himself later described it:

Working on my farm, I used to see aeroplanes flying around and pilots being taught to fly. It interested me, so I went along and found that the Scottish Flying Club could teach me to fly at a cost of two pounds per hour. I applied to 602 and was granted a commission in 1933. The Auxiliary Air Force was really the same as the Territorial Army. The officers and the airmen were recruited from Glasgow. Among the officers were bank clerks, stockbrokers, a plasterer, a miner and a couple of lawyers. The airmen came from all walks of life, many of them out of engineering works and technical jobs in the city. When it got really serious we were mobilized on 24 August. We were put on a war footing and we armed the aeroplanes and manned them. Of course we were proud of our ability to operate and we didn't really think the regulars had anything on us at all.[2]

So it was that on that afternoon, Blue Section of No. 602 Squadron, led by Pinkerton, and Yellow Section of No. 603

Squadron, under Flt. Lt. George Lovell Denholm, were up when unidentified aircraft were reported heading toward the Forth Bridge. Pinkerton was at ten thousand feet altitude when he spotted the Ju 88s dive-bombing warships off Rosyth. The Germans had spotted *Hood*, but she was in the harbor and too close to a civilian area. In accordance with his rules of engagement, Pohle went after secondary targets outside the harbor instead. One of the Germans' bombs struck the light cruiser *Southampton* amidships but failed to explode, while the light cruiser *Edinburgh* and destroyer *Mohawk* suffered light damage.

Pinkerton and his section attacked the leading bomber, which pulled out of its dive and fled over May Island. Pinkerton described what followed:

> We got fleeting glimpses of him but not sufficient to be able to fire at him. And when we got to the edge of a bank of cloud, he emerged—with me sitting on his starboard side and my number two sitting on his port side. We came up astern of him and carried out an attack. His aeroplane lifted up in the air and then went down into a dive and I think I had probably injured him in some way. My number two came in and gave him some more, and then I came back to finish him off.

The Spitfires took turns attacking Pohle's Junkers, killing his flight engineer and rear gunner, mortally wounding his navigator, and then disabling one of his engines. When the second engine was demolished in a hail of bullets, Pohle made for a trawler and crash-landed in the water nearby. "We watched how he flew over a merchant ship and then crashed down in the sea about a mile away," wrote Pinkerton. "We were quite glad to see him in the sea because the last

thing I wanted was to go back to my unit and say I hadn't shot him down!—considering that I had eight machine-guns that could fire ammunition at the rate of 2,400 rounds a minute."

Pohle suffered facial injuries and a concussion, but minutes later the trawler crew rescued him from his sinking plane, and he regained consciousness in a naval hospital at Port Edgar five days later. Pinkerton visited him afterward and said that, "I sent him some sweets and cigarettes and he wrote a letter thanking me."

"I thank you for the friendly conduct," Pohle wrote, "wish you the best and greet with you likewise, the other pilot. To all airmen, comradeship."[3]

Pinkerton's "other pilot" to whom Pohle referred was Flying Officer Archibald Ashmore McKellar, a diminutive five-foot-four plasterer from Glasgow who had been serving in 602 Squadron since 1936. Called into active service upon the declaration of war, he had just had his first fight and had scored the first of three shared victories. He would go on to become one of the heroes of the Battle of Britain, scoring an additional seventeen individual victories before being killed in action on November 1, 1940.

Meanwhile No. 603's commander, Squadron Leader Ernest H. Stevens, led Red Section against an He 111P of the *Stab* (Headquarters Squadron) of KG 26, which was observing the Ju 88s' mission, over Dalkeith. Another section, led by Flt. Lt. Pat Gifford, also joined the chase, which ended in the He 111's pilot being wounded and forced to crash-land—the first German aircraft to fall on British soil since 1918.

Three Spitfires of No. 603 Squadron, led by its commander, Squadron Leader Andrew Douglas Farquhar (a Glasgow stockbroker before the war), arrived in time to pursue a Ju 88 retiring off Aberdour. Joined by two of No. 602 Squadron's Spitfires, they sent it crashing into the sea off Crail, its entire crew being killed.

Four other Ju 88s were damaged in the fight. The pilot of one, *Ltn.* Horst von Riesen of 1st *Staffel*, KG 30, survived to describe how his right radiator was hit and erupted in a cloud of steam, forcing him to turn off the engine before it burst into flames. His speed went down to 112 miles per hour and he could barely keep his Ju 88 above the waves, but at that point the Spitfires, at the limit of their range, broke off their pursuit and headed for home. With a four-hour flight ahead, one German crewman suggested they turn back and crash-land in Scotland, but Riesen and the rest of his men decided to take their chances rather than face those Spitfires again. As the Junkers' fuel was expended, it became lighter and gained altitude, allowing Riesen to reach his air base at Westerland. He would somehow survive subsequent unpleasant encounters with Spitfires during the Battle of Britain and over the Mediterranean Sea.

An hour later, Farquhar and two other pilots damaged an He 111P over Rosyth, and Spitfires of his squadron sent another one crashing into the sea off Port Seton. One of the downed Heinkel's crew was killed, but three others were rescued, shaken but miraculously uninjured.

The Auxiliary Air Force had proven its worth—and so had the Spitfire. Not to be outdone, a Canadian Spitfire pilot of No. 41 Squadron RAF, Flying Officer Howard Peter Blatchford, downed an He 111 twenty or thirty miles east of Whitby the next day.

Over in France, the Hurricane's first chance to fight came at 1100 hours on October 15, when Squadron Leader Patrick J. H. Halahan led five sections of No. 1 Squadron, comprised of three fighters each, from the airfield at Étain and commenced a patrol of the Saarlautern sector. The British flew forty miles into German territory, came under heavy antiaircraft fire and saw their first Me 109s. But instead of accepting combat, the enemy dived away.

As with the Spitfire, the Hurricane first fired its guns in anger against coastal raiders. On October 21, nine Heinkel He 115A-1 twin-engine floatplanes of 1st *Staffel*, *Küstenfliegergruppe* 106 went on a sweep of the Channel looking for merchant ships to attack. It was they, however, not the ships, that would end up running the gauntlet. The Heinkels were about fifteen miles from Spurn Head off the Yorkshire coast when they encountered two Spitfires of No. 72 Squadron, flown by Flt. Lts. Thomas A. F. Elsdon and Desmond F. B. Sheen, who raked the formation and claimed to have inflicted fatal damage on two of the floatplanes before their fuel ran low and they had to break off the action. By then six Hurricanes of A Flight, No. 46 Squadron, had entered the fray, and in a fifteen-mile chase they claimed five more of the fleeing He 115s, a share in the destruction of one going to Pilot Officer Peter W. Lefevre, who went on to score five personal and four more shared victories before being killed in action on February 6, 1944.

The Hawkers on the Continent finally opened their account on October 30, when a Dornier Do 17P of 2. *Staffel* (*Fernaufklärung*)/*Aufklärungsgruppe* 123 (2.(F)/123) crossed into French territory and Sgt. Peter William Olbert Mould, a resident of Hallaton, Uppingham, serving in No. 1 Squadron, scrambled up to intercept the intruder in Hurricane L1842. call number "G." He caught the Dornier ten miles west of Toul and shot it down in flames, killing *Hptmn.* Baldüin von Normann and his crew, for the first official RAF fighter victory over France since 1918. "Boy" Mould was awarded the Distinguished Flying Cross for his success and would later add six more German aircraft to his score and a share in one more, before the French armistice on June 22, 1940.

There would be other run-ins with the Germans over the next month, almost invariably involving quick dashes into Allied air space by their reconnaissance planes. On November 2, Flt. Lt. Robert Voase-Jeff of No. 87 Squadron downed an He 111H of 2.(F)/122 near Hazebrouck, while one of his squadron mates, Flying Officer William Dennis David, damaged another half an hour later. David's quarry gave as good as it got, however—not only did its gunners damage his Hurricane, but they shot up the fuel and oil lines of Pilot Officer Christopher C. D. Mackworth's plane, forcing him to land at Seclin, after which the Heinkel succeeded in returning to its airfield at Münster-Handorf.

A somewhat significant victory was scored on November 8, when a Hurricane of No. 73 Squadron intercepted and shot down a Do 17P of 1.(F)/123 near Lubey, northwest of Metz, killing *Obltn.* Hans Kutter and his crew. The pilot was Flying Officer Edgar James Kain from Hastings, New Zealand, and it was the first of at least sixteen enemy planes that he would destroy over France, making him the leading RAF ace prior to the Battle of Britain. With tragic irony, just after taking off for Britain on June 7, 1940, "Cobber" Kain made a last slow roll over the field, his wingtip struck the ground and his Hurricane crashed, killing him.

German probes began to intensify later in November. On the twenty-first, Flt. Lt. Richard H. A. Lee of No. 85 Squadron caught an He 111 of *Stab./KG 4* ten miles north of Cap Gris Nez and sent it crashing into the sea. Flight Sergeant Frederick S. Brown and Flying Officer James W. E. Davies of No. 79 Squadron teamed up to bring down a Do 17P of 3.(F)/122 east of Deal. No less than eight reconnaissance planes were downed on November 23— including a Do 17 of *Stab./KG 2* by Flt. Lt. John E. Scoular and a squadron mate of No. 73 Squadron—and a Do 17P of 3.(F)/22 crashed near Conflans for Cobber Kain's second victory. Principal honors of the day went to No. 1 Squadron, with an He 111 falling to Squadron Leader Arthur V. Clowes northeast of Saarbrücken; an He 111 of 2.(F)/122 credited to Flying Officer Newell Orton near Königsmacher; another from the same unit brought down by Flying Officer George H. F. Plinston and two squadron mates; and the demise of yet another hapless Heinkel south of Boulay being shared between "Fanny" Orton, three other Hurricanes of No. 1 Squadron, and three French Morane-Saulnier MS.406s of *Groupe de Chasse* II/5. Flying Officer John Ignatius Kilmartin also teamed up with two fellow Hurricane pilots of No. 1 Squadron to down a Do 17 over Saint-Menehould.

The most unusual action of the day took place when Flying Officers Cyril D. Palmer, Francis Soper, and Mark H. Brown of No. 1 Squadron took off from the field at Vassincourt to intercept yet another intruder. They caught the German, a Do 17P of 4.(F)/122, as it was racing back toward the border, and three sets of eight machine guns soon set one of its engines on fire, after which "Pussy" Palmer saw its navigator and gunner take

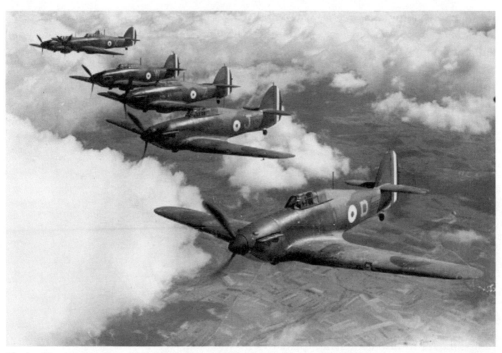

Hawker Hurricane Mark Is of No. 73 Squadron RAF fly over England in 1939. The unit was among those deployed to France, where the tight formations its pilots were trained to maintain proved to be a handicap against the Germans' more flexible "finger four" elements. (*Imperial War Museum*)

to their parachutes. Palmer's guns had jammed at that point, but he noticed that the Dornier pilot was slumped over the controls, so he brought his Hurricane alongside the stricken machine to make sure that it was indeed going down for good. At that point, however, *Uffz.* Arno Frankenberger, who in fact had only been "playing possum," suddenly came to life and throttled down both engines. Then, as a startled Palmer sped ahead of his plane, Frankenberger abandoned his controls, climbed into the empty navigator's seat, grabbed the MG 15 machine gun and fired a long burst into the Hurricane.

Palmer instinctively ducked and pushed his stick forward, but before his Hurricane nosed down a bullet whizzed past his head and smashed his windscreen, while other slugs struck his engine and his radiator. The engine stopped and, trailing a white plume of glycol, Palmer brought

his plane down—only to find that his undercarriage was also damaged and would not extend. He belly-landed in a field and saw Frankenberger—who was unwilling to press his luck with two other Hurricanes closing on his tail—jump from his doomed plane, which crashed not far from Palmer's.

After parachuting to earth, Frankenberger was taken prisoner by a squad of French soldiers and spent the night in jail; but on the following day, pilots of No. 1 Squadron came and had the German released into their custody for the evening, over the objections of French officials. They then took Frankenberger to the upstairs room of a small inn near Vassincourt, which served as their officers' mess, and treated him to a tankard of beer and the best dinner locally available. The airmen compared notes on the action, and Frankenberger showed off photographs of his wife and baby, and presented his hosts with an autographed picture of himself. It was well past midnight when, after a final toast to the universal fraternity of the air, the British wandered off to bed, while their German guest was returned to French custody.

There was more to the incident than the chivalry of an earlier war, however. There were things that the British could learn from an amiable reliving of the day's events with their captive (and somewhat inebriated) guest. For one thing, Palmer had been shot down and nearly killed, largely because the Hurricane, though equipped with an armored fuel tank and a bulletproof windscreen, had no rear protection for the pilot. "Bull" Halahan petitioned the Air Ministry for armor plate to be added behind the pilot's seat, only to have his request rejected on the grounds that it would upset the center of gravity. Undaunted, he removed armored plating from a Fairey Battle bomber, installed it behind the seat of one of his squadron's Hurricanes, and then had one of his pilots demonstrate its effects on the plane's flying characteristics to visiting experts with a brilliant aerobatic show. Shortly afterward, armored plating behind the pilot's seat became a standard item for both Hurricanes and Spitfires.

Thus far, the Hurricane pilots had managed to avoid any combat fatalities over France, but that run of good fortune came to its inevitable end on December 22, when a patrolling trio from 73 Squadron was bounced by Me 109Es fifteen kilometers northeast of Metz. Moments later, Sgt. Reginald M. Perry crashed to his death at Budange, the seventeenth victory for *Hptm.* Werner Mölders, commander of III *Gruppe*, *Jagdgeschwader* 53 (who misidentified Hurricane N2385 as a Morane-Saulnier in

his combat report). Hurricane L1967 likewise fell at Altroff, killing Sgt. John Winn, the first of an eventual thirty-one victories for the leader of JG53's 8th *Staffel*, *Oberleutnant* Hans von Hahn.

Such, then, were the beginnings of two legends. In the years to come, Hurricane and Spitfire would be familiar names to friend and foe alike.

Nine

Zero Hour

JAPANESE FIGHTERS, 1937–41

Although Japan was quick to appreciate the airplane's value as a weapon, Japanese aircraft of the 1920s and early 1930s were invariably dismissed by European observers as either outright imitations or second-rate variations on French or British designs. A case in point could be seen in an incident over the Chinese port of Shanghai on February 22, 1932, in which three Mitsubishi B1M3 three-seat biplane bombers from the aircraft carrier *Kaga* came under attack by a Boeing 212—a unique all-metal version of the P-12 that was being demonstrated for the Chinese by American pilot Robert Short. Short killed the lead plane's pilot, Lt. Susumu Kotani, and wounded his observer, Airman 1st Class Setsuro Sasaki, but the third crewman, Lt. j.g. Yoshiro Sakinaga, managed to land the stricken B1M at Shanghai. Besides coming under fire from the other B1M gunners, however, Short was attacked and sent to his death in flames by three fighters flown by Lt. Nokiji Ikuta, Petty Officer 3rd Class Toshio Kuroiwa, and Seaman 1st Class Kazuo Takeo. The three Nakajima A1N2s that shared in the Japanese navy's first aerial victory were, in fact, license-built

copies of the Gloster Gamecock biplane.

By the mid-1930s, the backward nature of Japanese aviation and the inferiority of Japanese airmen had become virtually ingrained in European and American strategic thinking. That complacency was rudely shaken on December 7, 1941, when the carrier-launched Japanese aircraft which staged a surprise attack on the US Navy and Army bases at Pearl Harbor on the Hawaiian island of Oahu, and the army aircraft that supported the Japanese invasion of British-held Malaya, proved themselves to be not only equal but superior to their Western counterparts. In the vanguard of that first devastating Japanese offensive was a fighter which would come to symbolize Japanese air supremacy in the first six months of the Pacific War: the Mitsubishi A6M2 Zero.

The Zero's performance seemed nothing short of phenomenal in 1941. Only a year earlier, Messerschmitt Me 109Es were handicapped by their inability to fight Supermarine Spitfires or Hawker Hurricanes for more than twenty minutes after crossing the English Channel. The Zero, in contrast, could fly as far as 1,930 miles with the aid of a drop-tank, vastly outdistancing an Me 109, Spitfire, or Hurricane. At a time when the structural reinforcement, deck landing gear, and navigational aids necessary for a carrier fighter intrinsically handicapped its performance, the A6M2, with its maximum speed of 331 miles per hour, made history as the first carrier-based fighter capable of outperforming its land-based opponents.

At Allied headquarters, intelligence officers wondered how the Japanese could have kept such a wonder weapon secret for so long. The truth was that they had not—the Zero had been in combat over China since August 1940. Such was the mindset in the British and American high commands about Japanese inferiority that even reports from their own military observers in China had been dismissed as fiction. Thousands of American, Dutch, and British Commonwealth airmen would pay with their lives for that self-imposed ignorance.

While they had been manufacturing Western designs under license in the 1920s, Japanese aircraft builders had been catching up on the fine points of aeronautical design. In the early 1930s, they were ready to proceed with designs of their own.

In February 1934, Imperial Navy air headquarters issued a 9-*shi* specification calling for a new single-seat fighter. Although carrier-based capability was virtually implicit in a navy requirement, one of its chief architects, Lt. Cmdr. Hideo Sawai, had deliberately avoided specific reference to such characteristics, concerned that it would inhibit the designers. In contrast to British and American naval aviation planners, Sawai wanted the manufacturers to produce a high-performance fighter that would

bring Japan up to world standard—the matter of adapting and equipping it for operations from carrier decks would be dealt with afterward. The specification called for a maximum speed of 350 kilometers per hour (217 miles per hour) at 3,000 meters (9,840 feet) and the ability to climb to 5,000 meters (16,405 feet) within 6.5 minutes. Armament was to consist of two 7.7mm machine guns, wingspan would not exceed 11 meters (36 feet), and length no more than 8 meters (26 1/2 feet).

Among those who sought to meet the requirement was Mitsubishi's chief designer, Jiro Horikoshi. After giving cursory consideration to other configurations, Horikoshi determined that the fighter must be a low-wing monoplane, with attention paid to the cleanest possible aerodynamics and to the minimum possible weight. The airframe he designed used stressed-skin aluminum over a two-spar box-type wing structure of inverted gull form, which was later altered into a flat center section with dihedral for the outer wing panels. After considering the possibility of incorporating retractable landing gear into his design, Horikoshi rejected it on the grounds that the 10 percent decrease in drag would result in only a 3 percent increase in speed, not enough to justify the system's greater weight and complexity. Instead, his monoplane would have fixed undercarriage, with streamlined fairings over the wheels.

The first Ka-14 prototype, powered by a 550-horsepower Nakajima Kotobuki 5 nine-cylinder radial engine, was completed in January 1935, just eleven months after Mitsubishi received the 9-*shi* specification. During its first flight tests in February, it reached a speed of 444 kilometers per hour (276 miles per hour) at 10,500 feet, exceeding both the navy's requirement and Horikoshi's own expectations. Nevertheless, the Ka-14 had a great obstacle to overcome when it competed with the Nakajima A4N1 biplane in the autumn of 1935, because of the almost insurmountable prejudice in favor of dogfighting capabilities over all others among senior JNAF officers. The Ka-14's low wing loading made it extremely maneuverable for a monoplane, but it still could not match the maneuverability of the A4N1—until the rules of mock combat were altered to include climb and dive tactics. That change gave the Ka-14 such an overall edge that even Minoru Genda, one of the most ardent biplane partisans, was won over to the new type.

After further development—and an unsuccessful attempt to interest the army in the design—Mitsubishi was able to put its monoplane into production as the A5M1 Model 96 carrier fighter in the autumn of 1936. An improved version with a 610-horsepower Kotobuki 2 KAI 3*ko* engine, a longer chord cowling, and a three-blade propeller in place of the two-blade airscrew entered production in the late spring of 1937 as the A5M2-*ko*. The outbreak of war with China on July 7, 1937, lent urgency to the JNAF's efforts to hasten the A5M to operational units, and the newly formed 13th *Kokutai* (Air Group) got its first monoplanes just four days later.

Initially, the Japanese army advanced quickly, taking Beijing and Tientsin; but resistance stiffened as it approached Shanghai, and at the same time Japan's regular arsenal of biplanes was proving unable to achieve a decisive degree of air superiority over the mixed bag of Chinese aircraft that opposed them. The need for such an edge became critical on August 14, when nine Mitsubishi G3M2s of the Kanoya *Kokutai*, led by Lt. Cmdr. Shinichi

A Curtiss Hawk III in Chinese Nationalist markings awaits the application of squadron numbers. These curious biplanes gave the Japanese a sobering reminder of the need for air superiority in August 1937. (*Chinese Aviation Historical Society via D.Y. Louie*)

Nitta, left Matsuyama (Sung Shan) air base near Taipei, Taiwan, crossed the Formosa Strait and tried to bomb Jienqiao training field near Hangzhou, while another nine G3Ms led by Lt. Cmdr. Nantaro Asano attacked Guangde airfield. Chinese intelligence had learned of the coming mission and Colonel Kao Zihang, commander of the 4th Pursuit Group at Zhoujiakao, had ordered his 21st, 22nd, and 23rd Squadrons to Jienqiao, where their Curtiss Hawks would refuel and take off to intercept the raiders.

Kao's fighters were the last of a long line of classic biplanes that had begun in 1925 with the P-1 Hawk (the first Curtiss to bear the name), and which culminated in the F11C-2 Goshawk. Final refinements in the Goshawk were metal wings and hand-cranked retractable landing gear on the XF11C-3, inspired by that of the Grumman XFF-1 two-seat fighter. Entering US Navy service in May 1933, the F11C-3—later redesignated as the BF2C-1 fighter-bomber—joined the Grumman F3F and Polikarpov I-153 as the only single-seat biplane fighters with retractable undercarriage to achieve quantity production.

Export versions of both Goshawk variants reverted to wooden wing structures, the fixed-gear plane being called the

Hawk II and the retractable model the Hawk III. Powered by a 740-horsepower Wright R-1820-F53 radial engine, the Hawk III had a top speed of 240 miles per hour, mounted two .30-caliber Browning machine guns, and served in the air arms of Thailand and Argentina as well as China, which imported its first twelve in March 1936. Another ninety arrived in crates and were assembled at the Central Manufacturing Company at Hangzhou. The Chinese had seventy-two operational when the Japanese invaded, as well as fifty Hawk IIs.

August 14 began with Chinese air attacks against the Japanese army in Shanghai and against supporting naval units offshore, including Hawks of the 5th Pursuit Group carrying one 250-kilogram bomb each. Lieutenant Liang Hongwen in plane No. 2401 of the 24th Pursuit Squadron was credited with hitting the stern of the Japanese light cruiser *Idzumo* off Nantong, but it turned out to be a near miss—which was fortunate, since his target was in fact the British heavy cruiser *Cumberland*. Two other Chinese bombs fell near the American heavy cruiser *Augusta*. Another "international incident" occurred that afternoon when three Hawks from the 24th attacked the Japanese-occupied Kungda Textile Factory, but one of their bombs fell on the Nanking Road in the International Settlement. Meanwhile, Liang came under attack by an aggressively flown Nakajima E8N1 floatplane from the light cruiser *Sendai*, and was so badly shot about that he had to force land and died of his injuries soon after. Next, five Hawk IIs and nine Hawk IIIs made another attempt to bomb the Kungda factory, and some also attacked the floatplane, scoring fifteen hits before it lost them in a cloud.

Poor though the Chinese bombing performance was, it spurred Vice Adm. Kiyoshi Hasegawa, commander of the Japanese Third Fleet, to order attacks against their airfields. Due to typhoon weather conditions, however, the only unit game to defy them was the Kanoya *Kokutai* with its modern G3Ms—all-metal twin-engine monoplanes capable of 232 miles per hour and of mounting three 7.7mm machine guns. They took off at 1305 hours Formosa time, but storm clouds and poor visibility limited their altitude to 1,641 feet. They would also have to carry out their mission without fighter escort, but the Japanese were not overly concerned, arrogantly assuming that they would not need it against the inept Chinese.

The 4th Group's Hawk IIIs were still in the process of refueling at Jienqiao when Nitto's bombers arrived, strung out due to the weather, which made it hazardous to fly in formation. Kao, boarding his personal Hawk III marked IV-I, ordered all fighters to scramble up, ready or not. This had grave consequences for two of them—Hawk III 2105 of the 21st Pursuit Squadron ran out of fuel and crashed, mortally injuring Lt. Liu Shufan, and 2106 force landed, injuring Lt. Chin Anyi. The rest laid into the Japanese, and Colonel Kao, after silencing two gunners, closed to twenty meters to shoot a G3M, piloted by Petty Officer 3rd Class Iyoshio Momosaki, down in flames over Banshan. He then attacked Warrant Officer Fujio Yamashita's bomber, crippling its left engine with fourteen hits, along with twenty-one in its right wing, and thirty-eight to the fuselage before his fuel gave out, compelling him to disengage and force land at Jienqiao. A third G3M, piloted by Petty Officer 1st Class Yanase Mitsui, and which was swarmed by 21st Squadron Hawk IIIs flown by Capt. Lee Guitan, Lt. Wang Wenhua, and Liu Chisheng, went down near Chiaosi.

At 1800 hours Asano's nine G3Ms arrived over Guangde in V formation but only scored one bomb hit. As they made for home, they encountered Capt. Chow Tingfong, commander of the 34th Provisional Pursuit Squadron, who had been test-flying a Hawk III. Although he had no ammunition, Chow attacked and managed to break up the formation. As the Japanese flew on, they ran into the 22nd Pursuit Squadron, and one of its pilots, Lt. Cheng Hsiaoyu, hit the right engine and wing tank of Warrant Officer Hitoshi Ogawa's G3M. Ogawa pressed on for Taiwan but ran out of fuel short of his goal and ditched near the Keelung harbor lighthouse, where he and his crew were rescued. Yamashita managed to get his riddled bomber back to Matsuyama on one engine, but its landing gear collapsed on landing, and it was written off.

Though overlooked in the West, the "Battle of Shanghai" had multiple significances in aviation annals. In 1940 the Republic of China pronounced August 14 as "Air Force Day." Indeed, Colonel Kao had scored his air arm's first aerial victories. To his shared kill that day, Liu Chisheng would add ten more flying Hawks, I-16s, and I-152s, to become China's ace of aces and a major general in the Republic of China Air Force. In terms of air strategy, the Chinese fighters' success struck a serious blow to Italian general Giulio Douhet's widely accepted theory that bombers alone could win wars. Conversely, the Japanese were rudely awakened to the folly of sending their bombers in unescorted, and to the critical need for a fighter capable of seizing control of the sky.

Land-based Japanese navy units, augmented by the air groups of the light carrier *Hosho* off Shanghai, were joined on August 15 by the large carrier *Kaga*, and both vessels achieved a marginal degree of local air superiority with their aging Nakajima A2N1 fighters. Then, two A5M2s landed on *Kaga* and flew their first combat sortie on August 22.

The A5Ms' first chance to test their mettle came on September 4, when Lt. Tadashi Nakajima and a companion encountered nine Hawk IIIs of the 23rd Pursuit Squadron over Lo Dan near Lake Taihu. In the ensuing fight, the Japanese claimed three Hawks and returned to *Kaga* undamaged. The Chinese, however, reported that the outnumbered Japanese inflicted no casualties on them before they escaped into thick cloud cover.

Five days later, the 13th *Kokutai*, equipped with twelve A5M2 fighters and eighteen Aichi D1A1 and Yokosuka B4Y1 bombers, arrived at Gongda airfield near Shanghai and commenced operations. Little contact was made with Chinese air elements until September 19, when all twelve A5Ms, led by Lt. Shichio Yamashita, escorted seventeen carrier bombers on a mission to Nanking and encountered a swarm of twenty Hawks and Boeing 248s (the latter being export versions of the P-26 "Peashooter" monoplane).

The results, as they appeared in the *Asahi Shimbun* on the following morning, were spectacular. In what the Tokyo newspaper dubbed the A5M's debut, the new Japanese fighters destroyed thirty-three enemy planes within fifteen minutes without loss to themselves. The account was considerably exaggerated, though the truth could have stood on its own merits. The large-scale engagement had not marked the operational debut of the A5M2—*that* had already occurred more than two weeks before—but in a swirling dogfight which lasted more than fifteen minutes, the Japanese fighters did record

Mitsubishi A5M2s of the 13th Kokutai are readied for a mission from Kunda airfield outside Shanghai. The air group's tail designator was changed from "T" to "4" in October 1937. (*US Navy*)

fifteen victories, including three probables. A second mission that same day brought Japanese claims up to twenty-six for the loss of three bombers, one reconnaissance plane, and no fighters.

According to Chinese combat records, at 0830 hours that day Capt. Mao Yingchu of the 23rd Pursuit Squadron, flying Colonel Kao's Hawk IV-I, led a mixed flight of eight Hawk IIIs against thirty fighters and bombers, which were flying in two tiers over Nanking. Mao attacked the three leading bombers at eleven thousand feet, striking all three before damage from their debris and depleted ammunition forced him to return to base. Wang Danbi also scored hits on a bomber, but his wingman, Dai Guanjin, was hit by return fire and crashed in Hawk 2509. Yang Munqing and Chen Huaimin damaged a floatplane, but by then the Japanese fighters, which had been flying at fifteen thousand feet, descended on the Chinese. Chen's Hawk 2405 was driven down for a forced landing at the Po River.

At 0840 eight more Hawk IIIs took off from Nanking and were splitting into three elements at ten thousand feet when they were jumped by ten A5Ms. Captain Hu Zhongru and Su Enru fought three of them for five minutes before breaking away. Three other Hawks engaged six A5Ms, during which Dung Mingde's 2404 took two hits in the cowling and Liu Yichin's Hawk 2512 was so badly damaged by fragments from a smoking Japanese fighter that he had to bail out. Zhou Gengzu belly-landed his damaged Hawk when its landing gear failed to extend. Engaging the next Chinese element, the same six A5Ms set the engine of Yang Jien's Hawk 2306 on fire, compelling him to bail out. Liu Zhongwu's 2101 took hits in the upper and lower fuel tanks and windshield, but in spite of a wounded right foot, he nursed his plane to a landing at Juyung. Wu Dingzhen, hit in both legs, brought 2102 in for an emergency landing at Yangzhou.

Meanwhile, at Juyung, Capt. John "Buffalo" Wong Sunshui, an American volunteer serving in the 17th Pursuit Squadron, led five Boeing P-12s up at 0840. At 0900 they intercepted eight Japanese floatplanes heading for Nanking, but again the A5Ms intervened. In the ensuing twenty-minute fight, Ho Jolung in P-12 1706 claimed a floatplane, but Wong had to crash-land in shot-up P-12 1703, and Liu Lanqing bailed out of his fighter, only to be mortally wounded in his parachute by three of the A5Ms.

Departing Nanking at 0840, two Fiat CR.32s of the 8th Pursuit Squadron joined the fight, but both were shot down by the A5Ms, and their pilots, Capt. Liu Chuiwei and Huang Zhuju, were killed.

At 1630 that afternoon eight Boeing P-12s had a five-minute engagement with at least ten A5Ms. Hu Jolung's P-12 1706 suffered damage to its left control cable. Squadron Leader Jiang Quiyan was shot down in 1705 and, like Liu Lanqing, was strafed in his parachute by the A5Ms, surviving with wounds to his back and both legs.

The Chinese air force wrote off a total of eleven aircraft in the course of September 19, while claiming one Japanese floatplane. Whatever the statistics, the A5M2 had had an auspicious debut, and in the months to come it would all but drive the Chinese air force from the sky. Lieutenant Yamashita would not be around to see it, however. Just a week after leading the 13th *Kokutai* to instant glory over Nanking, he was compelled to make a forced landing in Chinese lines and was taken prisoner, dying shortly thereafter.

Curiously, the Japanese Army Air Force (JAAF) was slower than the navy in adopting more innovative and potent fighter designs. One reason was its longer adherence to the Japanese penchant for comparing aerial combat to the swordsman's art of kendo, stressing dexterity of maneuver. Finally, in June 1935, the *Koku Hombu* (Army Headquarters) issued a requirement for an "advanced fighter" to replace the Kawasaki Ki.10 biplane. The new fighter was to be capable of 280 miles per hour in level flight and of reaching 16,405 feet (5,000 meters) in less than six minutes, but would still mount two 7.7mm Type 89 machine guns.

Nakajima's design team, led by Tei Koyama and assisted by Minoru Ota and Hideo Itokawa, had already had some experience with externally braced monoplane designs when they learned of the new army requirement. They went on to design a small cantilever monoplane around the Nakajima Kotobuki II-Kai nine-cylinder radial—a license-built version of the Bristol Jupiter—using all-metal stressed-skin construction. As with the A5M, fixed landing gear enclosed within two streamlined fairings was selected over the heavier and more complex retractable undercarriage, but an enclosed canopy was incorporated into the second prototype. Increased wing area and the incorporation of wing flaps kept the fighter maneuverable enough to satisfy the *Koku Hombu*, which chose it over its faster competitor, the Kawasaki Ki.28, following trials in the spring of 1937. Powered by an improved 710-horsepower Nakajima Ha-I-*Otsu* engine driving a two-bladed all-metal two-pitch (ground-adjustable) Sumitomo PE propeller, the production version of the Nakajima Ki.27-*Otsu* Type 97 fighter had a maximum speed of 291 miles per hour at 13,125 feet.

So confident was Nakajima in its new fighter that it was making production preparations at its new plant at Ota even before the army's order became official on December 27, 1937. Even the most conservative pilots were delighted with the new fighter's maneuverability, as well as with its higher performance.

Eager to blood the new fighter in combat, the army dispatched three of the first planes to leave the assembly line to the 1st *Chutai* of the 2nd *Hiko Daitai* (Air Battalion), which was then flying Ki.10s from Yangzhou, northern China. Just one week after their arrival, on April 10, 1938, Capt. Tateo Kato, commander of the 1st *Chutai*, was leading the trio along with twelve Ki.10s on patrol near Ma Muchi when they encountered eighteen Polikarpov I-152s (also called I-15bis, which were upgraded I-15s with cabane struts in place of the gulled upper wings) returning from an attack on Japanese headquarters at Chao Chuang. The Japanese promptly attacked the lower formation of eleven I-152s of the 4th Pursuit Group at 4,500 meters altitude, and in the dogfight that followed, Kato's Ki.27s zeroed in on Lt. Chang Guangming, who tried to escape with a climbing half-roll before he was hit. Bailing out just before his plane caught fire, Chang survived a Japanese attempt to strafe him in his parachute before alighting with an injured back.

At that point the seven I-152s of the 3rd Pursuit Group's 8th Squadron, which were five hundred meters above the 4th Group's flight, dived on the Japanese. The 8th's deputy commander, Capt. Zhu Jiaxun, shot Sgt. Maj. Risaburo Saito off the tail of a 23rd Pursuit Squadron I-152, only to see the Ki.27 veer away to collide with Lt. Chen Huimin's Polikarpov. Chen managed to bail out before both planes crashed, surviving with a leg injury. Meanwhile, the two remaining Ki.27s shot the cowling off Zhu's plane, but he managed to evade them and force landed in a wheat field. Zhu's and a second force-landed plane were subsequently recovered for repair. He was credited with Saito as the second of three victories he would score in I-152s; he would later add two more in Gloster Gladiators.

By the time the two sides finally disengaged, the Japanese had claimed twenty-four Chinese planes—more than they had been fighting—including two for Kato, two for Warrant Officer Morita, and a posthumous credit for Saito. Among the Ki.10 pilots, Sgt. Tokuya Sudo was credit-

ed with two and 1st Lt. Iori Sakai with three, but Lt. Yonesuke Fukuyama returned to base with wounds from which he died in hospital shortly after. Two Ki.10s force landed short of Yangzhou, and two others crash-landed on the airfield. Actual Chinese losses were two 4th Pursuit Group planes shot down, with Sun Jinjen killed, and three others, flown by Lin Yuexing, Wang Denbi, and Li Tingcai, force landing due to fuel exhaustion, but which were later recovered; a 3rd Pursuit Group plane lost with its pilot, Liang Zihang; and two 3rd Pursuit planes returning with their pilots, Liu Tianlung and Huang Yin, wounded.

Over the next few weeks, the 2nd *Daitai* supported the army's Hsuchow campaign, intended to give the Japanese complete control of the Peking-Nanking railway. Chinese aerial opposition became infrequent, but on May 20, three Ki.27s of the *Daitai*'s headquarters flight took on ten Chinese, resulting in three credited to Capt. Mitsugu Sawada, one to Lt. Katsumi Anma, and one to Sgt. H. Wada, as well as seven more by the Ki.10s. In actuality, four I-152s of the 17th Pursuit Squadron were shot down, Zhu Qunqui and Qiu Guo bailing out of their flaming planes. In addition, Zhang Sheungren and William Tang force landed their shot-up fighters, and that of their commander, Sen Jiuliu, returned damaged, while two of the 22nd Pursuit Group's Hawk IIIs, 2201 flown by Feng Yuhe and Chao Mosheng's 2205, never returned.

The 2nd *Hiko Daitai* was still in the process of replacing its Ki.10s with Ki.27s in August 1938 when a restructuring of the JAAF resulted in its being redesignated the 64th *Koku Sentai* (Air Regiment). Adoption of the Ki.27 proceeded rapidly thereafter, and the fighter went on to its most celebrated period during the undeclared conflict between

Japan and the Soviet Union in the Nomonhan region of Mongolia and Manchuria in the summer of 1939.

Meanwhile the Imperial Navy, encouraged by the A5M's success, but also mindful of the rapid advances being made in Europe, issued a new specification for an even more advanced carrier fighter to representatives of the Mitsubishi and Nakajima firms during a meeting at the Yokosuka Naval Air Arsenal on January 17, 1938. Nakajima, convinced that it would be almost impossible to meet the navy's requirements, immediately withdrew from the competition.

Mitsubishi's design team again accepted the challenge, but to meet the 12-*shi* fighter requirement, it had to literally build its airplane around an 875-horsepower, fourteen-cylinder Mitsubishi Zuisei-13 twin-row radial engine. Again, weight reduction became the paramount consideration to compensate for every innovation incorporated in the new fighter. The wing was built around a one-piece spar of extrasuper duralumin—the first use of that alloy in the main spar construction of an airplane—with the fuselage center section integral with the wing. Horikoshi's team designed the fuselage to be as long, slim, and streamlined as its radial engine would allow. The relatively long, narrow wing provided a high aspect ratio and low enough wing loading to meet the navy requirement for maneuverability equal to the A5M. Underwing flaps gave the plane a landing speed of less than sixty-six miles per hour.

The most noteworthy advances in the 12-*shi* fighter's design were the use of retractable landing gear and considerably heavier armament. In addition to two fuselage-mounted Type 97 7.7mm machine guns, the plane featured two wing-mounted Type 99 20mm cannon—license-built versions of the Swiss Oerlikon gun manufactured by the Dai-Nihon Heiki Company—which were reliable and accurate, although were somewhat handicapped by a slow rate of fire and low muzzle velocity.

The 12-*shi* prototype emerged from the Mitsubishi plant in southern Nagoya on March 16, 1939, and testing commenced at Kagamigahara airfield on April 1. Teething troubles were minor. Engine vibration problems were eliminated after Horikoshi replaced the prototype's two-bladed airscrew with a constant-speed three-bladed propeller—the first of its kind to be used on a Japanese fighter. By September 14, Mitsubishi's new design had once again matched or exceeded the performance requirements that inspired its creation, and it was assigned the military designation A6M1 Type 0 carrier-borne fighter—the "0" referring to the year 1940, when it was expected to enter service, and which on traditional Japanese calendars would be the year 2600. In Japanese parlance, it was known as *Rei Shiki Sento Ki*, but pilots soon abbreviated that to *Rei-sen*. It would eventually achieve even wider notoriety by the English translation of that nickname: Zero Fighter.

The third Zero prototype was modified to use the more powerful 950-horsepower Nakajima Sakae-12 engine, and was placed in production as the A6M2 Model 11, the first example leaving Mitsubishi's Nagoya plant in December 1939. First tested on January 18, 1940, the Sakae-12-powered Zero exceeded all of the original navy specifications. By then, word of the new fighter had reached pilots in China, and they began clamoring for it. Overriding objections from Mitsubishi's engineers, the navy ordered six preproduction A6M2s to be transferred from the Yokosuka *Kokutai* to Wuhan, China, on

July 21, 1940—ten days before the type was to be officially accepted into regular service. Those Zeros and nine more that followed them shortly thereafter were assigned to the 12th *Kokutai*, a new unit that had been formed the previous June under the command of Capt. Kiichi Hasegawa. On August 19, Lt. Tamotsu Yokoyama led twelve Zeros on their first operational mission, escorting fifty-four G3M2s on a 1,150-mile round trip to bomb Chunking. The Japanese encountered no aerial opposition, nor did they on another mission the following day.

The first aerial contact occurred on September 13, when thirteen Zeros led by Lt. Saburo Shindo escorted twenty-seven bombers and a Mitsubishi C5M1 reconnaissance plane to Hankow. After the bombing raid was carried out, the Japanese were about to head home when they spotted Chinese fighters converging over the smoking target area. Eager to test their new mounts, the Zero pilots climbed for height, then dived on the enemy—ten I-152 biplanes of the 21st Pursuit Squadron led by Cheng Xiaoyu, nine of Wang Yuqun's 23rd, and six from the 28th led by Louie Yimqun, flying at 3,500 meters with nine I-16 monoplanes of Yang Monqing's 24th Pursuit Squadron flying top cover for them at 4,500 meters. What the Chinese would call the Air Battle at Bishan would begin at high noon, with results that revived the A5M's heyday with a vengeance.

Petty Officer First Class Saburo Kitahata, No. 2 in Shindo's 1st *Shotai* (Flight) of the 1st *Chutai* (Squadron), claimed two enemy planes (his eventual score would be more than ten). He then strafed the Paishih Railroad Station. Another Chinese fighter was claimed by Petty Officer 2nd Class Kihei Fujiwara, while Petty Officer 3rd Class Yoshio Oki scored no less than four.

The leader of the 2nd *Chutai*'s 1st *Shotai*, Lt. j.g. Ayao Shirane, was having his first combat, but he managed to down one of his adversaries. He would eventually be credited with eight more. The number two man in his *shotai*, Petty Officer 1st Class Masayuki Mitsumasa, accounted for two, as did the number three man, Petty Officer 2nd Class Tsutomu Iwai; Iwai's final wartime total would be eleven.

Petty Officer First Class Toraichi Takatsuka, leader of the 2nd *Chutai*'s 2nd *Shotai*, claimed three enemy planes, while Petty Officer 2nd Class Kazuki Mikami got two, and Petty Officer 3rd Class Masaharu Miramoto accounted for one.

Among the three Zeros of the 1st *Chutai*'s 2nd *Shotai*, Petty Officer 2nd Class Toshiyuki Sueda downed one enemy plane—the first of an eventual total of nine—and Petty Officer 3rd Class Hatsuyama Yamaya shot down two. The star performance of the day, however, was by their *shotai* leader, Warrant Officer Koshiro Yamashita, who downed five Chinese fighters, the last of which he fought down to an altitude of fifty meters before finally sending it crashing in a rice paddy. He then rejoined his two wingmen, performed a victory loop, and led them in a strafing attack on the Paishih Railroad Station.

The Japanese claimed twenty-seven victories in all. Not one Zero had been touched. Chinese records showed that on this occasion, the Japanese claims, though typically optimistic, were not that far off the mark so far as perception went. Thirteen Chinese fighters had actually been destroyed, and another eleven had come down damaged. Ten pilots were killed, including the 24th Squadron's Yang Monqing, shot after parachuting from his stricken I-16, and also Zhou Yanxun, who was missing, as well as I-152 pilots Cao

Mitsubishi A6M2 Model 11 Zeros of the 12th *Kokutai* conduct a sweep on May 26, 1941. The plane in the background was flown by Lt. Minoru Suzuki. Petty Officer Third Class Kunimori Nakakariya, flying 3-136 in foreground, shot down two Chinese aircraft during this mission, his first of an eventual sixteen victories. (*US Navy*)

Fei, He Jemin, Hwang Donchen, Yu Bifeng, Louie Tingzi, Liu Yinyi, Kang Baozhong, Zhang Hunjiao, and Sito Gen. The eight wounded included Major Louie, who force landed at Siuning airfield with forty-eight bullet holes in his I-152, and Col. Cheng Xiaoyu, commander of the 4th Pursuit Group. Admiral Shigetaro Shimada, commander of the China Area Fleet, issued a special unit commendation to mark what he regarded—more rightly than he may have imagined—as a historic event.

By the end of 1940, the Zeros would account for a total of eighty-seven Chinese aircraft. The 12th *Kokutai*'s own losses remained nil until May 20, 1941, when Petty Officer 1st Class Eichi Kimura was shot down and killed by anti-aircraft fire over Central China. Petty Officer First Class Kichiro Kobayashi suffered a similar fate over Lanchow on June 23. Those would be the only Zeros destroyed in action prior to December 7. Their phenomenal performance did not go entirely unnoticed. Colonel Claire Lee Chennault, then engaged in organizing a

force of American mercenary pilots for the Chinese air force, examined the remains of shot-down Zeros, prepared a report on the fighter's capabilities and sent it to Washington, DC—where it was completely ignored.

Then came December 7, 1941, and the Zero's first appearance over Pearl Harbor and the Philippines, taking the Americans completely by surprise.

At the same time, surviving Allied airmen over Malaya and China were also recording their shock at the sudden appearance and astonishing performance of the Zero. In most cases there, however, the Zero's name and mystique were being passed on to another Japanese "mystery" fighter—an army type that was even newer, yet not as good as the Mitsubishi A6M2.

Although the JAAF's Nakajima Ki.27 had been a great success, it was fundamentally little more than a refinement of Mitsubishi's A5M formula. The *Koku Hombu* was so pleased with it, however, that essentially the only concession to technological progress that it put in its specification for a Ki.27 successor in December 1937 was for retractable landing gear, which was expected to bring the new fighter's speed up to 311 miles per hour (an unambitious 7 percent increase over the Ki.27's speed). The *Koku Hombu*'s complacency was also reflected in its decision not to open its requirement to competition, leaving it to Nakajima designers Hideo Itokawa and Yasumi Koyama to improve on the Ki.27 formula without sacrificing its maneuverability. They responded, roughly as Horikoshi had done, by designing a slim, lightweight fuselage with a three-spar one-piece wing around the new 990-horsepower Nakajima Ha-25 four-cylinder radial engine. The result, designated Ki.43, was completed on December 12, 1938, but during test flights in January 1939 it proved to be barely faster and far less nimble than its predecessor, while displaying unsatisfactory takeoff and landing characteristics. Only the provision for overload tanks for increased range offered any real improvement over the Ki.27.

Disappointed, the *Koku Hombu* was about to suspend further development, but Itokawa won a reprieve by promising to redesign the plane. New, finer tail surfaces were incorporated with a fuselage of reduced cross section on the first preproduction aircraft, which was completed in 1939 and displayed a promising improvement in speed and climb rate. Subsequent preproduction airframes were used in various engine and armament experiments; but it was the eighth—which introduced Fowler-type wing flaps boosting lift and maneuverability—that revived enthusiasm in the army. The final preproduction Ki.43, completed in September 1940, incorporated the "butterfly" combat flaps with an uprated Ha-

105 engine driving a two-bladed two-pitch metal propeller, and which featured twin Ho-103 12.7mm machine guns in the cowling. Its performance actually exceeded the *Koku Hombu*'s expectations, and Nakajima was urged to begin production on the Type 1 Fighter Model 1-*Ko Hayabusa* (Peregrine Falcon) as soon as possible.

Although it was even smaller and lighter than the Zero, the *Hayabusa*'s performance was inferior to that of the carrier fighter, its maximum speed being no more than 308 miles per hour. Moreover, the first Ki.43-I-*Ko* fighters retained the puny armament of two synchronized 7.7mm Type 89 machine guns, primarily because the 12.7mm Ho-103, classed as a small-caliber cannon by the Japanese because it fired an explosive shell capable of penetrating light armor, had exhibited instances of early production rounds jamming, or exploding before or just after leaving the muzzle. In spite of those problems, pilots were eager to use the newer, more powerful weapon, so as an interim compromise Nakajima also produced the Ki.43-I-*Otsu*, fitted with one 7.7mm and one 12.7mm gun, until the Ho-103 could be made fully reliable. The Ki.43-I-*Otsu* would in fact represent the majority of the first production batch, because by the time the Ho-103 was dependable enough for twin machine cannon to be mounted in the cowl, Nakajima was gearing up for its second variant, the Ki.43-II with a three-bladed propeller, on which twin 12.7mm guns would be standard.

The Ki.43 possessed one trait that a Zero pilot might envy—its fuel tanks were in the wings and under the cockpit, whereas the A6M2's were in front of the cockpit so that in the event of a hit it would burst into flames or explode in his face. Having been designed with a 900-kilometer (559-mile) range requirement

Nakajima Ki.43-I of the 64th *Sentai*, whose first use of the Japanese army's most ubiquitous fighter was by no means an unqualified success. (*US Navy*)

in mind, the *Hayabusa* compared fairly well alongside its naval stablemate—for example, it proved capable of flying from Formosa to the Philippines, engaging the Americans for up to thirty minutes and returning to base with fuel to spare.

In August 1941, the first Ki.43s began replacing Ki.27s in the 59th and 64th *Sentais*. During their training period in the new fighters, however, new cause for alarm arose during a mock dogfight on October 19, when Sgt. Maj. Saburo Seki of the 64th *Sentai* failed to pull out of a steep dive and crashed to his death. Ground crewmen checked the rest of the 64th's Ki.43s and found cracks around wheel wells and wrinkling of the skin covering the wing roots. Meanwhile, the 59th *Sentai* flew the first operational sortie to involve Ki.43s from Hankow, China, on October 24. This too proved to be a false start, as pilots found wrinkling on the metal skinning after they had engaged in aerobatic combat maneuvers. The *Koku Hombu* promptly recalled all the fighters and sent them to the Nakajima factory at Ojima airfield, Ota Prefecture, where their wing roots were strengthened. While it was at it, Nakajima installed fuel lines that would allow the

planes to use auxiliary drop-tanks to extend their range further.

After some hasty testing the Ki.43s were reissued to the 64th *Sentai*, whose commander, Maj. Tateo Kato, insisted upon receiving one with twin 12.7mm guns, with provision made for his other pilots to receive fully equipped planes if his own guns proved to be fully reliable. Kato also stressed oversea navigation, while he trained his men on their new mounts. There was little time for such practice, however, as war clouds loomed in the Pacific. By December 7, 1941, the 64th *Sentai* had thirty-five Ki.43s on strength and the 59th had twenty-one, as they and the Ki.27-equipped 1st, 11th, and 77th *Sentais* deployed for the coming invasion of British-held Malaya.

Stationed at Duong Dong airfield on the island of Phu Quoc off the Cambodian coast, the 64th *Sentai* covered the landings of the Japanese Twenty-Fifth Army at Kota Bharu, Singora, and Patani on December 8, 1941 (which, because of the international date line, actually occurred slightly before the Pearl Harbor raid). On the previous evening, Lt. Takeo Takayama led the 2nd *Chutai* of the 64th *Sentai* over the invasion fleet. The patrol was relieved at 1730 hours by six *Hayabusas* led by the regiment's veteran commander, Major Kato, but three planes of that second flight were lost when their pilots became disorientated amid darkness, bad weather, and low clouds. Eventually the Ki.43s ran out of fuel and went down in the Gulf of Siam, along with their pilots—1st Lt. Saburo Takahashi, Warrant Officer Kazuo Nakamichi, and Sgt. Maj. Masayoshi Tsuzuki.

The Ki.43's first combat mission took place when aircraft of the 64th *Sentai* departed Duong Dong at 0950 hours to escort Mitsubishi Ki.21 bombers of the 12th and 98th *Sentais* on their second raid against Sungei Patani airfield, where the Brewster Buffalos of No. 21 Squadron, Royal Australian Air Force, were based. Two Buffalos had been destroyed and five damaged in the earlier raid, and the bombers' second strike was followed up by *Hayabusas* strafing the field, leaving the Australians only four serviceable fighters. Two of the Ki.43s also claimed their first aerial victory when they attacked a Bristol Blenheim Mark IV of No. 34 Squadron that had been trying to bomb the Japanese beachhead at Ayer Tawar and compelled it to force land at Machang airstrip. One of the victorious pilots, Lt. Yohei Hinoki, would survive the war with twelve victories, the last of which was scored after he had lost a leg. On the way home, a Ki.43 of the 3rd *Chutai* had to ditch in the sea off Idu Phu Quoc after being hit by gunfire from a Japanese destroyer, and all the others landed at Duong Dong with their tanks nearly dry.

As other Blenheims were returning to Butterworth airfield to refuel, they encountered Ki.43s of the 59th *Sentai*. Warrant Officer Takeomi Hayashi claimed two Blenheims, but in reality only brought down one piloted by New Zealander Sgt. J. E. Smith, who belly-landed his damaged bomber and suffered slight injuries, as did one of his crewmen, Australian sergeant E. H. Brown. A second Blenheim, flown by Flying Officer N. N. H. Dunlop, was chased for twenty miles at treetop level, while his Australian gunner, Sgt. K. R. Burrill, stuck to his gun in spite of being wounded in the lower jaw. Although he was in great pain, Burrill's doggedness was rewarded by the sight of his attacker taking telling hits just as it was closing in for the kill, after which it crashed in the jungle. Dunlop landed safely at Butterworth, while the 59th *Sentai* recorded one pilot, 1st Lt. Takeshi Nakajima, missing. Bad weather and disorientation, however, were responsible for the loss of two other 59th *Sentai* planes—twice as many as were lost in combat that day.

Seventeen Ki.43s of the 64th *Sentai* escorted seventy-nine Ki.21s on a mission to the British airfield at Victoria Point on the following day, but when thick cloud cover forced the bombers to turn back, the fighters landed at the newly captured field at Singora to refuel. They then went off on their own to strafe the airfields at Penang Island and Butterworth.

After the Japanese occupation of Thailand, JAAF units were moved to air bases there, giving the British a few days' grace to reform their ravaged squadrons to the south. The 64th resumed its duties on December 11, escorting forty-one Ki.21s of the 12th and 60th *Sentais* on a raid against Penang. During the return flight, two of the 64th's Ki.43s landed at Kota Bharu, where the Japanese found stocks of fuel in the tanks of abandoned British aircraft. The rest of the unit moved there on the thirteenth, in preparation to support operations over Kuantan.

Thus far, the Ki.43s had encountered little aerial opposition, the more numerous Ki.27s proving to be sufficient to establish air superiority. However, when Buffalos of No. 453 Squadron RAAF attacked Japanese bombers over Penang, shooting down two Kawasaki Ki.48s and two Mitsubishi Ki.51s; five *Hayabusas* of the 59th *Sentai* were dispatched to save the bombers. They arrived thirty minutes later to find the Buffalos already gone and contented themselves with strafing Butterworth. On the return trip, however, two of the Ki.43s collided, resulting in the deaths of 1st Lt. Tomoichiro Fujisaki and the 59th's commander, Maj. Reinosuke Tanimura.

By December 16, the JAAF had lost a total of ten Ki.43s, almost all to noncombat-related causes. As three *Hayabusas* of the 59th *Sentai* were strafing Ipoh airfield that morning, they were dived upon by three Buffalos of 453 Squadron. A lively dogfight followed, in which the Australians found themselves being quite handily outfought. One *Hayabusa* got on Sgt. V. A. Collyer's tail, and he escaped only by jinking away at treetop height. Sergeants A. W. B. Clare and J. Summerton also survived, but both of their planes were badly shot up. Upon returning to base, the Japanese reported tackling six Buffalos and claimed two shot down, as well as seven bombers destroyed or damaged on the airfield.

The Japanese advance continued with such rapidity that they often found the airfields they occupied to have vast amounts of ammunition, fuel, equipment, and provisions. One such "Churchill aerodrome," as the Japanese wryly dubbed them, was Sungei Patani, which still had 1,500 drums of aviation fuel when the 27th and 59th *Sentais* arrived on December 20.

A dozen of the 59th *Sentai*'s Ki.43s were escorting fourteen Ki.48 and Ki.51 bombers to Kuala Lumpur on December 21 when they encountered two Buffalos of 453 Squadron—the only Allied fighter squadron left on the Malayan mainland. Sergeant Eric A. Peterson went after the bombers, claiming one shot down, one probable, and one damaged before breaking off and landing safely. Sergeant K. Ross Leys took on the fighter escort but was quickly overwhelmed, although he succeeded in bailing out and parachuting to earth in spite of the Japanese firing on him. The 59th pilots claimed four victories in the fight—two more than the total engaged—including one credited to 2nd Lt. Hiroshi Onozaki, who would go on to be the 59th *Sentai*'s leading ace with fourteen victories.

The *Hayabusas* encountered their first sizeable aerial opposition on the following day, when Major Kato led eighteen Ki.43s of the 64th *Sentai* on a mission to Kuala Lumpur. At 1140 hours the regiment's 2nd *Chutai*, under Lieutenant Takayama, spotted a formation of fifteen Buffalos, dived on them and subsequently claimed a total of eleven destroyed—three of which were credited to Takayama—as well as four damaged. Lieutenant Oizumi was credited with one Buffalo, and it is believed that others were credited to Major Kato, Lieutenants Hinoki and Shogo Takeuchi, and Sgts. Yoshito Yasuda and Miyoshi Watanabe.

In actuality, 453 Squadron only had twelve planes up at that time, of which five were lost and four others damaged. After being shot up by six Ki.43s, Pilot Officer Thomas W. Livesey belly-landed at Kuala Lumpur with shrapnel wounds in both calves and one ankle. Pilot Officer Robert W. Drury also crash-landed but hit the embankment at the end of the strip and died of his injuries minutes later. Sergeant Harry H. Griffiths made it down with a wound to his left hand. Sergeant Malcolm N. Read was killed, and was reported to have rammed or crashed into one of his opponents. Collyer was wounded in the right foot but managed to land at Sembawang, where he was taken to hospital. Sergeant Stanley George Scrimgeour was involved in a head-on duel with a *Hayabusa* when his Buffalo burst into flame, burning his face, wrists, and fingers before he was able to bail out. Although his parachute was shot up by his victorious adversary, Scrimgeour landed without further mishap.

Sergeant Gregory Richmond Board later recalled: "Spotting a 'Zero' below, I half rolled . . . before I could set up the fighter for the kill, all hell broke loose behind me . . . the instrument panel exploded and blew apart . . . brilliant fire gushed from the fuel tanks." Board managed to bail out.[1]

The Australians credited enemy fighters to Sergeants Griffiths, Read, and Clare, the last mentioned also claimed two Japanese probably shot down. Flight Lieutenant Richard D. Vanderfield was also credited with a "probable," as were Sgts. Keith Gorringe, "Bill" Collyer, and George Scrimgeour. In fact, the only Japanese loss was Lieutenant Takayama. The tail of his plane, which was found by Indian troops, revealed that he had taken some devastating hits from large-caliber bullets, but some of his squadron mates swore that they had seen his right wing collapse as he was pulling out of a dive. Moreover, an examination of the 64th's *Hayabusas* upon their return to Kota Bharu revealed cracks in the wings of six other aircraft—clearly the alterations they had undergone at Ojima had not been enough to cope with the stresses of aerial combat. More fundamental strengthening would be necessary on production Ki.43s, but the 59th and 64th *Sentais* could not wait—for the time being, their ground crews spent the night

improvising their own field modifications to reinforce the wing structures with which the *Hayabusa* pilots courageously made do.

The Australians had little time to rest before another air raid was called in. As the Buffalos taxied to take off, four more Ki.43s, this time from the 59th *Sentai*, arrived and began shooting up the airfield. All of the Buffalo pilots stopped their planes and abandoned their cockpits except one—Eric Peterson, who got airborne, only to find a *Hayabusa* on his tail, guns blazing. The Buffalo spun into the ground from seven hundred feet, killing Peterson. With only three serviceable aircraft, three others barely flyable, and six fit pilots left, what was left of No. 453 Squadron retired to Sembawang to be amalgamated with No. 21 Squadron RAAF.

In spite of its somewhat premature introduction to combat, the Ki.43 did well in the early months of the Pacific War—rather better, in fact, than its intrinsic design deserved—partly due to the skill and élan of its experienced pilots, and partly by riding on the coattails of the reputation of its navy stablemate, the Zero. Even so, the *Hayabusa*'s fighting debut cannot be called an unqualified success, and it was fortunate for its pilots that the Allies, in their stunned ignorance, did not learn of the early Ki.43's weaknesses and take fuller advantage of them.

While the fighting continued, the *Hayabusa*'s flaws were ironed out, more powerful engines developed and more potent armament installed. It became as much the workhorse fighter of the JAAF as the Zero was for the JNAF, and for the same reason—the Japanese committed themselves to it for too long, and their industry was unable to produce enough improved successors to completely replace it in frontline service. As with the Zero, a great many *Hayabusa* pilots became aces in spite of its rapidly advancing obsolescence, getting by on individual skill and their sheer familiarity with its flying characteristics. For every one of those extraordinary paladins, however, there would be scores of other JAAF airmen who died in the cockpits of the fighter that the Allies came to code-name Oscar.

On December 25, twenty-five Ki.43s of the 64th *Sentai*, led by Kato, escorted twenty-seven Ki.21s from the 12th *Sentai* and thirty-six Ki.21s from the 60th *Sentai* to Rangoon. After the target was bombed, the Japanese were attacked by Buffalos of No. 67 Squadron RAF and Curtiss P-40s of Chennault's American Volunteer Group or AVG. The 64th *Sentai* claimed ten victories but lost two fighters and their pilots, while the 12th *Sentai* lost three bombers, in addition to which a fourth damaged Ki.21 had to force land upon its return to base. Two Ki.27s were also lost by the 77th *Sentai*, which in turn claimed seven victories. The Allies, who lost two P-40s and four Buffalos destroyed and two Buffalos damaged, claimed a total of sixteen bombers and twelve fighters that day.

While that sprawling, confused Christmas dogfight raged over Rangoon, four other Japanese fighters remained at the 12th and 64th *Sentais'* base at Don Muang. The only flyable examples left of a nine-plane detachment, their sole duty that day was to patrol over the airfield and prevent the other aircraft from being attacked while taking off or landing. Minor though their role was, however, the arrival of these fighters at the front was not without significance. Bearing little familial resemblance to the Ki.27 or to the Ki.43, they were prototypes of a new Nakajima fighter that represented a radi-

cal conceptual change in the *Koku Hombu's* thinking.

The specification that brought the new fighter into existence stemmed from the undeclared "Nomonhan Incident" between Japan and the Soviet Union in the summer of 1939. At first, the Ki.27 pilots had enjoyed a field day against Soviet Polikarpov I-152 and I-153 biplanes, which were barely able to match the Japanese in speed or maneuverability. The I-16 monoplane proved to be a different matter, however—it was faster than the Ki.27, and Soviet pilots soon learned to eschew dogfighting the Ki.27s in favor of hit-and-run tactics. In response to that disturbing trend, the *Koku Hombu* began to reconsider its earlier dismissal of the fast, heavily armed fighters which were being developed in the West. The specification that it finally issued in 1939 called for a fighter capable of 373 miles per hour (600 kilometers per hour) at 13,125 feet (4,000 meters) and a climb rate of 16,405 feet in less than five minutes. Armament was to consist of two synchronized 7.7mm Type 89 machine guns in the cowling and two wing-mounted 12.7mm Ho-103 guns—less potent than the navy Zero's arsenal, but heavy by army standards.

Nakajima had already been experimenting with a more advanced fighter concept with retractable landing gear. Tohru Koyama's design team responded to the *Koku Hombu's* specification by designing the smallest possible fighter around a 1,185-horsepower Nakajima Ha-42 fourteen-cylinder twin-row radial engine. The short-span three-section two-spar wing, designed largely at the suggestion of Professor Itokawa, had an area of only 161.46 feet, resulting in a wing loading of thirty pounds per square foot—

very high by Japanese standards. The reduction in stability due to the wing's relatively small dihedral was to be compensated for by the substantial side area of the oval-section fuselage, and both climb and maneuverability in combat were to be enhanced by "butterfly" wing flaps similar to those used on the Ki.43.

The first prototype of the Nakajima Ki.44 first flew from Ojima airfield in August 1940, leaving Koyama and his team less than satisfied. Flying characteristics were acceptable, but its weight was 16 percent higher than the design team had originally calculated, raising the wing loading to 34.8 pounds per square foot and which consequently increased the landing speed. Drag was also greater than anticipated, resulting in a maximum speed of only 342 miles per hour at 13,125 feet and a climb rate of 16,405 feet in 5 minutes and 54 seconds. Increasing the rigidity of the engine mount, recontouring the supercharger air intake, and modifying the cowl flaps eventually brought the maximum speed up to 389 miles per hour and the diving speed to 528 miles per hour in 1941. At that point, it was estimated that a fully armed Ki.44 would be capable of 360 miles per hour, and a satisfied *Koku Hombu* ordered the second and third prototypes to be modified accordingly, along with seven more for evaluation purposes.

Veteran Ki.27 pilots assigned to evaluate the new *Shoki*, or "Demon Queller," as the Ki.44 was named, inevitably complained of poor forward visibility on the ground, the high landing speed and the excessive 3,280-foot landing run required for a safe landing, and, of course, the high wing loading and consequent loss of agility. As they became more accustomed to the plane, however, they came to appreciate its fast roll rate, high diving speed, and steadiness as a gun plat-

form. Those features and its armament would have rated it as a first-class interceptor, if the Japanese could bring themselves to use such a term; but their Bushido code was loath to even imply a defensive posture in their terminology, and its role was instead categorized as "aerial exterminating action."

On September 5, 1941, all nine of the first Ki.44s were delivered to Tachikawa airfield and assigned to a specially formed evaluation unit, the 47th *Dokuritsu Hiko Chutai* (Independent Air Squadron), unofficially dubbed the *Kawasemi Butai* (Kingfisher Force) because its use of the fighters was expected to be similar to the hit-and-run tactics of its avian namesake. Led by thirty-one-year-old Maj. Toshio Sakagawa, the unit's number was allegedly chosen in reference to the legend of the forty-seven ronin, former samurai who took revenge on the slayers of their daimyo, or lord. The pilots were also informally referred to as *Shinsengumi*, or "chosen ones."

On December 3, the 47th *Chutai* was shipped off to Canton, China, to evaluate its fighters under combat conditions. When war broke out a week later, three *hentais* (flights) of the 47th were ordered to Saigon, where they were attached to the Southern Area Army. Hastily daubed in brown camouflage, each plane was also marked with a small emblem under the cockpit representing the *Yamada-ryu* drum, which had signaled the attack of the forty-seven ronin. One, two, or three stripes, in descending order of rank, slanted their way down the front of the vertical stabilizer—in white for the 1st *Hentai*, led by Sakagawa and attached to the 1st *Sentai*; yellow for Capt. Susumu Jinbo's 2nd *Hentai*, assigned to the 11th *Sentai*; and red for the 3rd *Hentai* under Capt. Yasuhiko Kuroe, assigned to the 12th *Sentai*.

In late December, the JAAF diverted its efforts from Malaya to Burma in response to reconnaissance reports that large numbers of Allied aircraft were gathering at the airfields of Mingaladon and Toungoo. On December 24, the 47th joined the 12th and 64th *Sentais* in a move from Saigon to Don Muang airfield, in the course of which mechanical problems forced three of the Ki.44s, including the *chutai* commander's, to force land. Consequently, only six *Shokis* were combat ready the next day, and they were kept close to home, four flying their first sortie to do nothing more than cover the airfield. There was little time to lose in restoring the 47th to full operational status, however, as Sakagawa and his pilots learned of the stinging setback the JAAF had suffered over Rangoon at the hands of Buffalos of No. 67 Squadron RAF and P-40s of the AVG.

Captain Yasuhiko Kuroe, leader of the 3rd *Hentai*, 47th *Dokuritsu Hiko Chutai* (signified by the single red rudder band), taxis the eighth prototype of the Nakajima Ki.44 Shoki prior to takeoff from Saigon airport early in 1942. (*US Navy*)

In January 1942, while Nakajima was putting the Ki.44-I into production, the nine prototype and evaluation aircraft were transferred to Kuantan to support operations over Singapore. On the fifteenth, sixteen bombers attacked Tengah and Sembawang, while the 59th *Sentai's Hayabusas* claimed six Buffalos that tried to interfere, and Ki.27s of the 1st *Sentai* claimed seven more Buffalos.

Accompanying the raid were four Ki.44s. Spotting a flight of Buffalos approaching the bombers over Tengah, Kuroe dived on the leader, and when his first pass failed to bring it down, he pulled up while Jinbo engaged it. Both Japanese had scored hits, and while Jinbo climbed away, Kuroe came at the Buffalo again, reporting afterward that "after five bursts it fell to the ground." Pilot Officer Greville L. Hesketh, a twenty-six-year-old New Zealander in No. 243 Squadron,

had been leading a section of Buffalos from No. 488 Squadron when the *Shokis* attacked him, but while he managed to crash-land beside the oil tanks near Alexandria Hospital, he died of his wounds before medical personnel reached the wreck.[2]

The Ki.44 had been blooded, and it would happen again on January 18, when members of the 47th *Chutai* claimed another Buffalo over Singapore. On the night of the twentieth, however, Lockheed Hudsons of No. 8 Squadron RAAF bombed Kuantan and damaged four of the Ki.44s.

On January 26, a British force of Hudson bombers and Vickers Vildebeest biplane torpedo bombers, escorted by Buffalos and Hawker Hurricanes, left Singapore to attack a Japanese invasion force approaching Endau when they were set upon by nineteen Nakajima Ki.27s of the 1st and 11th *Sentais*. Five Vildebeests were shot down, while four Hudsons and several Buffalos returned with damage. Although the Commonwealth pilots claimed numerous "Zeros" and "Navy Type 0" and "Navy Type 96" fighters shot down, only one Ki.27 of the 1st *Sentai* was actually lost—possibly the one claimed in flames by Hurricane pilot Flt. Sgt. Ronald L. Dovell of No. 232

Squadron—and its pilot bailed out safely.

Also present during the action was Kuroe in his Ki.44, initially staying discretely above the fight while observing and reporting results. He was still there when the British launched a second strike from Singapore, comprised of nine Vildebeests and three Fairey Albacore biplane torpedo bombers, escorted by seven Hurricanes and four Buffalos. The motley force arrived over Endau at 1750 hours and immediately came under attack by ten Ki.27s of the 1st *Sentai*, which shot down six Vildebeests and two Albacores. Kuroe, now joined by Jinbo, also attacked the Vildebeests, two of which were credited to the Kingfisher pilots. When 232 Squadron tried to intervene, the *Shoki* pilots plunged through its formation and were subsequently credited with a Hurricane each, although only one was actually shot down, with Jinbo being the more likely victor. Its pilot, Canadian-born Sgt. John P. Fleming, reported:

> On this last trip of the day I immediately attacked three fixed-undercarriage Japs attacking a Vildebeest who was making a bomb run on one of the ships—I think a freighter. The three scattered and I saw a bomb strike at the waterline of the ship—as I broke away from this attack I was struck by fire from a 'Zero' I think. Oil pressure collapsed—managed to fly south for about 20 miles before the engine seized, abandoning the aircraft at low level over the beach just north of Mersing.[3]

The Hurricane pilots claimed three or four Ki.27s in the fight, corresponding to the 1st *Sentai*'s logging several aircraft damaged, though none were lost.

On January 28, four Boeing B-17Es of the American 19th Bomb Group operating from Java bombed Kuala Lumpur airfield, setting one Japanese fighter on fire,

badly damaging three others, and causing lesser damage to another six. The 11th *Sentai* was reduced to only one combat-ready *Chutai*, while the 47th *Dokuritsu Hiko Chutai*'s local detachment was left with only one operational Ki.44.

By February 8, four Ki.44s were serviceable again, providing air cover for the landings on Singapore Island without encountering any opposition. Moving on to Mudon in southeastern Burma, the 47th *Chutai* lost a plane on February 24 when Allied fighters strafed the field and burned one of the Ki.44s. The three remaining *Shokis* participated in a large fighter sweep over Mingaladon in Burma the next day, claiming two victories during a wild fight with Hurricanes of No. 17 Squadron RAF and with three P-40s of the AVG. Both sides, in fact, made extravagant claims that day, but one of the Ki.44s' misperceived victims may have been Sgt. John F. Barrick, an American serving in 17 Squadron, who described his lone fight with fifteen enemy fighters:

> I attacked and shot down one 'Army 97' [Ki.27] and was then jumped from above by a 'Zero' [undoubtedly a Ki.44]. I went into a tight turn which caused one of the gun panels to fly open. This made the aircraft 'flick' and probably saved my life, because the 'Zero' was in an excellent position behind me. As it was[,] my plane was not hit.[4]

During a return strike on Mingaladon on the twenty-sixth, Sakagawa was leading his two wingmen in a line-astern formation when they were jumped by a Tomahawk. Skillfully evading his assailant, Sakagawa got on the tail of another P-40 and Warrant Officer Etsugi Mitsumoto attacked the first, only to be himself beset by a third AVG fighter. "I went after him," said Kuroe, who had been flying too far in the rear to assist effectively, "but I was not fast enough. The enemy fired at Mitsumoto, then turned upside down and got out of the battlefield by diving fast." Mitsumoto was wounded but got his damaged plane back to base.[5]

Four of the 47th's Ki.44s may also have been the so-called "Zeros" that attacked four Blenheims over the Sittang area on March 4, shooting down one plane from No. 45 Squadron. Rangoon fell on March 8, and on the twentieth the unit's quartet of Ki.44s had joined the thirty-two Ki.27s of Headquarters *Shotai*, 12th *Hiko Daitai* and of the 1st and 11th *Sentais* at the newly taken airfield at Pegu as components of the 5th Air Division during the continuing Japanese offensive into Burma.

On March 21, ten 17 Squadron Hurricanes strafed Mingaladon airfield, destroying two Japanese planes in flames and badly damaging two others. The Japanese retaliated with a massive strike on Magwe field, to be opposed by six Hurricanes and three P-40s. The 11th *Sentai* lost its commander, Maj. Tadashi Okabe, to one of the Hurricanes. Elsewhere, Parker Dupouy and Kenneth Jernstedt of the 3rd Pursuit Squadron, AVG, were attacking Ki.21 bombers when one of the Japanese gunners put a bullet through Ken Jernstedt's windscreen and a shard of glass entered an eye, forcing him to disengage. Dupouy was then attacked by seven "Model O" fighters, to which he responded in characteristic "Flying Tiger" form by turning and coming at the nearest one, resulting in .30-caliber hits to his plane, and his hand being laid open by shrapnel, but his opponent catching fire and crashing thirty miles northwest of Toungoo. Both P-40s force landed, damaged beyond repair, along with six equally shot-up Hurricanes; but it seems likely that Dupouy's victim was not a Ki.43, but the

first Ki.44 to be destroyed in air-to-air combat along with the first *Shoki* fatality, 1st Lt. Shunji Sugiyama.

On April 21, Captain Kuroe left the 47th *Chutai* to take command of the 3rd *Chutai* of the 64th *Sentai*, succeeding Capt. Katsumi Anma, who had been killed on April 8. Four days later, in reaction to Lt. Col. James H. Doolittle's carrier-launched B-25 raid on the home islands on April 18, the 47th *Chutai* was recalled to Japan and turned its Ki.44s over to the 64th *Sentai*. In spite of the three victories credited to him while flying the *Shoki*, Kuroe reported that it did not live up to his expectations. By then, however, Nakajima had committed the plane to production, and it was accepted for JAAF service.

Eventually identified by Allied intelligence and given the code name of Tojo, the Ki.44 was nevertheless to be frequently misidentified as a "Zero" in Allied combat reports in China for years to come. Its speed and rate of climb were soon overtaken by newer Allied fighters, but the Ki.44 *Shoki* would soldier on throughout the war, its pilots often managing to distinguish themselves as late as 1945 defending their homeland against Boeing B-29s. More important, it set a precedent that roused the JAAF from its obsession with maneuverability for its own sake, leading to a generation of more potent fighters. Unfortunately for Japan, those new fighters still came out too little and too late; even as the Pacific War neared its end in 1945, the Japanese never fully replaced the navy Zeros and army *Hayabusas* with which they had started it.

Ten

Under Foreign Management

LICENSED AND LEASED FIGHTERS, 1939–42

Not all of the famous fighters of World War II started out serving the countries that built them. The most obvious examples came from the United States, which did not officially enter the conflict until December 7, 1941, but which provided aircraft to several of the combatants prior to that date. Two other cases, however, can be found in two of the war's most famous flying anachronisms, built by Britain and Italy but first used in combat by two other air arms entirely.

Incredibly, the Gloster Gladiator was still in first-line service with several Royal Air Force squadrons when Britain entered the war, even though low-wing monoplane fighters with retractable landing gear had been making their way into the RAF since 1936. Developed by Harry P. Folland in 1934, the Gloster S.S. 37 was first flown by Flt. Lt. P. E. G. Sayer on September 12, 1934, and entered production, with an 830-horsepower Bristol Mercury IX engine, on July 1, 1935, under the official designation of Gladiator. A compact but conventional biplane, the Gladiator looked more modern than its Italian contemporary, the Fiat CR.42, only in that it had an enclosed sliding canopy and two .303-inch machine guns under the wings in addition to the usual two in the fuselage—touches that, to the retrospective eye, only look quaintly out of place on an aircraft that should have been retired within a year of being committed to production.

Tangmere-based No. 72 Squadron began replacing its Bristol Bulldogs with Gladiators on February 22, 1937, the first of an eventual nine squadrons to get the type. Rough running of the Mercury IX engine with a two-bladed airscrew led to the decision to use a three-bladed Fairey-Reed propeller in July 1937. The resultant improvement was standardized, along with a Mercury VIII or VIIIAS engine, as the Gladiator Mark II.

The 1938 Munich Crisis led to increased production of all available fighter types for the RAF, and the Gladiator II was among those built for use by the Auxiliary Air Force. At the same time, foreign orders for the Gladiator I came from Latvia, Sweden, Norway, Greece, and China. The Royal Navy, too, wanted a replacement for its aging Hawker Nimrods, for which Gloster developed the Sea Gladiator, equipped with a naval

T.R.9 radio, an airspeed indicator calibrated in knots, and provision for an arrestor hook.

By September 1939, Gladiators and Sea Gladiators were serving actively with the RAF and thirteen other air arms. Long before World War II, however, it was the Chinese who had the distinction of first blooding the type in combat.

In January 1938, the Chinese central government ordered thirty-six Gladiators, which were accompanied to Kai Tak airfield, Hong Kong, by a Gloster test pilot and a small party of riggers and technicians. From there, the crated fighters had to be transported by rail and then by junk up the Pearl River to Canton. As each plane was assembled at Tien Ho airfield, it was flown first by the test pilot and then by a Chinese pilot—who more often than not would prang the tricky biplane while taxiing or while in a heavy-handed stall upon takeoff. Nevertheless, by February Gladiators were flying operationally with the 28th and 29th Pursuit Squadrons of the 5th Pursuit Group, and the 32nd Squadron of the 3rd Pursuit Group.

Some of the Gladiators' earliest combats were flown by American citizens of Chinese descent who had volunteered to defend their ancestral homeland from the Japanese invaders. Wong Sunshui, also known as John Wong, was born on March 15, 1914, in Los Angeles, California, which was also where he earned his pilot's license. He subsequently went to China, and after training at the Guangdong Air Force Academy, became deputy commander of the 17th Squadron, which was originally equipped with Boeing 248s—export versions of the P-26 Peashooter monoplane. "Buffalo" Wong, as he was also nicknamed, was credited with two victories in the Boeings before being wounded over Nanking during a disastrous encounter with new Mitsubishi A5M2s of the Japanese 13th *Kokutai* on September 19, 1937, which cost the Chinese eleven planes.

When Wong returned to combat, he learned that the 17th Squadron's Boeings had been virtually annihilated while defending Nanking in December. He was given command of the 29th Squadron at Nan Hsiung, newly equipped with Gladiators, and it was in the British biplane's first action that Wong reopened his account. On February 24, 1938, eight Nakajima E8N1 floatplanes from the auxiliary seaplane carrier *Notoro* and five from *Kinugasa Maru* set out to bomb Nan Hsiung. In response to the threat, three Gladiators of the 28th and nine from the 29th scrambled up, and upon spotting the enemy formations at 6,000 feet in the nine o'clock position, Wong waggled his wings and signaled his men to follow him down. In spite of jamming guns, which plagued all the Gladiators, Wong quickly set one of the floatplanes on fire and saw it dive away, while he sent another down streaming fuel. He and Lts. Huang Guangching, Chou Linghsu, and Huang Nengrong then shot up another E8N, which was last seen emitting black smoke as it fled the scene.

The Chinese found the E8Ns to be much feistier than they looked. As Lt. Yang Rutong pursued one of the floatplanes, its pilot turned to engage, fired and shot Gladiator No. 2902 down, killing Yang. Lieutenant Chen Chiwei was chasing another jinking E8N at 300 feet altitude when he lost control of Gladiator No. 2808, went into a tailspin and crashed to his death. Lieutenant Chou Linghsu came back with Gladiator 2810 badly damaged by the rear gunner of the E8N he had fought. To top off a less than glorious sortie, brake failure caused Gladiator

2901, flown by the 29th Squadron's leader, Capt. Xieh Chuangwo, to career off the runway, damaging a wing.

The Chinese credited Wong with one victory, along with one plane damaged, and shared credit in another. The Japanese noted that one of their planes failed to return and a second, from *Notoro*, returned with 138 hits, the observer dead, and the pilot wounded in the right leg, only to crash-land and be subsequently written off. Two other Japanese aircrewmen had been killed in the action.

On the following day Wong, flying Gladiator No. 2905, downed a Nakajima A4N1 biplane fighter over Guangzhou, his late opponent being either Lt. Shigeo Takuma or Petty Officer 1st Class Hisao Ochi of the 13th *Kokutai*, both of whom were killed in action that day. Other sources, however, cite his victim as having been one of four E8Ns that he attacked above the Kowloon-Guangzhou railroad on February 28.

Wong scored three more victories during a battle with aircraft from the carrier *Kaga* on April 13, bringing his total up to eight, but was himself wounded by A5M pilot Petty Officer 1st Class Jiro Chono and had to bail out of his burning Gladiator 2913. He was later promoted to the rank of major, and on November 15, 1940, was put in command of the 5th Fighter Group, with which unit he flew Polikarpov I-152s. On March 14, 1941, Wong was hit during a dogfight with Mitsubishi A6M2 Zeros and died of his wounds two days later.

Another early exponent of the Gladiator was Chin Shuitin, a.k.a. Arthur Chin from Portland, Oregon. Born on October 23, 1913, Chin became one of fifteen Sino-American volunteers who trained at Al Greenwood's flying school in Portland before joining the Cantonese air force on December 1, 1932. In 1936,

Major Wong Panyang, commander of the 5th Pursuit Group, was flying Gloster Gladiator Mark I 2909 on June 16, 1938, when he downed a Mitsubishi "Ki.21" (actually one of three G3M2s lost by the Takao *Kokutai*) and shared in three others. (*Andrew Thomas*)

Chin was sent to Germany for additional training and then returned to become executive officer (XO) of the 28th Squadron of the 5th Group, Chinese air force. First flying the Curtiss Hawk III biplane fighter, Chin opened his account with a Mitsubishi G3M2 on August 16, 1937. The squadron reequipped with Gladiators in the spring of 1938, and on April 13 another American volunteer, Louie Yim-Qun (a.k.a. Clifford "Long-legged" Louie from Seattle, Washington), shot down an A4N and shared in downing another over Guangzhou (Petty Officer 1st Class Naoshi Eitaku and Petty Officers 3rd Class Yukio Miyasato and Yuji Mori from *Kaga*'s fighter squadron were killed that day). Chin used a Gladiator to down an E8N floatplane over Hukou, Kiangsi, on May 31, killing Seaman 1st Class Hiromitsu Takahara of the 12th *Kokutai*. On June 1, 1938, he was promoted to captain and placed in command of the 28th Squadron, with Louie as his XO.

On June 16, Chin downed a Mitsubishi Ki.21, as well as damaging another bomber over Shaokuan, near Canton, while Louie and a Chinese Gladiator pilot teamed up to bring down another. Also active that day was Wong Panyang of the

17th Squadron, who downed a Ki.21 and shared in the destruction of three others over Nan Hsiung.

Born in the village of Cheung, Guangdong Province, on July 29, 1913, Wong Panyang had accompanied his father to Seattle in July 1920 and subsequently became a US citizen under the name of John Wong. After earning a private pilot's license in Portland, he went to Shanghai in September 1932 and joined the Chinese air force, scoring two victories in Boeing 248s with the 17th Squadron in August 1937.

That the Gladiator's days were numbered became apparent on August 3, 1938, when Chin's 28th Squadron tangled with A5M2s over Liang-Chiatien in Hupei Province. Chin, flying Gladiator No. 2809, managed to damage one of the swift monoplane fighters and then, out of sheer desperation, rammed a second and succeeded in bailing out. Louie and a Chinese comrade also claimed an A5M, and at least one more was also forced down, for the 15th *Kokutai* recorded the deaths of Lt. j.g. Shinjo Naoshisa and Petty Officer 2nd Class Hitoshi Fukosawa, while Petty Officer 3rd Class Namitaro Matsushima was brought down as a POW. Chin went on to destroy two more A5Ms and share in the destruction of two more before being badly burned in combat on December 27, 1940. Returning to the United States, he died in 1997.

After the war John Wong Panyang flew for the Central Air Transport Corporation until 1947. He then formed the Tan Yuen Industrial Corporation in Bangkok, running it until his death in 2005.

The Gladiator was still in Chinese use in 1939, but by then it was hopelessly outclassed by the A5M. Had anyone in Europe been paying serious attention to what was going on in China, the Gladiator

would probably have been retired before May 10, 1940, when the Germans launched their offensive in the West. As it was, desperate circumstances often brought the biplanes into the forefront over Finland, Norway, Belgium, France, the Western Desert, Greece, and Malta, to name but a few, and heroic deeds were performed in the Gladiators before the last of them were withdrawn from RAF service in 1942.

If ever there were an airplane that was obsolete before its wheels left the runway, the Italian Fiat CR.42 *Falco* (Falcon) would be the all-time champion. The last of a successful series of biplane fighters designed by Celestino Rosatelli, the CR.42 had a robust airframe of light alloy and steel, and its wings were braced by the same sturdy Warren-truss structure as its predecessors. Its engine was a reliable 840-horsepower Fiat A74R.1C.38, neatly cowled and driving a three-bladed, two-speed Fiat 3D41 propeller. Armament was comprised of one 7.7mm and one 12.7mm Breda-SAFAT machine gun in the upper fuselage decking, which was later wisely upgraded to two 12.7mm weapons. The prototype displayed the same excellent maneuverability of the earlier CR.32, but could go much faster at 274 miles per hour. Moreover, its development proceeded so smoothly that it went from flight testing into production and began squadron service all in the same year. Unfortunately for Italy, that year was 1939.

When Italy entered the war on June 10, 1940, the CR.42 was numerically the *Regia Aeronautica*'s principal fighter. Arguably, it represented the ultimate development of the old World War I formula of a biplane fighter with fixed landing gear and an open cockpit, but that could not make up for the fact that that configuration was long out of date.

The CR.42 joined the Italian army in its fourteen-day invasion of southern France, beginning with strafing attacks on the air bases at Fayence and Hyères, and escorting Fiat BR.20 bombers over Toulon harbor on June 13. More than a month before Italy's entry into the conflict, however, its fighter had already seen some action—and drawn a few drops of blood—in the hands of Belgian pilots.

Belgium had undergone a fitful rearmament in the face of the growing threat from a resurgent Nazi Germany during the late 1930s. Fully roused by the German invasion of Poland in September 1939, the Belgian military began making desperate feelers abroad to acquire more modern aircraft for its woefully inadequate *Aéronautique Militaire*. One of the most desperate of such actions was the purchase of forty Fiat CR.42s that month. The first of them was delivered in March 1940, and by May there were thirty-four delivered and thirty assembled to equip the 3ème and 4ème *escadrilles* of *Groupe de Chasse* II. While Italian fighter pilots may have still favored the seat-of-the-pants atmosphere of the Fiat's open cockpit, the Belgians were less nostalgic. So low was their opinion of their new mounts that the two *escadrille* commanders were said to have flipped a coin to see who would test-fly the biplane first—with that distinction going to the loser.

By May 1940, Belgium had a total of 180 aircraft at its disposal, of which only eleven Hawker Hurricanes of the 2ème *Escadrille* could be considered up to contemporary standards. In addition to the two CR.42 squadrons, the 1ère *Escadrille* was operating the Gladiator, which in the coming year would be better known for dueling the Fiats, rather than fighting alongside them. Still, the CR.42s and Gladiators seemed more likely to survive in the air than the fighters with which the 5ème and 6ème *Escadrilles* were equipped. Those two squadrons were still using the Fairey Fox VIC, a Belgian-built Hispano-Suiza-engine single-seat version of the British reconnaissance and training two-seater, pending the arrival of forty replacement Brewster B-239s (export versions of the F2A Buffalo).

Such was the situation when the CR.42's true combat debut came on May 10, 1940, as Luftwaffe aircraft heralded the end of the Phony War and the beginning of the German offensive in the West—with Belgium and the Netherlands among their initial stepping-stones on the road to Paris.

Among the few Belgian aircraft to get into the air on that disastrous first day was a CR.42 flown by *Sgt.* Jean Henri Marie Offenberg. Born at Laeken, near Brussels, on July 3, 1916, "Peike" Offenberg had joined the 4éme *Escadrille* at Nivelles in March 1939, just as it was exchanging its Foxes for the barely more up-to-date Fiats.

In his autobiography, Offenberg described how he was awakened at 0100 hours on May 10 by his squadron mate, *Premier Sergent* (First Sergeant) Alexis Jottard, who told him that war had broken out with Germany. Anticipating trouble, at 0415 the II *Groupe* leader, *Commandant* Jacques Lamarche, took off in CR.42 R-43 and led the twenty-four flyable fighters from Nivelles airfield to the emergency airstrip at Brustem, near Saint-Trond. Junkers Ju 87Bs arrived just as the last CR.42s were taking off and damaged two of them.

Just before landing, *Capitaine* Jean de Callataÿ, leading a three-plane flight, spotted Ju 52/3ms of 17th *Staffel*, *Kampfgeschwader zur besonderen Verwendung* dropping dummy parachutists over Alken

Fiat CR.42 No. 21 belonged to the 4*e Escadrille* of *Groupe de Chasse* II, based at Nivelles. One of three planes left there on May 10, 1940, it was damaged in a German air raid and unable to participate in Belgium's air defense that day. (*J. L. Roba*)

as a diversionary tactic and attacked one of them. It subsequently crash-landed east of Maastricht at 0530. Moments after scoring the first CR.42 victory, Callataÿ's flight was attacked by Me 109Es of I *Gruppe*, *Jagdgeschwader* 1, but he managed to disengage and, believing Brustem to be the enemy's next target, led his pilots back to Nivelles.

At 0605 five CR.42s took off to protect Brustem. Two of them clashed with Me 109Es of 2nd *Staffel*, JG 1 over Waremme, and *Lt.* Charles Goffin shot one down, *Ltn.* Erwin Dutel parachuting to earth near Aix-la-Chappelle. The Belgians then came under attack from ten Me 109s of 3./JG 27, *Ltn.* Erwin Axthelm and *Uffz.* Heinrich Becher each claiming a "Gladiator" over Tirelemont, one of whose pilots took to his parachute. In fact, Goffin's Fiat was damaged and his wingman, *Premier Sergent* Roger Delannay, bailed out but was hit in a lung by ground fire and died before he reached the hospital.

Also taking off at 0605 were Offenberg, who had asked *Commandant* Lamarche to fly a patrol, if only to avoid being caught on the ground, and *Premiers Sergents* Jottard and Jean Maes. As the trio circled up over Saint-Trond, Offenberg noticed a twin-engine airplane with twin tails, decked in black crosses—his first view of a German. He signaled to his wingmen, but they too had spotted it, and Maes had already peeled off after it. Offenberg was pondering why Maes hadn't waited for orders, in accordance with Belgian fighting instructions, when Jottard gestured toward a ragged formation of aircraft on the horizon, closing fast. Offenberg instinctively went into a spin, dived between two Messerschmitts, and approached a Dornier Do 17 whose rear gunner fired at him. He too fired, only to see his right machine gun jam.

Pulling out of his dive, Offenberg found himself alone over Diest, having descended 5,000 feet. As he prepared to set course for Saint-Trond, he spotted another twin-engine bomber below him. Offenberg pursued it in a shallow dive but was unable to catch up with the German.

After glancing down at Saint-Trond airfield, Offenberg followed the main road toward Brussels, and then spotted another Do 17 approaching from his left. Having learned a lot in the past two combats, he dived—the only way to build up the speed to catch up with the swift bomber—then performed an Immelmann turn and positioned himself above it. After an exchange of gunfire, Offenberg disengaged and last saw the Do 17 below him, rapidly losing altitude with black smoke curling from an engine. At 0748 Offenberg returned at Brustem, where he claimed one Dornier destroyed and one damaged. The *Aéronautique Militaire* did not confirm his claims or Maes's Dornier claim, all of which may have been made on the same plane—a Do 17P of 3rd *Staffel* (*Fernaufklärung*) *Aufklärungsgruppe* 11 (3.(F)/11), which had been reconnoitering the Saint-Trond-Tongeren area for the German XVI *Panzer Korps*, and which, in fact, made it back to base.

The last CR.42 success for May 10 occurred at 1440 hours, when *Lt.* Prince Werner de Mérode attacked a Do 17P over Waremme and claimed it in flames. *Oberleutnant* von Schaezler of 2.(F)/123 managed to belly-land his Dornier at Mönchengladbach.

At 1500 hours the Germans found Brustem, and five Me 109Es strafed the Fiats. Although the fighters were not well dispersed, the Messerschmitts accomplished little, but even as Offenberg was suggesting that they get airborne again before the enemy came back, more Germans did arrive—this time Ju 87Bs of 4th *Staffel*, *Sturzkampfgeschwader* 2. The Belgians took what cover they could as bombs fell on the CR.42s. When the Stukas departed, not one of the 3*ème Escadrille*'s fifteen fighters remained operational. In contrast, none of the 4*ème Escadrille*'s Fiats had been seriously damaged.

As night fell, Offenberg felt like weeping. On the first day of the German invasion, all of the Belgian airfields had been bombed or strafed, and Fort Eben Emael had been seized by an audacious assault by glider-borne German paratroopers. After that, the remaining seven Fiats of *Escadrille* 4/II redeployed to Nieukerk, where they claimed two more enemy planes for the loss of another CR.42 before May 16 when, with Belgium about to be overrun, Offenberg and his comrades were ordered to withdraw to Chartres, France.

When France fell a month later, Offenberg and his squadron mate, *Sergent* Jottard, commandeered two Caudron Simouns and flew to Montpelier, then made their way across the Channel to England. There, Offenberg enlisted in the RAF, joining No. 145 Squadron on August 17. After flying Hurricanes in the Battle of Britain and later switching to Supermarine Spitfire IIBs, "Pyker" Offenberg, as he was known to his British comrades, became the first foreign volunteer in the RAF to be awarded the Distinguished Flying Cross, in June 1941, and on the twenty-seventh of that month he joined No. 609 Squadron. He subsequently commanded a Belgian flight within the squadron, but during a training flight on January 22, 1942, he was killed in a midair collision with a Spitfire from No. 92 Squadron. At the time of his death, Offenberg had added four victories, plus two shared, to the Do 17 he had claimed while flying the Fiat CR.42, rejected by his own air arm but accepted by the RAF. At that same time, the *Falco*, though finally eclipsed by monoplanes as a first-line fighter, was still soldiering on as a fighter-bomber and night fighter in the *Regia Aeronautica*, as well as in the Royal Hungarian Air Force (*Magyar Királyi Honvéd Légierö*), and later even in the Luftwaffe.

At about the same time Belgium was vainly struggling to slow the German onslaught, the Netherlands was also fighting a losing battle against the Luftwaffe. The principal Dutch fighter at that time was the Fokker D.XXI, a low-wing monoplane with an enclosed cockpit and fixed, spatted landing gear. Before the Dutch commenced their gallant but doomed fight against the Germans, however, Fokker D.XXIs had been blooded in another war against equally daunting odds, yet with far greater success.

Designed in 1936, the Fokker D.XXI was a reasonably advanced, if unexceptional, plane for its time, with a 760-horsepower Bristol Mercury VIII nine-cylinder radial engine that gave it a maximum speed of 286 miles per hour, and with an armament of four 7.7mm Vickers machine guns. The D.XXI gave Fokker's Nederlandsche Vliegtuigenfabriek, which had suffered during the economic depression and recession of the 1930s, a new lease on life. In addition to orders from the Royal Netherlands Air Force, the firm sold rights for the license manufacture of ten D.XXIs by Denmark and thirty-eight by Finland in 1937. By May 1940, however, the D.XXI was becoming obsolescent, and only thirty-nine were operational in the Royal Netherlands Air Force. It is small wonder, then, that the Germans overran the Netherlands in a matter of days.

Long before that, however, the Finnish Fokkers had already drawn blood—and plenty of it.

Taking advantage of the Nazi-Soviet pact of August 1939 while Adolf Hitler's attention was on Britain and France to the west, Josef Stalin sought to secure his country's northern flank by occupying part of Finland's Karelian Isthmus. When the Finns refused to give up that territory, the Soviets commenced hostilities with a bombing attack on Helsinki on November 30, 1939.

The Finns were astonished that war had actually broken out in so sudden a fashion, but they got over their shock quickly. The army was mobilized and soon managed to stall the Red Army's invasion in the Finnish forests while inflicting tremendous casualties on its personnel.

Although numerically no match for the Soviet air forces, the *Ilmavoimat* had done its best to compensate through intense training. "The younger pilots got additional training in aerial combat and gunnery," recalled then-*Lentomestari* (Warrant Officer) Eino Ilmari Juutilainen of *Lentolaivue* (Flying Squadron) 24. "During bad weather, we indulged in sports, pistol shooting and discussions about fighter tactics. Our esprit de corps was high despite the fact that we would be up against heavy odds. We were ready."[1]

The first clash occurred at 1145 on the morning of December 1, when two Finnish Bristol Bulldog Mark IVAs—minor refinements of a British biplane design that had entered RAF service in 1929—were attacked by six I-16s of the 4th *Eskadrila*, 7th *Istrebitelny Aviatsionny Polk*. One of the Finns, *Ylikersantti* Toivo Uuttu of the 3rd *Lentueessa* (Flight) of *LeLv* 26, reported that three I-16s circled his Bulldog BU-64, and that whenever the Soviet flight leader fired his guns, his two wingmen would also, whether they were aiming at him or not. Uuttu scored some hits and reported one I-16 going down emitting black smoke from its engine

Newly delivered in Dutch camouflage, Fokker D.XXI FR-108 bore a blue 6 on its rudder when Kersantti Eino Ilmari Juutilainan destroyed an SB-2 bomber on December 19, 1939. *Luutnantti* Eino A. Luukkanen, leader of 3/LeLv 24, used it to share in destroying a Polikarpov R-5 on December 23, and to down an SB-2 on January 6, 1940. (*Aviation Museum of Central Finland, Tikkakoski, via Mikko Veijalainen*)

before his Bulldog took hits in the fuse-lage that jammed the controls and sent it down out of control over the Kannas forest. Uuttu managed to regain control in time to crash-land in a clearing. Although strafed by the victors and temporarily taken prisoner by a Finnish patrol that assumed him to be Soviet (an error which he played along with for awhile), Uuttu survived his ordeal with only a lump on his head.

Uuttu, credited with Finland's first-ever air-to-air victory, may in turn have been the "reconnaissance aircraft" jointly credited to *Starshy Leitenant* Fyodor Shinkarenko (the 4th *Eskadrila*'s com-mander), *Starshy Politruk* (Senior Political Officer) Gabriel Didenko, Boris Aleksandrovich Grigoriev, and *Lt.* Piotr I. Pokryshev, the last mentioned of whom force landed his I-16 due to damage that he attributed to antiaircraft fire. Pokryshev would go on to be credited with a total of thirty-eight solo and eight shared victories during World War II, for which he received two Gold Stars of a Hero of the Soviet Union.

Meanwhile, two hundred fifty Soviet bombers were on their way to Helsinki and other ports, as well as to airfields in southwestern Finland. This time, howev-er, the Fokker pilots of *LeLv* 24, led by *Kapteeni* Gustaf Erik Magnusson, took off in pairs, and by the time the first bombers arrived at 1200, the Finns were waiting for them over Kannas. Over the next three hours, *Luutnantti* Jaakko Vuorela downed two bombers and *Luutnantti* Eino A. Luukkanen downed an Ilyushin DB-3, while Tupolev SB-2s of the 24th and 41st *Bombardirovochny Aviatsionny Polki* fell to "Eka" Magnusson; *Luutnantti* Jussi Räty; *Vänrikki* (Second Lieutenant) Pekka J. Kokko; and *Kersantti* (Sergeants) Lauri V. Nissinen, Lasse Heikinaro, Leo Rauta-korpi, and Kelpo Virta.

It was only the beginning of a disastrous mismatch between the overwhelming quantity of Soviet aircraft and the out-standing quality of Finnish pilots. Fighting alongside more aged types such as Bulldogs and Gladiators, the Fokkers were credited with 120 victories by the time Finland capitulated on March 13,

1940, making *LeLv* 24 the top-scoring unit of the Winter War. Five of the squadron's pilots became aces, and a good many future ones opened their accounts in the D.XXI. Juutilainen, who shot down a DB-3 and a Polikarpov I-16 and joined five other pilots in downing an SB-2, compared the Fokker D.XXI with its principal adversary:

> It was our best fighter in 1939, but the Soviet Polikarpov I-16 was faster, had better agility and also had protective armor for the pilot. I flew a war booty I-16, and it did 215 knots at low level and turned around on a dime. I liked that plane. In comparison, the Fokker could make about 175 [knots]. The D.XXI also lacked armor, but it had good diving characteristics and it was a steady shooting platform. I think that our gunnery training made the Fokker such a winner in the Winter War.[2]

The Fokker D.XXI's most successful exponent was a relative latecomer, *Luutnantti* Jorma Sarvanto, who opened his account on December 23 with two SB-2s over Kannas. Flying the same plane, FR-97, on January 6, he encountered a formation of DB-3s over Kannas and shot down six of them in about four minutes. He went on to down four more Soviet planes and share in the destruction of two others before the Winter War ended, making him the leading ace of the conflict.

In addition to seven D.XXIs bought from Fokker and the thirty-eight that it was licensed to produce, the Finnish State Aircraft Factory (*Valtion Lentokonetehdas*, or VL) built sixty more, incorporating such modifications as a 1,050-horsepower Pratt & Whitney R-1535 Twin Wasp Junior engine, enlarged vertical control surfaces, the transfer of the two fuselage guns to the wings, and more canopy glaz-

ing for better visibility. VL also built one experimental version with inward-retracting landing gear. Although the careers of the Dutch and Danish Fokkers were extremely brief, the Fokker D.XXI steadfastly served Finland throughout the Continuation War until September 1944.

By 1940, the Gladiator had been eclipsed in RAF service by a number of British-built monoplane fighters, supplemented amid the exigencies of war by several types imported from the United States through the Lend-Lease program. As a result of that arrangement, three famous American fighters first fired their guns in anger in British or Commonwealth service.

Arguably the best-known fighter in the US Army Air Forces (USAAF) in the first year of direct American participation in the war was the Curtiss P-40. Designed by Donovan Reese Berlin, the prototype XP-40 was more evolutionary than revolutionary in conception, being nothing more than the tenth production airframe of Berlin's P-36A Hawk fighter with an Allison V-1710 liquid-cooled inline engine substituted for the P-36's 875-horsepower Wright GR-1820-G3 Cyclone air-cooled radial.

Donovan's original monoplane design dated to 1934. He designated it the Curtiss Model 75, reflecting his obsession with that number, while the emotive nickname of "Hawk," already famous from an earlier generation of Curtiss biplane fighters, was revived for the new-generation monoplane. Although the Hawk 75 first flew in 1935, engine problems delayed its development until July 7, 1937, when the Army Air Corps gave Curtiss the largest American peacetime production order up to that time—210 P-36s, as the Army called them, for

$4,113,550. Hawk 75s—built with retractable landing gear or with fixed, spatted undercarriage, depending on whether speed or simplicity was the customer's main priority—were also sold to numerous foreign air arms.

The first to employ them in combat were the Chinese. The first Hawk 75 to arrive was purchased by Madame Chiang Kai-shek for $35,000 as a present for Col. Claire Lee Chennault in July 1937. In July 1938, the first official Chinese air force consignment of Hawk 75Ms with fixed undercarriage was assigned to the veteran 25th Pursuit Squadron, whose pilots began training under Chennault's tutelage. Its baptism of fire came on August 18, when twenty-seven Mitsubishi G3M2s came in three waves to bomb Hengyang, and Squadron Commander Tang Pusheng led three of the new Hawks up along with seven I-152s. Intercepting the first flight of nine, Tang shot one down and damaged another, then attacked the second wave of bombers, only to be caught in a cross fire from the gunners, who shot him down in flames. The two other Hawk 75Ms crashed on landing.

On October 1, nine more Hawk 75Ms were assigned to the 16th Pursuit Squadron at Zhiguang, which had just been redesignated from a light bomber unit that had previously flown Vought V-92 Corsairs. Similarly, the next unit to get nine Hawk 75Ms, the 18th Pursuit Squadron, had previously flown Douglas O-1MCs. Training continued apace but suffered a setback on January 2, 1939, when five pilots of the 25th Squadron were killed in crashes while flying from Kunming to Sichuan. Curtiss Hawk III biplanes were assigned to the squadron to supplement its reduced numbers. Chinese confidence in the new plane remained so low three months later that when twenty-three Japanese bombers

Delivered to China in August 1937, Curtiss Hawk 75H (civil registration NR 1276) was flown by Lt. Col. Claire L. Chennault. In spite of the improvements he recommended that went into the Hawk 75M, the Curtiss proved no match for the Japanese Zero in 1940. (*Library of Congress*)

raided Kunming on April 8, no Hawks rose to challenge them, and the nine were bombed to destruction on the ground.

In August 1939 the 25th Squadron was disbanded, followed by the 16th within the following month, leaving only the 18th to be attached to the newly formed 11th Pursuit Group, which operated six Hawk 75Ms alongside I-152s and I-16s. Success continued to elude the monoplane Hawks, and the ultimate blow came on October 4, 1940, when twenty-seven G4Ms attacked Chengdu, escorted by eight new A6M2 Zeros of the 12th *Kokutai*, led by Lt. Tamotsu Yokoyama and Lt. j.g. Ayao Shirane. Far from intercepting the enemy, the 18th Pursuit Squadron was ordered by the Third Army air staff to disperse to Guangxi, but as it did some of its planes were caught en route by the aggressive Zeros. Hawk 75M No. 5044 was shot down in flames and its pilot, Shi Ganzhen, killed when his parachute failed to open. Additionally, an I-152 and a Gladiator were sent crashing with their respective pilots, Gin Wei and Liu Jon, wounded. Two other I-152s returning to base with engine trouble, piloted by Zing Ziaoxi and Liang Zhengsheng, were strafed as they landed and were apparently counted among the five "I-16s" the Zero pilots claimed to have shot down, along with an "SB-2 bomber," which was in fact an Ilyushin DB-3 of the 6th Bomber Squadron, returning on the mistaken assumption that the raid was over, only to be destroyed with its three crewmen.

At Taipingzhi airfield the Japanese reported destroying nineteen aircraft on the ground, four of their pilots audaciously landing to set fire to some of them. The Chinese acknowledged the loss of two I-16s, two I-152s, two Hawk IIIs, six Fleet trainers, one Beech UC-43 Traveler, a Dewoitine 510, and a Hawk 75M. The Hawk 75M's inauspicious combat debut in China officially came to an end when the virtually impotent 18th Pursuit Group was disbanded in January 1941.

Although its performance was eclipsed by other types by 1940, the Hawk 75 was fondly remembered by a lot of others who flew it for its easy handling and excellent maneuverability. Export Hawk 75As, powered by 950-horsepower Pratt & Whitney Twin Wasp engines, became the best fighters available in quantity to the French in May 1940, and were credited with more enemy planes shot down than any other French fighter. Thailand purchased Hawk 75Ns with fixed landing gear and used them during its war against the French in Indochina in January 1941, as well as during its brief resistance to invading Japanese forces that December. The US Army Air Corps still had P-36Cs on strength when Pearl Harbor was attacked on December 7, 1941, and they scored some of the first American air-to-air victories that day. The exiled Royal Netherlands Air Force used Cyclone-engine P-36s in the East Indies, where they became easy prey for the Japanese Zero. The RAF also used P-36s, which it called Mohawks, as stopgap fighters, most notably in Burma in 1942. Vichy French pilots flew Hawk 75As against the British

and Americans during their invasion of North Africa in November 1942, and other French H-75As, shipped to Finland by the Germans, fought the Soviets until September 1944. Although neither as modern nor as famous as the P-40 that succeeded it, the P-36 can lay claim to greater ubiquity and the rare—if somewhat dubious—distinction of having fought on both sides during World War II.

The Allison-engine XP-40 first flew on October 14, 1938, and competed against the Lockheed XP-38, Bell XP-39, and Seversky AP-4 at Wright Field near Dayton, Ohio, on January 25, 1939. The XP-40 emerged the winner, and in April Curtiss was rewarded with what was touted as the largest contract since World War I, for 543 P-40s.

Although outclassed by the Zero, P-40s soldiered on as best they could, starting with the very first Japanese attack on Pearl Harbor. In the months that followed, the P-40 pilots fought desperately and often heroically, with decidedly mixed fortunes. They were ultimately annihilated in the Philippines and the Dutch East Indies, but in the hands of master tactician Col. Claire Chennault and his American Volunteer Group in China, the Hawk 81 (as the export version of the P-40 was known) did a disproportionate amount of damage and became one of the war's legendary fighters.

More than half a year before the United States entered the war, however, P-40s had already fought over terrain that could not have been farther removed from Pearl Harbor, the mountains of China, or the jungles of Southeast Asia.

Even while it was filling the Army Air Corps orders for P-36s and P-40s, Curtiss was marketing both the Hawk 75 and 81 to overseas customers. Although not used quite as widely as the Hawk 75, the Hawk 81 saw considerable Chinese use—first by the American mercenaries of the AVG and later by Chinese pilots— as well as service in the Soviet army and naval air arms, the RAF, and the Royal Australian Air Force (RAAF). It was, in fact, in British service that the "Hawk" designator probably got more use in reference to the P-40 than it ever did by the fighter's American pilots. The earliest models were called Tomahawks by the British, while the later ones were christened Kittyhawks.

The first major action in which Tomahawks figured prominently was a sideshow that nevertheless serves as a reminder of just how global a conflict World War II was. On May 2, 1941, British forces in Iraq came under attack by Iraqi forces directed by the anti-British, pro-German chief of the National Defense Government, Rashid Ali El Ghailani. The revolt was quickly crushed and Rashid Ali fled the country on May 30, but not before a number of Axis aircraft had been committed to his cause. Sixty-two German transport aircraft carried matériel to the Iraqis, making refueling stops at airfields in Syria and Lebanon, then mandates of nominally neutral Vichy France. The Germans and Italians had also sent fighters and bombers which, hastily adorned in Iraqi markings, had made their way into the country from the French air bases. Amid the British counterattack, at 1650 hours on May 14, two Tomahawk Mark IIbs of No. 250 Squadron RAF, flown by Flying Officers Gordon A. Wolsey and Frederick J. S. Aldridge, carried out the first P-40 combat mission when they escorted three Bristol Blenheims in an attack on suspected German and Italian aircraft at Palmyra in Syria.

A Curtiss Tomahawk Mark IIB of No. 250 Squadron RAF taxis for takeoff in May 1941. Its first combats were primarily against the Vichy French in the Levant. (*Andrew Thomas*)

In allowing Axis planes to stage from their territory, the French-mandated Levant presented a threat to British security in Palestine and Egypt at a time when *Generalleutnant* (Major General) Erwin Rommel's Afrika Korps was starting to make its presence felt in North Africa. On May 15, British aircraft attacked French air bases at Palmyra, Rayack, Damascus, Homs, Tripoli, and Beirut. The Vichy government responded by dispatching *Groupe de Chasse* III/6 to Rayak, with a complement of top-of-the-line Dewoitine D.520 fighters. In addition to those, *Général de Division* (Major General) Jean-François Jannekeyn, commander of the *Armée de l'Air* in the Levant, had GC I/7, equipped with Morane-Saulnier MS.406 fighters; *Groupe de Bombardment* II/39, with American-built Glenn Martin 167F twin-engine bombers; GB III/39, with antiquated Bloch 200s; *Groupe de Reconnaissance* (GR) II/39 and Flight GAO 583, both with Potez 63-IIs; and a myriad of less effective army and navy planes at his disposal for a total of ninety aircraft.

By the end of May, the British had decided to seize the Levant. General Archibald Wavell organized an invasion force by pulling the Australian 7th

Division, less one brigade, from its defensive position at Mersa Matruh and combined it with the 5th Brigade of the Indian 4th Division, elements of the 1st Cavalry Division, a commando unit from Cyprus, a squadron of armored cars, and a cavalry regiment. The scratch force was to receive some offshore support from the Royal Navy, and air support from sixty aircraft of the RAF and the RAAF. The Aussie contingent included Tomahawk Mark IIbs newly delivered to replace the Gladiators of No. 3 Squadron RAAF at Lydda, Palestine. Meanwhile, on May 22, No. 250 Squadron had been reassigned to the defense of Alexandria, Egypt.

The three-pronged invasion, dubbed Operation Exporter, commenced from Palestine and Trans-Jordan at 0200 hours on June 8. The British and accompanying Gaullist French forces had hoped that the 35,000 Vichy troops in Syria would be loath to fight their former allies, but they were in for a disappointment. Anticipating Allied propaganda appeals based on the idea of saving Syria from German domination, Vichy High Commissioner Gen. Henri-Fernand Dentz saw to it that all signs of German presence were removed throughout the country. The Luftwaffe, too, had discretely evacuated all of its planes, aircrews, and technicians from the Syrian airfields forty-eight hours ahead of the expected invasion. In consequence the Allied propaganda fell on deaf ears, the French officers defending Syria refused to deal with their Gaullist compatriots, and the Allied invasion force found itself with a fight on its hands.

On the day of the invasion, June 8, Hurricanes of Nos. 80, 108, and 260 Squadrons, and five Tomahawks of No. 3 Squadron, RAAF, made preemptive attacks on the French airfields. Among other targets, the Hurricanes and Tomahawks strafed GC III/6's fighters on the ground at Rayak, burning a D.520 and damaging seven others. On this occasion, the Tomahawks drew relatively little fire from French ground gunners, who mistook the unfamiliar new fighters for their own D.520s. On the same day, two Tomahawks of 250 Squadron, flown by Flying Officer Jack Hamlyn and Flt. Sgt. Thomas G. Paxton, together with shore batteries, shot down an intruding Cant Z.1007bis reconnaissance bomber of the Italian 211*a Squadriglia* five miles northwest of Alexandria.

As Vichy France rushed reinforcements to the Levant, the Germans and Italians put airstrips in newly conquered Greece at their disposal, allowing the swift ferrying of Lioré et Olivier LéO 451 bombers of GB I/31, I/14, and I/25; D.520s of GC II/3; and Martin 167s of the 4*e Flotille* of the *Aéronavale* to Syria by June 17.

French bombers attacked Adm. Sir Andrew Cunningham's naval force off Saida on June 9, damaging two ships. Eight German Junkers Ju 88A-5s of II *Gruppe*, *Lehrgeschwader* 1, operating from Crete, also turned up to harass the fleet on June 12, but they were intercepted by Tomahawks of No. 3 Squadron. Squadron Leader Peter Jeffrey shot down one of the attackers, Flt. Lts. John R. "Jock" Perrin and John H. Saunders claimed two others, while Flt. Lt. Robert H. Gibbes caught a fourth bomber right over the fleet and claimed it as a "probable." In fact, two Ju 88s failed to return—one from the 4th *Staffel*, piloted by *Ltn.* Heinrich Diekjobst, and one from the 5th *Staffel*, flown by *Ltn.* Rolf Bennewitz.

Three days later, the Aussies turned their attention to the Vichy French, as Jeffrey and Flt. Lt. Peter St. George B. Turnbull each accounted for a Martin 167F of GB I/39 in the area of Sheik

Meskine. On June 19, Turnbull damaged another Martin bomber over Saida, and Pilot Officer Alan C. Rawlinson damaged two others near Jezzine. Damascus fell to the Allied forces on June 21. In an encounter between Tomahawks and D.520s of GC III/6 two days later, *Capitaine* Léon Richard, commander of the *6e Escadrille*, was credited with shooting down a Tomahawk south of Zahle— probably Turnbull, who crashed his damaged Tomahawk upon returning to Jenin. The starboard wing of Sgt. Frank B. Reid's Tomahawk was also damaged by 20mm shells, but the French took the worst of the fight. *Sous-Lieutenant* Pierre Le Gloan was forced to beat a hasty retreat when his D.520 began to burn, probably after being hit by Flying Officer Percival Roy Bothwell, who also sent *Lt.* Marcel Steunou, a five-victory ace, down in flames near Zahle, and killed *Sgt.* Maurice Savinel between Ablah and Malakaa. Flying Officer Lindsey E. S. Knowles claimed to have shot the wingtip off another Dewoitine, which was credited to him as damaged.

Among the toughest Vichy strongpoints was Palmyra, a fortified air base surrounded by concrete pillboxes, antitank ditches, observation posts, and snipers' nests. When Habforce, a composite invasion group drawn from British and Arab troops occupying Iraq, crossed the border and moved on Palmyra on June 20, it came under heavy and very effective bombing and strafing attacks by the base's aircraft. Habforce's advance ground to a halt for about a week. Then, on June 25, appeals for air support by Habforce's commander, Maj. Gen. J. George W. Clark, were finally answered as Commonwealth aircraft arrived, including No. 3 Squadron's Tomahawks. In an aerial engagement fifteen miles southwest of Palmyra, Saunders, Flying Officers

John F. Jackson and Wallace E. Jewell, and Sgt. Alan C. Cameron each claimed a LéO 451 of GB I/12—though only three such bombers were in fact present, and all were lost along with the lives of five of their twelve crewmen. When six Martin 167s of *Flotille* 4F sallied out of Palmyra to attack Habforce again on June 28, they were intercepted by No. 3 Squadron's Tomahawks, and six were promptly shot down in sight of the British ground troops, Rawlinson accounting for three of them, Turnbull downing one, and Sgt. Rex K. Wilson destroying another. The action heralded a pivotal turn in Habforce's fortunes, the day's only sour note being struck after the Tomahawks refueled and returned to Jenin, where Sergeant Randall's plane suffered engine failure, and he was killed in the crash.

Ultimately, the Allied forces prevailed, launching their final thrust on Beirut on July 7. Amid fierce but hopeless resistance, High Commissioner Dentz passed a note to Cornelius Engert, the US consul general in Beirut, expressing his willingness to discuss surrender terms with the British, but absolutely not with the Gaullist French.

Negotiations dragged on for sixty hours, during which fighting continued, and the Tomahawks had one more occasion to test their mettle against the D.520 in the air. On July 10, seven Tomahawks of No. 3 Squadron RAAF went to cover twelve Bristol Blenheims of No. 45 Squadron, which were to bomb an ammunition dump near Hamana, south of Beirut. The Blenheims bombed the target, but the explosions drew the attention of five D.520s of *Aéronavale* Flight 1AC, which had transferred to Lebanon six days earlier and were escorting Martin bombers on a mission. Attacking from head on and below, the French quickly shot down three Blenheims, riddled a fourth so badly that it subsequently had to crash-land, and damaged six others.

Diving to the belated rescue, the Tomahawk pilots claimed all five of the Dewoitines—two by Turnbull, and one each by Jackson, Pilot Officer Eric Lane, and Sgt. Geoffrey E. Hiller. In actuality, only two French fighters were lost: *Second Maître* (Petty Officer Second Class) Pierre Ancion was mortally wounded, dying in Beirut hospital two days later, while *Premier Maître* (Petty Officer First Class) Paul Goffeny bailed out of his burning D.520 over the Bakaa Valley with slight wounds, subsequently claiming that his pursuer had crashed into a mountain while trying to follow his evasive maneuvers. The other three French pilots returned—*Enseigne de Vaisseau de 1ère Classe* (Lieutenant Junior Grade) Jacques du Merle being credited with two bombers, *Premier Maître* Jean Bénézet with another, and *Lieutenant de Vaisseau* (Lieutenant) Edouard Pirel sharing in the destruction of the fourth Blenheim with Goffeny, who was also credited (wrongly) with the "crashed" Tomahawk.

On the following day, *Lt.* René Lèté of GC II/3, lagging behind his formation due to engine trouble, spotted three Tomahawks and attacked, shooting down Flying Officer Frank Fischer. Lèté was one of only two Frenchmen to shoot down a Tomahawk during the campaign, but he had little time to exalt in the distinction, for moments later Gibbes and Jackson got on his tail and claimed to have sent him down in flames—Bobby Gibbes subsequently getting full credit for his first of an eventual ten victories by winning the coin toss. In fact, Lèté survived his crash landing—as did Fischer, who, after hiding from the French in an Arab village, rejoined 3 Squadron after hostilities ceased. The squadron lost one other Tomahawk to antiaircraft fire over Djebel Mazar that day, Flying Officer Lin

Knowles crash-landing near Yafour, ten miles from Damascus.

At 1201 hours on July 12, a cease-fire finally went into effect, ending the thirty-four-day Syrian campaign. A formal armistice was signed at Saint-Jean-d'Acre on July 14—Bastille Day, ironically enough. The Tomahawks of No. 3 Squadron RAAF were transferred to the Western Desert, where their colleagues in RAF and South African air squadrons were already battling the Luftwaffe and *Regia Aeronautica*. Among other early successes over the Western Desert, Pilot Officer Thomas G. Paxton of 250 Squadron, who had shared in the Cant Z.1007bis on June 8, added an Me 109E to his growing score south of Tobruk on June 26, while on June 30 one of Paxton's squadron mates, Sgt. Robert J. C. Whittle, shared in the destruction of an Me 110, damaged an Me 109, and probably downed an Italian aircraft.

And so, the Curtiss P-40 had its baptism of fire in the Middle East. Although its principal virtue in 1941 was its ready availability, great things would be done in the P-40, and a series of improved models kept the Curtiss fighter in production for five years, a total of 13,737 being produced.

Another American workhorse that gave good, if less extensively publicized, service for the British was the Grumman F4F, dubbed Wildcat by its builder but known in Royal Navy circles—at least initially—as the Martlet. Originally conceived as a biplane successor to Grumman's F3F series, the F4F prototype had evolved into a midwing monoplane with retractable landing gear by the time the XF4F-2 appeared in 1937. In the course of development, the rounded wings and control surfaces became more square in

At left, Lt. Rodney H. P. Carver in Grumman Martlet Mark I BJ562 S-7l leads a flight of No. 804 Squadron. At right, Ju 88A-5 4N-AL of 3rd *Staffel* (*Fernaufklärung*)/*Aufklärungsgruppe* 22, which Carver and Flt. Sub-Lt. Thomas R. V. Parke shot down on Christmas 1940 after *Ltn*. Karl Schipp belly-landed at Sandrick, south of Loch of Skaill, Scotland. All four crewmen survived to be taken prisoner. (*Andrew Thomas*)

shape, and the machine guns were moved from the fuselage to the wings.

Soon after entering production, the F4F-3 attracted orders from the French and Royal Navies. The French ordered eighty-one G-36As, as the export F4F-3 was called, to be fitted upon arrival with two fuselage- and four wing-mounted 7.5mm Darne machine guns. The first of the French G-36As, with a Wright R-1820-G205A engine, was tested at the Grumman factory on May 11, 1940, but by then Germany had launched its offensive in the West. With the capitulation of France in June, the export order was transferred to Britain, augmenting a British order for 100 G-36Bs with 1,200-horsepower Pratt & Whitney S3C4-G single-stage supercharged engines. The first F4Fs were designated Martlet Mark Is by the British, and later arrivals with folding wings were called Martlet IIs. In all cases, the British armed their Grummans with four wing-mounted .50-inch Colt-Browning machine guns.

The first Martlet Is reached Prestwick in August 1940, replacing the Sea Gladiators of No. 804 Squadron, while the first Martlet II flew in October. The F4F's first aerial success was achieved under circumstances curiously similar to those of the Supermarine Spitfire's. On Christmas Day 1940, two of No. 804's Martlet Is, BJ562 and BJ561 flown respectively by Lt. Rodney H. P. Carver and Sub-Lt. T. R. V. Parke, intercepted a Ju 88A-5 as it tried to enter the Royal Navy's anchorage at Scapa Flow, and shot it down. It was a modest beginning for a corpulent little fighter that would soon become legendary in US Navy service at the battles of the Coral Sea, Midway, and Santa Cruz, and with the Marines at places like Wake Island and Guadalcanal.

One of World War II's enduring ironies was that the war's best all-around fighter owed its very creation to a British order—and its status among the war's immortals to the subsequent substitution of a British engine for the American one with which it was originally powered. First flown on October 26, 1940, the prototype North American NA-73X had been developed to satisfy an order from the British Purchasing Commission earlier that year in response to the imminent German invasion threat. By the time production on the Mustang Mark I began, however, the Battle of Britain had been won, and the need for a complementary fighter to the Hurricane and Spitfire was less pressing.

The first Mustang Mark Is reached Britain in October 1941. The Mustang impressed its RAF pilots with its strong construction, good flying characteristics, eight-gun armament (six in the wings and two alongside the engine crankcase), and, above all, its four-hour endurance (twice that of a Spitfire), but its unsupercharged Allison V-1710 engine performed poorly above twenty-five thousand feet. For the high-altitude combat that dominated the Battle of Britain, and the subsequent duels over the Channel, the Mustang was seriously outclassed by the Spitfire Mark V, as well as by its most likely opponents, the Messerschmitt Me 109F and Focke-Wulf Fw 190A. It was therefore relegated to the Army Cooperation Command, for which role the Mustang pilots spent most of their time training in England for the inevitable Allied counterinvasion of Hitler's *Festung Europa*.

The first Mustang I unit, No. 26 Squadron at Gatwick, Sussex, began converting to the type in January 1942. Among the tasks envisioned for the close-support Mustang was high-speed photoreconnaissance using a single F24 camera installed behind the pilot's seat. After some testing, it was determined that the Mustangs could get the best results from an altitude of 9,000 feet, holding the airplane in a bank and aligning the camera with a mark made on the wing trailing edge.

In May 1942, No. 26 Squadron began participating in Rhubarb raids, crossing the Channel to strafe German targets in France, at the same time carrying out reconnaissance sorties over targets of potential interest along the French coast. As the RAF and the newly arriving bomber squadrons of the United States Army Air Forces (USAAF) began planning operations against industrial centers in German-occupied Europe, the

North American Mustang Mark I AG427 of No. 414 Squadron RCAF flies an army cooperation mission in the summer of 1942. A plane from this unit scored the first Mustang victory on August 19. (*No. 414 Squadron records via Andrew Thomas*)

Mustangs were fully committed to the photoreconnaissance role. On July 27, 1942—three weeks before the American Eighth Air Force was to dispatch its first daylight attack with Boeing B-17s—sixteen RAF Mustangs flew their first photo mission over the Dortmund-Ems Canal, leading into the Ruhr Valley and its vital industrial centers.

The first enemy aircraft officially credited to a Mustang was brought down by a member of No. 414 Squadron, Royal Canadian Air Force. Curiously but appropriately, however, the pilot was an American. Born in Baxter, Iowa, on March 25, 1915, Hollis Harry Hills had been a civilian flyer during the 1930s before slipping across the border and enlisting in the RCAF on September 5, 1940. Flying Officer Hills was posted to No. 414 Squadron, then equipped with Tomahawks, on October 12, 1941.

Hills's first air-to-air fight took place during Operation Jubilee, the RAF's effort to support the Anglo-Canadian landing at Dieppe on August 19, 1942. By then, fifteen RAF squadrons were equipped with Mustang Mark Is, of which four—Nos. 26, 239, 400, and 414, making up No. 35 Wing at Gatwick—were committed to low-level reconnaissance missions over the Dieppe area, primarily to scout for German motorized activity.

Hills in Mustang AG470 RU-M took off early in the morning to accompany Flt. Lt. Frederick E. Clark in AG375 RU-F on a low-level road reconnaissance from Abbéville to Dieppe, checking for movements of German armor. The two lost one another in the darkness and returned to their base at Gatwick with nothing to report. They flew a second such mission later in the day, and Hills described what happened in an account in the summer 1990 edition of the US Navy's aviation journal, *The Hook*:

The weather was sunny, not a cloud in the sky. As we approached the French coast, the sky was full of fighters in one massive dogfight from sea level to the contrail level. In hurried glances, I counted seven parachutes in the air at one time. A couple of miles short of landfall I spotted four Fw 190s off to our right at about 1,500 ft. Their course and speed was going to put them directly overhead when we crossed the beach. I called Freddie twice with a 'tally ho' but there was no response. He did not hear the warnings and apparently did not see the Fw 190s. When Freddie turned right to intercept our recce road at Abbeville, we were in an ideal position for the FWs to attack. I swung very wide to Freddie's left during the turn, dusting the Abbeville chimney tops. That kept me beneath the FWs and I believe they lost sight of me.

My plan was to cut off the lead Fw 190 before he could open fire on Freddie, but my timing went to pot when a crashing Spitfire forced me to turn to avoid a collision. That gave the lead Fw pilot time to get into firing position and he hit Freddie's Mustang with the first burst. I got a long range shot at the Fw leader but

had to break right when his number two man had a go at me. Number two missed and made a big mistake of sliding to my left side ahead of me. It was an easy shot and I hit him hard. His engine caught fire, and soon after it started smoking the canopy came off. I hit him again and he was a goner, falling off to the right into the trees.[3]

The second *Rotte* of Fw 190s had disappeared, so Hills went looking for Clark and found him heading for Dieppe harbor at 1,000 feet, streaming glycol and still being pursued by the leading Fw 190. Hills fired a short high-deflection burst, just to distract the German from finishing Clark, and saw the enemy plane break into a tight left-hand turn to engage him. Hills found that he could outmaneuver the Fw 190, but whenever he tried to press his advantage, the Fw 190 pilot would break away, using his superior speed to get clear of the Mustang and fly inland, only to turn back and attack Hills when he tried to retire over the Channel. At one point in the long duel, Hills had to dodge a falling Me 109, and the Focke-Wulf pilot got his best chance at a shot but missed. "My opponent was a highly competent pilot and I was ready to call a draw as soon as I could," remarked Hills.[4]

Finally, during one of the Focke-Wulf's breakaways, Hills made a break toward Dieppe at maximum speed and was relieved to find the German fighter no longer in pursuit. As he crossed the Channel, Hills passed under some Ju 88s and Me 109s returning from a raid on Southampton, but they all kept going. The American returned without further incident, but was saddened to learn that Clark had not come back. His squadron was not alone in suffering a loss during Operation Jubilee—No. 26 Squadron had lost no less than five of its Mustangs to antiaircraft fire or enemy fighters, while No. 239 lost two and No. 400 lost one.

At about 0500 the following morning, Hills was rudely awakened as his door burst open. "I was grabbed in a bear hug by what smelled like a huge clump of seaweed," he recalled. "It was Freddie. He'd ditched his Mustang in the Dieppe harbor and was rescued unconscious by a brave soldier of the amphibious forces. With Freddie as witness, the Fw 190 shootdown was confirmed for me and RCAF No. 414 Sqn. as the first Mustang kill."[5]

Later, of course, the Mustang would have its Allison engine replaced with the Rolls-Royce Merlin and would reach its full potential as the best all-around fighter of the war. Scores of Allied pilots would become aces in Mustangs, but "Holly" Hills would not be one of them.

As the USAAF began establishing bases in Britain, Hills was pressured to transfer to his own country's air arm. He was willing to do so, provided he got to fly Spitfires with the 4th Fighter Group, then being created from the three volunteer Eagle squadrons of the RAF, Nos. 71, 121, and 133. Instead, he was offered a slot in a squadron equipped with twin-engine Lockheed P-38s. Hills refused to do so, transferring into the US Navy instead on November 9. After returning to the United States and training on Grumman fighters, Lieutenant Junior Grade Hills was assigned to fighter squadron VF-32 on May 28, 1943, and returned to action in the Pacific aboard the light carrier *Langley* in October. It was in that unit, flying Grumman F6F-3 Hellcats, that he added significantly to his tally by downing three Zeros and damaging a fourth over Truk on April 29, 1944. Another Zero over Manila on September 21 made Lieutenant Hills an ace. On the following day he was shot down by antiaircraft fire over Subic Bay, but like Freddie Clark, he was pulled from the drink by crewmen of the submarine *Haddo*. Holly Hills died in Palm Bay, Florida, on October 31, 2009.

There was one other American fighter that had an auspicious combat debut in the hands of foreign pilots during World War II—the Brewster B-239, an export version of the F2A-1 that was known to its RAF pilots as the Buffalo. Designed in 1935 and entering service with VF-3 aboard the carrier *Saratoga* late in 1938, the F2A-1 has a place in history as the US Navy's first monoplane fighter with retractable landing gear and an enclosed canopy. It also acquired a less than flattering reputation during the early months of Japanese expansion in Southeast Asia and in the Pacific, when Buffalos of the RAF and RAAF, B-239s of the Royal Netherlands East Indies Air Force, and US Marine F2A-1s based on Midway Island suffered terribly against nimbler, heavier armed, and usually more numerous Japanese fighters like the A6M2 Zero and Nakajima Ki.43 *Hayabusa*. Survivors of the debacles in Malaya, Singapore, Burma, and the East Indies would never have believed it, but almost half a year before their ordeals began in December 1941, Buffalos were enjoying incredible success—over Finland.

During the Winter War, Finland had desperately sought new aircraft to replace the obsolescent machines with which its air arm had to defend the country. In spite of the official ban on the export of all military matériel issued by the US government at the start of World War II, on December 16, 1939, Finnish representatives managed to order forty-four fighters from the Brewster Aeronautical Corporation at a cost of $55,404 each, since deliveries of the aircraft to the US

Lentomestari Eino Ilmari Juutilainen taxis his assigned Brewster 239, BW364 "Orange 4," through the snow at Solomanni in Soviet Karelia on December 18, 1941. Juutilainen scored 33 of his 94 1/6 victories in Brewsters—and he was only its second-most successful user. (*Aviation Museum of Central Finland, Tikkakoski, via Mikko Veijalainen*)

Navy had been delayed and the aircraft were still technically Brewster's property. The Finnish order was comprised of thirty-eight F2A-1s intended for the US Navy, and six B-239Bs slated for Belgium, all standardized to basic B-239 configuration. In the case of the F2A-1s, all US Navy property was removed, including guns, instruments, and carrier deck landing equipment, resulting in a significant reduction in weight. The F2A-1's 950-horsepower Wright Cyclone R-1820-34 nine-cylinder radial engine was replaced by a civilian version of equal output, the Cyclone R-1820-G5.

After being shipped to Stavanger, Norway, the Brewsters were transported by train to Trollhättan, Sweden, where they were assembled and armed with two synchronized machine guns—one 7.62mm and one 12.7mm—and two wing-mounted 12.7mm Colt-Browning weapons. From there, they were flown to Finland, the first four arriving on March 1, 1940.

The first six B-239s were hurriedly assigned to *LeLv* 22, commanded by *Kapteeni* Erkki Heinilä and based in south-ern Finland. Flying from a frozen lake at Säkylä in western Finland, they made one uneventful sortie before the Finns and Soviets agreed on a truce, ending hostilities on March 13, 1940. A month later, on April 19, the Brewsters were all reassigned to *Majuri* Gustaf E. Magnusson's *LeLv* 24, based at Helsinki-Malmi airport. Pleased to have a replacement for their aging Fokker D.XXIs, *LeLv* 24's experienced pilots enthusiastically strove to master the new Brewsters and to develop flexible tactics for using them in combat. The squadron lost two aircraft and pilots in the course of fourteen months, but that intense training would soon pay off.

When Germany launched *Unternehmen Barbarossa*, its invasion of the Soviet Union, on June 22, 1941, Finland tried to avoid becoming involved, though at the same time it kept its military forces on alert on the assumption that involvement might be unavoidable. The Soviets seem to have harbored similar notions, and on the morning of June 25 they chose to strike first, dispatching 150 unescorted bombers to neutralize Finnish defensive installations.

They should have known better. Alert Finnish patrols spotted the oncoming first wave, and soon fighters were scrambling up to intercept them. The Finns had not been as well prepared for trouble as they should have been—only 24 fighters of the 125 in the *Lentolaivues'* operational inventory actually engaged the bombers, but they wrought havoc aplenty on them. First blood was drawn when six Fokker D.XXIs of the 1st *Lentueessa* of *LeLv* 32 intercepted a wave of DB-3s near the railway junction at Riihimäki, and *Luutnantti* Veikko Arvid Evinen managed to shoot down two of them. *Lentolaivue* 26, flying its Fiat G.50s over central eastern Finland, shot down thirteen SB-2s, while *Kersantti* Antti Johannes Tani, flying a Morane-Saulnier MS.406 of *LeLv* 28, downed another in the same area.

Making their fighting debut alongside those fighters were seven B.239s of *LeLv* 24's 2nd *Lentueessa*, which took off at 0715, led by *Luutnantti* Jorma Sarvanto. The flight split into two divisions, both of which ran into waves of unescorted Tupolev SBs. *Lentomestari* Eero Aulis Kinnunen downed no less than four SB-2s and shared in the destruction of a fifth with *Kersantti* Heimo Olavi Lampi, who personally accounted for two more, while Sarvanto added one more to his Winter War tally. Meanwhile, a Brewster of *LeLv* 24's 1st *Lentueessa*, flown by *Lentomestari* Yrjö Olavi Turkka, was vectored to another sector and downed two more SB-2s. The second division landed, refueled, rearmed, and took off again at 1050. The Finns met more SB bombers, and *Lentomestari* Kinnunen destroyed another two of them.

In total, the twenty-four Finnish fighters that took off accounted for twenty-six bombers. The incident gave the Finnish government an excuse to declare war on the Soviet Union. The Continuation War was on.

By the time Finland accepted Soviet terms for an armistice on September 8, 1944, its Brewsters had accounted for 496 enemy aircraft for the loss of only 19 in aerial combat. Most of Finland's leading aces flew the type, including Sarvanto, who used it to add 4 victories to the 12 1/6 with which he had been credited during the Winter War, and also Ilmari Juutilainen, whose total of 94 1/6, 33 scored in Brewsters, was the highest confirmed fighter pilot's score outside of the Luftwaffe. Finland's second-ranking ace, Hans Henrik Wind, scored 39 of his 75-victory total while flying the Brewster—a record that none of its American, Dutch, or Commonwealth pilots would have found easy to believe.

Eleven

The Tide-Turning Generation
AMERICAN FIGHTERS, 1941–43

When the Japanese surprise attack on Pearl Harbor abruptly plunged the United States into the war in official earnest on December 7, 1941, the Americans had only one fighter that was intrinsically comparable to its opposition—the Grumman F4F-3 Wildcat. The Curtiss P-40E had arrived in the Philippines, and the Brewster F2A-1 Buffalo and Bell P-39D Airacobra were in full production, but they would all prove disappointing when matched against the astonishing speed, climb, maneuverability, range, and firepower of Japan's Mitsubishi A6M2 Zero.

Better designs were waiting in the wings. The twin-engine, twin-boom Lockheed P-38E Lightning and the inverted-gull-wing Vought F4U-1 Corsair had entered production, but they were not yet available for frontline use. The Wildcat was, and although its performance was generally inferior to the Zero's, it was the closest thing the Americans had to a fighter that could face it on even terms.

The Grumman Martlet, to use its British designation, had already seen action with the Fleet Air Arm a year before the first American Wildcat fired its

guns in anger. The Pearl Harbor strike caught half of Marine squadron VMF-211's complement of F4F-3s on the ground at Ewa Field on the island of Oahu, destroying nine out of twelve.

The other twelve Wildcats of VMF-211 had been delivered on December 4 to Wake Atoll, America's farthest Pacific outpost, lying 1,025 miles from Midway and 1,300 from Guam. It was also just 764 miles from Japanese-held Marcus Island and 620 miles from Roi and Namur in the Japanese-mandated Marshall Islands. Wake's garrison, too, was taken by surprise when thirty-six Mitsubishi G3M2s of the Chitose *Kokutai* departed Roi and Namur, slipped past a patrol under the fortuitous cover of a rainsquall and bombed the airfield, destroying seven Wildcats and demolishing the auxiliary fuel tank in the sole plane that remained in one piece. To make the day complete, one of the four returning F4Fs was damaged when it ran into a bomb crater on the runway.

In spite of that disastrous beginning, Wake's garrison put up a heroic defense in the air and on the ground—starting on December 9, when 2nd Lt. David Kliever and Technical Sgt. John Hamilton shot down a G3M for the first Marine aerial victory of the war. Cannibalizing parts from the wrecks to keep two to four Wildcats flying at a time, the Marines were credited with shooting down nine enemy aircraft in two weeks (the Japanese acknowledged the loss of seven). Two bombers were downed on December 10 by Capt. Henry T. Elrod, who also used an improvised rig to drop a hundred-pound bomb among the depth charges on the Japanese destroyer *Kisaragi* on December 11, the resultant explosive chain reaction sinking it with all hands.

After suffering the ignominy of having launched the only amphibious assault to be repulsed by shore batteries in World War II—with the loss of the destroyer *Hayate* to Wake's five-inch guns, as well as *Kisaragi* to Elrod's air attack—the Japanese assembled a more powerful invasion force at Kwajalein. The aircraft carriers *Soryu* and *Hiryu*, then en route home from the Pearl Harbor raid, were diverted to support the renewed effort. When a planned sortie by the US Navy to relieve Wake was canceled, the island's ultimate fate was sealed. The carrier planes attacked on December 21, and during another air strike the next day, Wake's last two airworthy F4Fs rose to challenge them. Attacking a formation of enemy planes head on, Capt. Herbert C. Freuler claimed two Zeros—the first such victories credited to the Wildcat—but his victims were, in fact, Nakajima B5N2 torpedo bombers from *Soryu*. One of *Hiryu*'s Zero pilots, Petty Officer 3rd Class Isao Tahara, then shot up Freuler's plane, which had to crash-land, and Tahara was subsequently credited with downing the other F4F, killing Lt. Carl R. Davidson.

With the last of their planes gone, the twenty surviving pilots and maintenance crewmen of VMF-211 joined the Marines on the ground as the Japanese landed on the morning of December 23. Wake's naval commander, Cmdr. Winfield Scott Cunningham, finally judged the situation hopeless and surrendered that afternoon. Elrod, who had died while manning a machine gun position on December 23, was posthumously awarded the Medal of Honor.

The F4F's Pacific combat debut in defense of Wake would have been enough in itself to ensure the rotund little scrapper a place in military aviation annals, but it had many more names still to add to its laurels—such as Coral Sea, Midway, Guadalcanal, the Eastern Solomons, and Santa Cruz. Seven more Navy and Marine Wildcat pilots would be awarded the Medal of Honor for courage above and beyond the call of duty.

As would be the case with the P-40—particularly as used by Col. Claire Chennault's American Volunteer Group over China—the Wildcat pilots had to develop tactics that would exploit the Zero's weaknesses to the fullest degree, while minimizing their own. Since the F4F-3 was almost as fast as an A6M2 in level flight and slightly faster in the dive, the best tactic was to start the engagement from a higher altitude, hit hard, and dive away. Another tactic by which the F4Fs were able to "hold the line" against the Zero in 1942 was the development of a US Navy variation on Werner Mölders's "finger-four" concept of flexible teamwork. Called the "Thach Weave" after Cmdr. John S. "Jimmy" Thach of VF-2 from the carrier *Lexington*, it involved constant coordinated weaving by two-man teams during a dogfight, with each pilot ready to cover his partner's tail if an enemy plane got on it. The Japanese airmen, individually skilled though they were, seldom became as proficient in that technique as their well-drilled US Navy and Marine counterparts—a factor that resulted in the deaths of numerous veteran Zero pilots to prowling Wildcats.

While the Wildcat had seen service with the British prior to Pearl Harbor, another Lend-Lease American fighter, the Bell P-39 Airacobra, likewise saw some action with the Royal Air Force. It remained for its own air arm to claim its first aerial successes, however, because in contrast to the Wildcat's, its British career was to be singularly short lived.

Conceived by Lawrence Dale Bell and made practicable by his gifted chief engineer, Robert J. Woods, the P-39 emerged from two ideas for fighters that sought to improve maneuverability by locating the engine near the center of gravity, using a ten-foot shaft to connect it to the propeller. The Bell Model 3, with the cockpit placed far aft behind the engine, afforded poor visibility for the pilot, so the Model 4, with the pilot sitting just ahead of the engine, was selected for development, using an Allison V-1710-E4 engine with a B5 turbosupercharger. Never a man to stop at one novel approach when a second or third would be even better, Bell also proposed installing a 25mm cannon, which would fire through the propeller shaft, and a tricycle landing gear arrangement. His proposal was approved on October 7, 1939, and the first XP-39 was completed in March 1939, with the cannon's bore increased to 37mm at the Army Air Corps' request, along with two synchronized .50-caliber machine guns placed in the nose.

The prototype was flown under a veil of secrecy on April 6, with James Taylor at the controls, and achieved a speed of 390 miles per hour at twenty thousand feet. Severe cooling problems were encoun-

tered, so the oil-cooler scoops on the fuselage sides were enlarged. As the promising design made the transition from testing to acceptance, the Army abandoned the supercharger, a measure that facilitated production and maintenance, but which sacrificed a critical amount of performance. The oil-cooler intakes were relocated from the fuselage sides to the wing roots, a carburetor intake was installed behind the canopy, and covers were added over the main wheels. Two additional .30-caliber machine guns were also installed in the fuselage.

While the turbosupercharger had been removed, the extensive modifications that the Army Air Corps had had done to the P-39 raised its empty weight from about 4,000 pounds to over 5,600 pounds. Its maximum speed was reduced to 375 miles per hour at 15,000 feet, but the Army Air Corps was satisfied and ordered 80 P-45s, as the revised fighters were initially called, although that designation was later changed back to P-39C. After 20 P-39Cs were built, a small dorsal fillet was added to the vertical stabilizer, and the gun arrangement was changed to one 37mm cannon and two .50-caliber machine guns in the nose, and four .30-caliber machine guns in the wings. In that form, the remaining 60 planes—followed by 369 in a follow-up order—were designated P-39D. In addition to the American order, on May 8, 1940, the British Purchasing Commission ordered 675 of the fighters under the name of Caribou, later changed to Airacobra Mark I. Export Airacobras were to use a 20mm cannon in place of the 37mm, and 175 of them were repossessed by the US Army Air Forces in December 1941 and given the designation P-400.

The only operational British unit equipped with Airacobras was No. 601

"County of London" Squadron of the Auxiliary Air Force, which received its new planes in August 1941. The squadron flew its first desultory low-level strafing mission, or "Rhubarb," on October 9, when two planes left Marston airfield, crossed the Channel and attacked a German trawler, although its ultimate fate went unrecorded. Two more Airacobras flew over the same area the next day, but found nothing and returned without firing a shot. On October 11, two Airacobras attacked German barges near Gravelines and Calais, while three planes scouted the area around Ostende.

Those four missions in three days constituted the entirety of the Airacobra's fighting career in the RAF. Problems with the plane's compass was the official reason for grounding 601 Squadron's fighters. But in spite of the superior maneuverability displayed by the Airacobra when pitted against a captured Messerschmitt Me 109E, its rate of climb was inferior to those of both the Me 109E and the Supermarine Spitfire Mark VB, and it was clearly no match for the new Me 109Fs and Focke-Wulf Fw 190As that it would be more likely to encounter. "Iron Dog" became the third British term for the P-400, courtesy of its disgusted pilots, as 601 Squadron stood down until it was reequipped with Spitfire VBs in March 1942.

While the Channel Front had stabilized enough for Britain to afford to hold off using its Airacobras in earnest, the situation in the South Pacific in early 1942 offered no such luxury. In March, the American 8th Pursuit Group was shipped to Australia. From there, in early April it moved to Port Moresby, New Guinea, which had been under increasing pressure from units of the Japanese Navy Air Force, operating from bases at Lae and Salamaua, since February 3.

Ground crewmen wax the finish on a Bell P-400 Airacobra of the 39th Squadron, 35th Fighter Group, flown at Port Moresby, New Guinea, in June 1942 by 1st Lt. Robert L. Faurot. The smoking skull and ace of spades motif also appeared on Faurot's Lockheed P-38F-5 when he was shot down and killed on March 3, 1943. (*US Air Force*)

Petty Officer First Class Saburo Sakai's memoirs refer to victories over P-39s as early as April 11, 1942, but these have since turned out to be Curtiss Kittyhawks of the Royal Australian Air Force. The 35th and 36th Pursuit Squadrons of the 8th Pursuit Group settled in at Port Moresby much later, on April 26, and the first encounter between the group's Airacobras and the vaunted Zeros actually occurred on April 30, when Lt. Col. Boyd D. Wagner, commander of V Fighter Command, led eleven drop-tank-equipped P-39Ds of the 35th and 36th Squadrons on their first major sweep. Crossing the Owen Stanley Mountains at twenty thousand feet and then descending to a hundred feet above Huon Gulf, they surprised the Japanese at Lae, with four Airacobras leading the pack to draw off any patrolling Japanese fighters they encountered; the rest of "Buzz" Wagner's force achieved complete surprise, heavily damaging nine bombers and three fighters on Lae airfield.

As Wagner led his pilots to carry out a similar strafe of Salamaua, the Tainan *Kokutai* scrambled up after the departing Airacobras, catching up with and attacking the last four in the formation as it was departing Salamaua. Seven other P-39 pilots turned to assist their comrades, and the resulting dogfight ranged thirty miles up the coast and back. Although a number of Americans claimed to have scored hits on their opponents, only Wagner's somewhat ambiguous claims were officially confirmed, adding three victories to the five already credited to him over the Philippines. The Tainan *Kokutai*'s only recorded loss in the action was Petty Officer 2nd Class Hideo Izumi, killed in action.

The Americans lost four planes, but only one pilot, 2nd Lt. Edwin D. Durand of the 35th Squadron, was killed; last seen going down twenty miles south of Salamaua, he was later reported to have been captured and executed by the Japanese. First Lieutenant Arthur E. Andres, his 35th Squadron P-39 hit by antiaircraft fire, force landed eighteen miles south of Buna, but with the help of local natives he made his way back to Port Moresby on May 27. In the 36th Squadron, 1st Lt. James J. Bevlock ran out of fuel and crash-landed on a beach, but natives helped him get back on May 2, while 1st Lt. Paul G. Brown went down due to coolant loss. He, too, returned after running into Australian soldiers, who sent him home with the added charge of a Japanese pilot they had captured.

All things considered, the P-39 had acquitted itself reasonably well in its first action, but the shoe was on the other foot on May 1, when Port Moresby's Seven-Mile Drome came under a strafing attack by seven Tainan *Kokutai* Zeros. Five P-39s of the 36th Pursuit Squadron intercepted them, and in the low-level melee that followed, 2nd Lt. Donald McGee chased a Zero that was on another Airacobra's tail, and after scoring hits in its fuselage, saw it veer off to the left and explode in the jun-

gle. He, in turn, was attacked by Zeros that shattered his canopy and damaged his plane before they departed, probably short of fuel. First Lieutenant David Campbell of the 36th also claimed a Zero, and the Americans claimed three others damaged. The only Japanese loss, however, was McGee's victim, Petty Officer 1st Class Yoshisuke Arita, whose body was later found about a mile from Seven-Mile Drome.

The Tainan *Kokutai*, including its most skilled ace, Petty Officer 1st Class Hiroyoshi Nishizawa, made the more extravagant claim of eight victories. The principal American losses were the P-39 of 1st Lt. John Mainwaring, who crash-landed, and McGee's, which was badly shot up but was restored to flyability using parts cannibalized from Mainwaring's wreck.

In the month that followed, the 8th Pursuit Group claimed forty Japanese planes destroyed, but at a cost of twenty-five of its own planes in combat, eight in forced landings, and three destroyed on the ground. The group was relieved by the 35th Pursuit Group shortly thereafter, but that outfit was to fare no better with its P-39Ds and P-400s, the latter of which was derisively referred to by its crews as "a P-39 with a Zero on its tail."[1]

By July 1942, the USAAF had issued orders that P-39 pilots were not to engage enemy aircraft in air-to-air combat over any front, unless circumstances compelled them to do so. In spite of the appalling losses they suffered, the Airacobra pilots did their best to hold the line in the Pacific until the arrival of better fighters made it possible for them to relegate their planes to the fighter-bomber and reconnaissance roles. The RAF had already abandoned the Airacobra, but Free French pilots flew P-39s over North Africa and the Mediterranean when there was nothing else available, and so did members of the *Regia Aeronautica* after Italy changed over to the Allied side in September 1943. Soviet pilots flew great numbers of Lend-Lease Airacobras at low and medium altitudes where their lack of superchargers were less of a critical factor—and, surprisingly, liked them very much. Second-ranking Soviet ace Aleksandr Pokryshkin accounted for a sizeable share of his fifty-nine credited victories at the controls of a P-39, as did Grigori Rechkalov and a good number of other notable Russian fighter pilots.

The most successful fighter design available to the US Army Air Forces in the first year of the Pacific War had its origin in a 1936 Army Air Corps requirement for an interceptor capable of 360 miles per hour at 20,000 feet and of staying at least an hour at that altitude. Lockheed designers H. L. Hibbard and Clarence L. Johnson worked to fill that unusual specification by going through an assortment of twin-engine arrangements before settling on twin booms attached to a long tail plane, with the pilot and armament located in a central nacelle. Using a pair of Allison V-1710-11/15 engines fitted with General Electric turbosuperchargers, the Model 22 had a projected top speed of 417 miles per hour at 20,000 feet—far exceeding the Army's expectations—and on June 23, 1937, Lockheed was awarded a contract to build a prototype.

The first XP-38, serial 37-457, made its first flight from March Field, California, on January 27, 1939. Aside from the failure of a flap linkage, the strange-looking plane functioned perfectly. As a result of that encouraging start, the XP-38 was flown by 1st Lt. Benjamin S. Kelsey from March to Mitchel Field, New York, with

the intention of breaking the transcontinental speed record. Unfortunately, Kelsey became overtired, and after he missed his first approach and then applied power too quickly to go around for a second try, one of the engines faltered and he had to crash-land. Nevertheless, the XP-38 had demonstrated its potential, and the Army Air Corps ordered thirteen pre-production YP-38s on April 27, 1939, followed by a full production order on August 10.

As with the Spitfire, the P-38 was a sophisticated, complex piece of machinery, resulting in a long process of tooling up for production. The first P-38, with a 37mm cannon and four .50-caliber machine guns in the nacelle, was not delivered until June 1941. By then, the type had been undergoing further development, and the first P-38D, with self-sealing fuel tanks and a 23mm Madsen cannon in place of the 37mm gun, came out in August, followed by the P-38E, with a 20mm cannon, in October. All three types were assigned to the 1st Pursuit Group at Selfridge Field, Michigan, followed by the 14th Pursuit Group at March Field.

The Lockheed P-38E was first deployed in a combat zone to defend American soil—with the 54th Fighter Squadron, which had flights stationed at Elmendorf, Thornborough, and Cape airfields in the territory of Alaska. The Lightning was a logical choice to patrol the treacherous skies over the Alaskan coast and the Aleutian Islands, since its twin 1,325-horsepower Allison V-1710-49 liquid-cooled engines gave its pilots a greater feeling of security than did single-engine aircraft. Its range of 1,500 to 2,000 miles also suited it for the long distances that it would have to fly.

The pilots, by and large, were as new as the planes they flew. As they searched for the Japanese carrier task force that had recently raided Dutch Harbor, Alaska, on June 5, one flight of Lightnings saw a ship below and promptly attacked it. Fortunately nobody aboard was hurt, for it turned out to be Russian. As one of the planes was landing at Cape Field, its pilot accidentally pressed the gun button on his control yoke and sent a stream of .50-caliber and 20mm shells down the runway.

The P-38 pilots had much to learn and little time in which to do so, for a more tangible threat was coming. On June 7, Cmdr. Sukemitsu Ito's Yokohama-based Toko *Kokutai* arrived at newly occupied Kiska Island with six four-engine Kawanishi H6K4 flying boats, code-named Mavis by the Allies, which the Japanese secured to buoys that the US Navy had originally installed there to handle Consolidated PBYs. Normally used for maritime reconnaissance, the H6K4s could also carry torpedoes and bombs, and on July 20 three of the big boats tried to bomb the seaplane tender *Gillis* in Kuluk Bay, Adak. Although *Gillis* was not hit, the Americans withdrew her from that advanced island base, and the P-38s were ordered to fly regularly over Nazan Bay in case the Mavises returned.

The first two such patrols were flown on August 4, with a bomber from Col. William O. Eareckson's 28th Composite Group providing navigation support to each two-plane P-38 flight. The third flight of the day was comprised of Lts. Kenneth Ambrose and Stanley Long, led by 1st Lt. Major H. McWilliams in a Boeing B-17E. Once the Lightnings arrived over Nazan Bay, they began circling over the two seaplane tenders anchored there, while McWilliams flew on toward Kiska. Soon afterward, McWilliams spotted three Mavises flying toward Atka Island and radioed that information to Ambrose and Long.

Lockheed P-38Es of the 54th Fighter Squadron undergo maintenance on an Alaskan air base, from which two drew first blood for the Lightning on August 4, 1942. (*National Archives*)

The fighters climbed to 22,000 feet in order to be in a better position to intercept the intruders, and Long was first to see the H6Ks flying along the south coast of Atka in a V formation, approaching Nazan Bay at 7,000 feet altitude. As the two P-38s dived on them, the Japanese broke formation and fled to the west, trying to seek cover in a sea fog that rose some 1,500 feet above the ocean. Overhauling their targets, Ambrose and Long began circling around to attack from the front or sides, since the H6K4 was armed with a 20mm cannon in the tail position. Long was the first to achieve success; as he made a firing pass at close range on two of the boats, he saw the canopy on one disintegrate. As the two Lightnings climbed away, Ambrose congratulated Long and then said, "Let's concentrate on the outside one." They then dived for another firing pass, during which Ambrose set the Mavis's left wing on fire. It was last seen descending out of control into the overcast, while the third H6K escaped.[2]

Since they still had enough fuel, Ambrose and Long headed for Kiska, hoping to catch another enemy plane. After flying about 150 miles, they were rewarded by the sight of another H6K emerging from a cloud. As the Japanese took evasive action, Ambrose made a frontal attack while Long attacked from the side and raked it from nose to tail before it disappeared into the clouds again.

Upon their return to Cape Field at 1300 hours, the B-17 and two Lightnings buzzed the runway at low altitude. As he taxied into his revetment, Ambrose held up two fingers and hollered, "We got two of them!" Long had scored the first P-38 victory of the war, but was nevertheless chided by one of his comrades for leaving his ejection door open and losing all his machine gun casings, which should have been saved for the war effort. "I really couldn't have cared less," Long remarked afterwards. "I was more interested in getting some strength back into my weakened legs, so that I could stand up after all the excitement."[3]

Long and Ambrose were credited with one Mavis apiece, and the effect of their interception was indeed felt on the other side of the Aleutian chain. Commander Ito launched no more offensive operations, and on August 17 he and the remainder of his flying boats withdrew to Paramushiro naval base in the Kurile Islands. Left with only fighter and reconnaissance floatplanes, Kiska had little or no further offensive capabilities.

Later in that same month of August 1942, another P-38 scored the first victory for the type against the Germans under similarly unlikely circumstances.

The unit involved, the 1st Pursuit Group, spent the first months of the war defending the continental United States from what some people thought to be an imminent Japanese invasion. After the initial panic subsided, the unit was redesignated the 1st Fighter Group and went on to record several "firsts," including the first fighter unit to fly the North Atlantic

ferry route and the first P-38 unit to arrive in England.

During Operation Bolero, as the transatlantic ferry flights to Britain were called, the 1st Group's 27th Fighter Squadron spent several boring weeks on Iceland, the midway point of the route. Flying weather was extremely rare, but on August 14, 1942, it was clear enough for the Allies to spot a Focke-Wulf Fw 200 Condor prowling around the vicinity. At 1015 hours Maj. John W. Weltman, commander of the 27th Squadron, scrambled up with 2nd Lt. Elza E. Shahan as his wingman. Also taking off in quest of the intruder were two P-40Cs and a P-39D of the 33rd Fighter Squadron.

The four-engine Fw 200C-4 maritime reconnaissance bomber, bearing the Werke Nr. 000125 and call markings FB-BB from 2nd *Staffel*, *Kampfgeschwader* 40, was being piloted by *Uffz.* Friedrich Kühn on a long-range reconnaissance mission from Vaernes, Norway, in search of Allied convoys departing Reykjavik. This time the Germans got more than they bargained for, because at 1040 the bomber emerged from a cloud to be spotted and attacked by Weltman and Shahan. Although the Americans scored hits, the German gunners returned fire and damaged one of Weltman's propellers, which, combined with a jammed 20mm cannon, forced him to disengage. Kühn took cover in the clouds once more, while Shahan continued to patrol the area, hoping the snooper would turn up again.

At 1113 the Condor reemerged from its cloud cover, only to be spotted by the 33rd Squadron and attacked by 2nd Lt. Joseph D. R. Shaffer in the P-39, who scored hits that started a fire in the bomber's left inboard engine. At about that point Shahan also joined the chase, and after dodging more defensive fire, closed for a deflection shot that struck the fuselage. As he turned and climbed to make a high-side attack, he saw the Fw 200's left wing give way and then explode. Flying through a cloud of bruising debris, Shahan returned to Patterson Field to share credit for the USAAF's first aerial victory against the Luftwaffe with Shaffer. A US Navy rescue boat rushed to the crash scene, but all that was found of the plane and its seven-man crew was an oil slick, a notebook, and half a boot.

Later, on September 1, the 1st Fighter Group flew the first P-38 combat sortie over German-occupied territory—an uneventful sweep over France by the 71st and 94th Squadrons—and moved on to become the first such unit to be deployed to North Africa, commencing operations from Tafraoui, Algeria, on November 13.

The P-38 went on to live up to much of its promise in North Africa and Europe, although its record there would be greatly outshone by its accomplishments over Asia and the Pacific. There, the Fifth Air Force's Lightning pilots would include the first- and second-ranking American aces, Richard I. Bong with forty victories and Thomas G. McGuire with thirty-eight. One of the many aces of the Thirteenth Air Force, Rex A. Barber, was credited with only five victories, but one of them was the G4M1 bomber transport carrying the strategic mastermind of the Japanese navy, Adm. Isoroku Yamamoto, when he was killed on April 18, 1943.

Unquestionably the most heavily armed single-engine fighter in the USAAF's arsenal was the Republic P-47 Thunderbolt. Weighing in at 13,800 pounds, the P-47D's bulk was compensated for with an equally monstrous 2,300-horsepower Pratt & Whitney R-2800-21 radial engine. Armament consisted of eight wing-mounted .50-caliber machine guns. Its size, weight, and the raw power required to yank it into the air lent the Thunderbolt the look of an oversized juggernaut, which its pilots soon abbreviated into the sobriquet "Jug."

The P-47's ancestor was the SEV-1XP, a single-seat fighter built by expatriate Russian World War I naval ace Alexander P. de Seversky to compete against the Curtiss Hawk 75 and the Northrop 3-A for a USAAC fighter contact in August 1935. Seversky was given an order for seventy-seven P-35s on June 16, 1936, but it was essentially meant to tide things over for the Army Air Corps until problems with the Hawk 75's engine could be worked out. Meanwhile Seversky's chief designer, a fellow Russian exile named Alexander Kartveli, set to work on improving the already less than state-of-the-art P-35. One of Kartveli's principal improvements was to replace the P-35's landing gear, which retracted backward behind two underwing fairings, with undercarriage that retracted inward, flush with the wing undersides. It was also powered by a 1,200-horsepower Pratt & Whitney R-1830-19 radial engine with a two-stage supercharger. Designated the XP-41, it produced a maximum speed of 313 miles per hour at 15,000 feet—better than the P-35, but not impressive enough to ensure acceptance.

In spite of some brisk export sales, Seversky was unable to overcome his financial difficulties, and his firm had to undergo a reorganization in 1938. In June 1939, the company was renamed the Republic Aviation Corporation. Seversky had left, but Kartveli remained as Republic's chief designer.

When the USAAC held a new fighter competition at Wright Field, Ohio, in January 1939, Kartveli entered not only his XP-41, but also an improved version with an R-1830-31 engine and a turbosupercharger in the central fuselage with an air intake in the left wing root, called the AP-4. The USAAC accepted the Curtiss XP-40, but kept its options open on May 12 by ordering thirteen test versions of the AP-4, which received the Army designation of YP-43 Lancer. Powered by a 1,200-horsepower R-1830-35 Twin Wasp engine, the Lancer had two .50-caliber machine guns in the cowling and two .30-caliber weapons in the wings, and could reach 351 miles per hour at 20,000 feet.

In spite of the fact that inline water-cooled engines seemed to be favored by the USAAC (as seen in the production contracts granted to the P-40, P-39, and P-38), Kartveli remained a believer in the power and easier installation of the radial and that its principal disadvantage, drag, could be overcome by careful design (a theory that, unknown to Kartveli, Japanese designer Jiro Horikoshi was already proving with his Zero fighter). He proposed refined versions of the P-43: the AP-4J, to be powered by a 1,400-horsepower R-1800-1; and the AP-4L, using a 2,000-horsepower R-2800-7 Double Wasp. The USAAC showed an interest in the former, and at the end of 1939 Kartveli set to work on a mock-up for what would be designated the P-44. In August, however, Republic submitted proposals for an AP-4 variant using a 1,150-horsepower Allison V-1710-39 inline engine. The USAAC was interested, but suggested that it be enlarged to carry an increased armament of four .30-caliber wing guns in addition to the .50-caliber machine guns in the fuselage. Republic got a contract to build two prototypes—the XP-47 built to its specifications and a lighter version, the XP-47A.

It soon became apparent to Kartveli that neither version of the XP-47 would be able to produce the performance he had originally envisioned while taking on the added weight of armament and internal protection that the Army Air Corps required. In June 1940, therefore, he proposed a further modification, inspired by the passed-over AP-4L, using the Pratt & Whitney XR-2800 engine with turbosupercharger, which could produce 2,000 horsepower at 27,800 feet. He promised that the higher power produced by the new engine would give his enlarged fighter a speed of 400 miles per hour, and the ability to reach 15,000 feet in five minutes while carrying eight machine guns and long-range fuel tanks with a total capacity of 305 US gallons. Within two months the USAAC agreed, and on September 6, 1940, the contract was altered to order a single XP-47B, using the new Double Wasp engine. Plans to produce the P-44 were amended on September 13, Republic instead being ordered to produce 54 P-43s, 80 P-43As, and 125 P-43A-1s. The Lancer order was made primarily to keep Republic's assembly lines active until the P-47B was ready to enter production—most were shipped off to serve in the Chinese air force.

The first prototype XP-47B flew on May 6, 1942, with Lowry L. Brabham in the cockpit. Poorly sealed joints in the exhaust ducts resulted in smoke and fumes filling the cockpit, but Brabham managed to make an emergency landing at Mitchel Field. The hinged canopy was subsequently replaced by a sliding canopy, and eventually the plane lived up to its promise, with a speed of 412 miles per hour at 25,800 feet and a service ceiling of 38,000 feet. An order for 171 P-47Bs was placed in March 1942, and Republic delivered the last of that order six months later. That was followed in September 1942 by the first of 602 P-47Cs, with an eight-inch extension of the fuselage forward of the firewall, improving the center of gravity and allowing the installation of a shackle under the fuselage for a drop-tank or a 500-pound bomb. The P-47C also replaced the P-47B's forward-sloping radio mast with an upright one and had revised elevator balance systems.

The first P-47Bs were allotted to the 56th Fighter Group, still being organized in the United States, and the 78th Fighter Group subsequently got P-47s after the P-38s originally assigned it were diverted to North Africa to replace losses there. The first group to fly combat missions in the new type, however, was the 4th Fighter Group, already operational with the Eighth Air Force, which started receiving its P-47Cs in January 1943.

On the face of it, the 4th was a logical choice to blood the new fighter: its three squadrons were directly formed from the three Eagle squadrons—Nos. 71, 121, and 133—all comprised of American volunteers in the RAF. The one fly in the ointment was that their experience was in the Spitfire, a finely balanced instrument of aerial warfare that hardly prepared them for a brute like the P-47. One of the flight leaders nominated to try out the new plane, Lt. Richard L. Alexander, spoke for most of his colleagues: "Sitting in a P-47 after having flown Spitfires was like coming out of a small office at the end of a gymnasium and walking into the playing area. It was awfully big."[4] Indeed, it was apparently the 4th Group pilots who first began to coin the then-derogatory sobriquet of "Juggernaut," or "Jug," for their new mount.

"The Repulsive Scatterbolt, we called it," added former Spitfire ace William R. Dunn. "It had a lot of nicknames, most of them nasty. . . . The first time I saw one of them, I asked, 'Where's the other engine?' This klunk weighed almost twice as much as the Spitfire."[5]

The RAF was not ready for the P-47, either. In February 1943, four of the 4th Group's new planes were fired upon by Spitfire and Hawker Typhoon pilots and by ground gunners, who mistook them for Focke-Wulf Fw 190As. Late in the month, the noses and tails of the P-47s were painted with white bands as an identification measure.

After accumulating about fifty flying hours in the Thunderbolts, eleven of the 4th Group's pilots left their air base at Debden on their first combat mission on March 10. They flew over Flushing in the Netherlands, then turned inland and swept over Dunkirk before returning home. In addition to complications with the turbosupercharger system, the pilots experienced problems with the radios, making interaircraft communication virtually impossible. Antiaircraft guns threw some shells their way, but the Jugs encountered no aerial opposition. "A lot of Germans were up there, but they just stood off and looked at us," recalled Lt. Don D. Nee. "They hadn't seen Thunderbolts before, and were wary."[6] Upon their return to Debden, Lt. Col. Chesley G. Peterson, one of the most prominent of the group's old Eagles, indicated that he was relieved at the German fighters' reticence. "I don't mind telling you," he said, "I was scared."[7]

A rash of accidents continued to plague the 4th Group before its pilots got the measure of their big fighters. On April 3, Lt. Frank Smolinsky's plane caught fire, and he was killed while attempting an emergency landing at Sawbridgeworth. A similar fate befell Lt. Oscar Coen, but he safely bailed out of his burning machine. The 334th Fighter Squadron, which was flying the bulk of the 4th Group's early P-47 missions, was soon being referred to as the Suicide Squadron.

A second sweep by twenty-four P-47Cs over the Pas de Calais on April 8 was uneventful, but during the next mission

Republic P-47C-5-RE 41-6538 QP-B was originally assigned to Capt. Stanley M. Anderson of the 334th Squadron, 4th Fighter Group, at Debden, but after he was killed in action flying another plane on April 15, 1943, it was reassigned to Hawaiian-born Capt. Walter P. Hollander and christened *Welakahao!* Hollander never flew it before being transferred to command the 6th Fighter Wing on May 12, 1943. (*4th FG Association*)

over Saint-Omer by thirty-six Thunderbolts of the 4th and 78th Groups on April 13, one plane developed engine trouble and its pilot had to bail out over the Channel, resulting in the first combat loss. On another patrol that evening, another Jug suffered from engine trouble, and when it passed by a British coastal battery off Deal, Kent, it was mistaken for an Fw 190 and hit, the pilot being forced to ditch his plane off the coast.

Anxiety about the P-47's prospects against the Fw 190 was not totally alleviated when a P-47C was flown in mock combat against a captured Fw 190A-4/U8. Although the Thunderbolt was able to hold its own at twenty thousand feet and higher, its supercharged engine gave it no advantage at fifteen thousand feet, where the Fw 190 proved to be the better fighter.

April 15 saw the first shooting encounter with Fw 190s. Major Donald J. M. Blakeslee was leading ten Thunderbolts of the 335th Squadron on a sweep over Ostende when they spotted three Fw 190As at twenty-three thousand feet.

Blakeslee led his men in a diving attack and fired into one of the enemy fighters. According to Blakeslee, the German pilot descended to five hundred feet, then tried to bail out, but was too low and fell to his death in the backyard of a house in Ostende. Later, Blakeslee was congratulated on his victory and for proving that the P-47 could outdive the Fw 190. "By God it ought to dive," he retorted. "It certainly won't climb."[8]

Lieutenant Colonel Peterson and Lt. Douglas Boock were also credited with downing Fw 190s. "It was our first tangle with the Fw 190," Peterson recalled. "I knew the P-47 could out-turn and out-dive the 190, and I knew it could not out-climb it. When we mixed it up, the 190 made the mistake of trying to dive away. It was really duck soup for me, and this became the first 190 to be shot down by a P-47."[9]

Postwar German records told a rather different story. The Focke-Wulfs that the Americans encountered were from the II *Gruppe* of *Jagdgeschwader* 1, patrolling rather farther west than it usually did. Its pilots claimed two of what they had by then identified as P-47s, and suffered no losses whatsoever. Their own claims, on the other hand, were substantiated by the deaths of Capts. Stanley M. Anderson and Richard D. McMinn. Chesley Peterson's plane was also lost in that mission under different circumstances which he at least lived to describe, although he later said that from the standpoint of terror, it was his worst day of flying:

> As I shot the 190 down and pulled away to get back up to altitude, the engine blew up. A couple or three cylinders in front just let go.
>
> I did my best to glide back across the Channel to England, but the engine finally froze up. I was determined not to be taken prisoner; that would have to be the last thing that would happen to me. I got about five miles off the coast and was at very low altitude. I thought about coming down in the water but knew the P-47 probably would not ditch successfully. I made my up mind to bail out.
>
> The altimeter registered 500 feet when I rolled the plane over on its back, with my straps undone and the canopy open, still losing altitude. I fell out at around 300 feet and pulled the ripcord. The chute just streamed; it never did open.
>
> I made a beautiful swan dive into the Channel—a dive so absolutely perfect that the impact did not knock me out. I went down in the water 35 or 40 feet. The silk kept me from going too deep. . . . Somehow I managed to get into my dinghy, and I sat in the water for a couple or three hours. Finally a wonderful old [Supermarine] Walrus airplane from the RAF Air-Sea Rescue practically flew into the Ostend harbor and picked me up.[10]

On May 4 the Jugs flew their first bomber escort mission, accompanying B-17s to Antwerp. By the end of May, P-47 pilots had logged a total of 2,279 sorties and claimed ten enemy planes destroyed, along with seven probables and eighteen damaged. In that same time, however, eighteen P-47s had been lost on combat operations, of which five were attributable to engine failure.

In 1943, the P-47C's deficiencies were ironed out, and the plane soon proved its worth. Not only could it overhaul both the Fw 190A and Me 109G in level flight, but it could outmaneuver them both at altitudes above fifteen thousand feet. And if worse came to worst, neither German fighter had a chance of catching a Thunderbolt if it resorted to diving away.

While the North American P-51 Mustang, with its greater range, was the better overall performer, the rugged and heavily armed P-47 gained a solid following, particularly in the 56th Fighter Group, the only one in the Eighth Air Force to fly Thunderbolts throughout the war. Nineteen of the top twenty-five aces in VIII Fighter Command flew P-47s at some time during their careers, and the two leading American aces over Europe, Francis S. Gabreski with twenty-eight victories and Robert S. Johnson with twenty-seven, both flew P-47s exclusively with the 56th Group. Fifty years later—and right up to his death on December 27, 1998—Bob Johnson still swore by the Jug:

> In the Thunderbolt, I could beat the crap out of a P-51 anytime I wanted to, and I did, many times, in mock dogfights. The P-47 would hold together no matter what you did, while the P-51 would break under many circumstances. We used to joke that in the P-47, if you couldn't outfly them, you'd run through them.[11]

Twelve

Illustrious Beginnings
FAIREY FULMAR AND FIREFLY, 1940–44

While the United States and Japan were developing single-seat shipboard fighters during the 1930s, Britain retained a fixation on the need for a navigator to help the pilot of a long-range carrier-based fighter find his way to his target and back to his ship. With that in mind, the Admiralty issued Specification 0.8/38, calling for a two-seat naval fighter. As was so often the case with carrier fighters of the time, that requirement carried built-in restrictions on the airplane's performance; yet the plane that was ultimately built to satisfy it, the Fairey Fulmar, proved to be remarkably successful.

The Fulmar really began with an attempt to improve the performance of the Fairey Battle, a single-engine monoplane bomber that was pleasant to fly but overweight, underpowered, and doomed to failure in the early months of World War II. Designed by Marcel Lobelle, the Fairey P.4/34 light bomber was much smaller and cleaner than the Battle, and was built to withstand the stresses of dive bombing. When Royal Navy Specification 0.8/38 came out, it proved fairly easy to modify the second prototype of the light bomber into a two-seat fighter capable of absorbing the punishment of carrier landings. Among the changes involved were provision for wing folding, catapult points, arrestor gear, an uprated 1,080-horsepower Rolls-Royce Merlin VIII twelve-cylinder engine, and eight .303-inch Browning machine guns mounted in the wings. Besides the navigator, a relatively large airplane was necessary to accommodate the fuel required for more than four hours' endurance.

After examining a mock-up based on the P.4/34, the Admiralty told Fairey to proceed with its carrier fighter on May 5, 1938. Just twenty months later, on January 4, 1940, Fairey test pilot Duncan Menzies took the first true Fulmar prototype up for its first test flight from Ringway airfield. By the time the second Fulmar flew at Boscombe Down on April 4, approval of the new fighter was all but a foregone conclusion—by June, ten Fulmars had been delivered to No. 806 Squadron, which began training on them for use aboard HMS *Illustrious*, a new carrier equipped with radar and an armor-plated flight deck.

With a maximum level speed of 256 miles per hour, the Fulmar was hardly a world-beater by 1940 standards; but it was pleasant to fly, easy to take off from and land on carrier decks, and capable of absorbing considerable punishment. For the Admiralty, it at least represented the long-overdue introduction of a modern monoplane fighter with retractable landing gear to the Royal Navy's arsenal. It was also a classic case of the right plane coming at the right time, for on June 10 Italy invaded France, thereby entering the war as a German ally and threatening such British holdings in the Mediterranean as Gibraltar and Malta. Indeed, British and Italian warships had had some indecisive engagements, and aircraft of the *Regia Aeronautica* had been attacking Malta-bound convoys when *Illustrious* arrived at Gibraltar with her complement of brand-new Fulmars on August 29.

On the following day *Illustrious*, with battleship *Valiant* and antiaircraft cruisers *Coventry* and *Calcutta*, departed to reinforce Adm. Sir Andrew Cunningham's fleet at Alexandria, Egypt, as part of a multipronged maneuver called Operation Hats. While naval units under Vice Adm. Sir James Somerville feinted and raided to divert Italian attention from Force F, as the reinforcing vessels were called, *Illustrious* kept six-plane patrols up to cover the little fleet.

The Fulmars fired their guns in anger for the first time on September 1, but tragically the airplane they attacked was not one of the enemy's. As Force F approached the Sicilian Narrows, some of *Illustrious*'s Blackburn Skua dive bombers spotted an unidentified aircraft approaching eighteen miles to the southeast and went to investigate. As they closed, the Skua leader recognized the plane as a Lockheed Hudson from Malta and exchanged signals with its pilot. At that point, three Fulmars came up from below and opened fire on the Hudson, causing

considerable damage before signal flares fired from the Skua and their victim convinced them to break off their attack. The Hudson's pilot, Flying Officer G. W. Vincent Davies of No. 233 Squadron, headed for the nearest landfall, the Cap Bon peninsula in Vichy French Tunisia, and managed to crash-land on El Aouina airfield at Tunis. Two of his crew were injured and taken to hospital, while Davies and his other crewman were interned at Le Kef. There they would remain until their liberation in the wake of the British and American invasions of North Africa in November 1942.

The Fulmars that attacked Davies were led by 806 Squadron's commander, Lt. Cmdr. Charles L. S. Evans, but the pilot who actually shot at the Hudson was forty-three-year-old Lt. Cmdr. Robin A. Kilroy, the former commander of the Fairey Swordfish-equipped No. 815 Squadron, who was on his way to take command of the Royal Naval Air Station at Dekheila outside Alexandria, Egypt. He had offered to assist in flight duties, but in his first combat patrol in the Fulmar, he mistook the twin-engine Hudson for a Junkers Ju 88 until he saw the roundel on the fuselage.

While Somerville approached Malta from the west, Admiral Cunningham, flying his flag from the battleship *Warspite*, left Alexandria to rendezvous with his reinforcements. There were some brushes with Italian aircraft, and search planes from the carrier *Eagle* reported an Italian battle fleet to be coming out to intercept Cunningham on August 30, but the next day, a Malta-based flying boat reported the Italian warships to be retiring into the Gulf of Taranto. At 0900 hours on September 2, Cunningham's lookouts spotted Force F.

The two naval units were also being observed by the crew of a Cant Z.501

Fairey Fulmar N1886 of No. 806 Naval Air Squadron was flown by Sub-Lt. Ivan Lowe when he and two other Fulmar pilots shared in the fighter's first victory, downing a Cant Z.501 off Rhodes on September 2, 1940. (*Andrew Thomas*)

Gabbiano flying boat, which spotted them thirty-five miles west of Malta. Evans was leading three Fulmars on patrol, however, and by means of some primitive Morse code transmission he, Kilroy, and Sub-Lt. Ivan L. F. Lowe were directed toward the shadower at 1120 hours. The Fulmars attacked the Z.501 from astern at five hundred feet altitude and sent it flaming into the sea, to loud cheers from the ships' companies.

The Italian aircrew had managed to get off a transmission before going down, however, and at 1415 hours, nine Savoia-Marchetti SM.79 *Sparviero* trimotor bombers of the 36*o Stormo Bombardamento Terrestre*, followed by nine more from the 41*o Stormo B.T.*, set out to destroy the British fleet. Only the 41*o Stormo*'s SM.79s found their target and made for *Illustrious*. Again, however, two sections of Fulmars, led by Lt. William L. Barnes and Sub-Lt. Stanley G. Orr, were up, and as they went to intercept the bombers, two more sections were catapulted from *Illustrious*'s flight deck. The Italians had been expecting nothing more potent than Sea Gladiators to bar their way, and two of the attackers jettisoned their bomb loads

and fled. The remaining seven were made of sterner stuff and pressed on.

Now the Fulmars truly earned their keep, riddling the bomber formation before it could reach *Illustrious*. Barnes downed an SM.79 in flames and damaged two others.

"This action, not lasting more than a few minutes, had been witnessed by the whole Fleet," recalled Lt. Desmond Vincent-Jones, 806 Squadron's senior observer who watched the action from the carrier deck. "As Admiral Cunningham said later, 'Seldom has a fleet had a greater boost to its morale, and from that moment on the whole picture changed.'"[1]

The 41*o Stormo* made another appearance later that afternoon, with six SM.79s of the 235*a Squadriglia*, 59*o Gruppo B.T.*, trying to strike at *Illustrious* when they were intercepted by Bill Barnes's section at 1715 hours. Vincent-Jones was watching from the deck as he reported: "Barnes, as he always did, pressed home his attack until he appeared almost to ram the enemy bomber and Orr wasn't far behind. After a few moments parachutes appeared out of the two rear S.79s and they fell out of formation. The two lead-ing Italians jettisoned their bombs and turned toward Sicily, hotly pursued by the Fulmars who claimed serious damage." Barnes downed his second SM.79 of the day and damaged another, which was sub-sequently finished off by antiaircraft fire from the ships. The Fulmar team of Sub-Lt. G. R. Golden and Leading Airman Harry Phillips claimed one SM.79 shot down and a second damaged, while Sub-Lt. Alfred Jack Sewell was credited with two more. The perspective of Sewell's observer, Leading Airman Denis J. Tribe, suggests the relative helplessness of the Fulmar's unarmed "back-seater" in aerial combat:

> We claimed two shot down but in the action caught up with fire from the enemy. From the rear seat it was difficult to see except for tracers coming at us—a row of holes in my perspex cover; the glycol coolant pouring into my cockpit. Also the intercom was damaged and out of action. I thought I was badly hit as whenever I moved I could see blood. I was greatly relieved to find it was com-ing from a flesh wound under my right forearm. Although Jackie Sewell had armour-plated glass in front of him I thought he must have been hit because the plane seemed out of control. I could not communicate owing to the damage and I could not open my sliding cover. Fortunately he regained control and with damaged engine we made quite a good landing at Hal Far—guided by Lt. Barnes, who also landed at Hal Far. I had first aid locally and in the early evening we were picked up by one of the Swordfish from *Illustrious*.[2]

Italian Junkers Ju 87Bs dive-bombed the fleet that evening, but little damage was done by them or to them by the car-rier fighters and antiaircraft guns which tried to stop them. Cunningham then steamed into the Aegean to raid the Isle of Rhodes, with two flights of six Fulmars providing fighter cover for the carriers, while Swordfish from *Illustrious* and *Eagle* bombed the airfields at Callato, Maritza, and Gadurra on the morning of September 4. The 163*a* *Squadriglia Autonoma*, comprised of Fiat CR.32s and CR.42s and Meridionali biplanes, scram-bled up from Maritza and shot down four of the Swordfish, while losing three of their own—including a CR.32 and CR.42 that collided during takeoff.

Four SM.79s of the 39*o* *Stormo* were able to get off the ground at Gadurra and approached the British fleet at 1030 hours, but they were intercepted, shot up and chased off by Barnes and Sewell. When two more SM.79s of the 201*a* *Squadriglia*, 92*o* *Gruppo B.T.* showed up, Barnes shot one down in flames, killing *Tenente* Nicola Dell'Olio and his crew, and damaged another. Sub-Lieutenant I. P. Godfrey and Leading Airman Phillips attacked another SM.79 and scored dam-aging hits on its port engine. Two other SM.79s were also damaged by the Fulmars, returning with two of their crewmen dead and five wounded.

Elsewhere, at 1005, Kilroy and Orr were vectored over to investigate a trimo-tor aircraft identified as a Caproni Ca.133 bomber transport that was about to land, and they sent it crashing into the sea. Two crewmen were seen to bail out of what apparently was in fact a Savoia-Marchetti SM.81 *Pipistrello* of the 233*a* *Squadriglia B.T.* At 1345 that afternoon, a Fulmar sec-tion led by Lt. O. J. Roger Nichols encountered two SM.79s fifty miles south of Castello Point and pounced. Nichols thought he saw pieces fly off the starboard wing of the plane he attacked, but neither one was shot down.

Three Fulmars intercepted another Cant Z.501 at 1100 hours on September 17. Stanley Orr attacked and thought he killed the pilot, because the flying boat started to climb before slowly spinning into the sea. On the twenty-ninth, Sub-Lt. Graham Angus Hogg's section caught another of the shadowing flying boats at 8,000 feet between Alexandria and Malta. Hogg sent it spiraling down in flames, but not before its return fire disabled Lowe's Fulmar. Lowe ditched five miles astern of the Australian destroyer *Stuart*, which turned about and rescued him and his observer, Leading Airman P. Douet. Three hours later, Orr's section caught another Z.501 at 10,000 feet, and he sent it down in flames as well. As the fleet headed back to Alexandria on October 1, it surprised the surfaced Italian submarine *Berille*, whose crew surrendered and were taken off, after which the vessel was sunk. At 1115, Evans caught a Cant Z.506B *Airone* trimotor floatplane at 4,000 feet and sent it crashing into the sea with its right engine on fire. The body of an Italian airman identified as A. Girondola was found by the British destroyer *Coral* five days later, suggesting that he may have been one of the Z.506's crewmen.

The Italians came out in force again as *Illustrious* was making her way back from Malta to Gibraltar on October 12, only to run afoul of 806 Squadron's Fulmars once more. Jackie Sewell and Sub-Lt. J. M. L. Roberts intercepted a shadowing Cant Z.501 at 1145 and pursued it from 3,000 feet until it finally ditched in the sea, after which they strafed it. Lieutenant Roger Nicolls's section encountered twelve SM.79s at 1230, and he sprayed their second flight with bullets before latching onto a lone plane. Attacking from astern, he reported white smoke pouring from its starboard engine, but it was not con-

firmed as shot down. Evans, Hogg, and Lowe ran into five more SM.79s of the 36*o Stormo* at 1350 hours and made beam attacks on them. Evans and Hogg sent one of the Italians down in flames and compelled a second to ditch, corresponding to the losses of aircraft piloted by *Tenente* Alberto Soldati of the 108*o Gruppo B.T.* and *Tenente* Francesco Tempra of the 109*o Gruppo*. Barnes claimed a third SM.79, which could either have been a plane that returned to Sicily damaged with one dead and two wounded crewmen aboard, or another damaged machine, flown by *Tenente* Giorgio Pieri of the 109*o Gruppo*, which flew into a mountainside and was destroyed.

By October 14, the Fulmars of No. 806 Squadron had accounted for a total of ten Italian bombers. *Illustrious* returned to the Mediterranean on November 11 to launch a raid against the Italian naval base at Taranto, in which its Swordfish torpedoed and sank the battleships *Caio Duilio*, *Italia*, and *Giulio Cesare*—the last mentioned for the duration of the war. The strike accomplished much in proportion to the type and number of aircraft involved, but the material damage the Swordfish meted out was far exceeded by the psychological crippling they inflicted on the *Regia Marina*.

While the Swordfish were so engaged, *Illustrious*'s Fulmars continued to protect the carrier, downing several shadowers before they could send accurate reports to their superiors. Sub-Lieutenants Orr and W. H. Clisby claimed a Z.501 in flames at 1155 hours; Evans and Lowe sent another *Gabbiano* down in flames ten minutes later. Half an hour later, Evans found and attacked a Z.506B at 7,000 feet and after two attacks saw it go down in flames. One of their victims reportedly fell alongside the battleship *Warspite*, flagship of the task force. As a result of the

Fulmars' efforts, no Italian counterattack materialized following the Taranto raid. Orr and Hogg teamed up against one more Z.506B at 1530 the following afternoon, expending all their ammunition on it and then seeing it catch fire and descend after they broke off their attack.

All in all, the Fairey Fulmar had truly gotten off to an illustrious start, and over the next few years it would perform yeoman service for the Fleet Air Arm as a fighter, reconnaissance plane, and night fighter. Although the Fulmar was intrinsically no match for the Me 109, the Macchi C.200, or the Mitsubishi A6M2 Zero, its pilots often gave a good account of themselves, even against those single-seat foes. Eventually the Fulmar's performance was eclipsed, but Fairey already had a successor on the way—curiously similar in basic layout, but destined for a more distinguished and much longer career.

Following its acceptance of the Fulmar, the Admiralty persisted with its rationale that two sets of eyes were better than one. In 1939, therefore, it issued two specifications around the new Rolls-Royce Griffon engine—N8/39, armed with eight .303-inch machine guns; and N9/39, to use a power-operated four-gun turret, as employed on the Boulton Paul Defiant and the Blackburn Roc. Fairey's design team, supervised by L. Massey Hilton and led by H. E. Champlin, concluded early on that no amount of ingenuity would compensate for the weight and drag of the power turret and therefore concentrated on satisfying Specification N8/39. The brutishly handsome airplane they came up with bore a clear family resemblance to the more elegant-looking Fulmar, and although slightly smaller in dimensions, its empty weight almost matched that of a fully loaded Fulmar. As with the Fulmar, the new reconnaissance fighter was quick to gain Admiralty approval—in fact, with the completion of the mock-up on June 6, 1940, it was given a new specification number, 5/40, and a two-hundred-plane order placed virtually off the drawing board. On December 22, 1941, C. S. Staniland flew the first hand-built Fairey Firefly off the grass airfield at Heston.

Besides its 1,735-horsepower Griffin IIB V-12 engine, the Firefly differed from the Fulmar in having a fuselage built in two halves, joined along the vertical center line, and deriving its strength from a series of heavy U-shaped frames and the outer skin, rather than from longitudinal stringers. The engine was mounted in a quickly detachable "power egg," and the wings featured innovative hydraulically operated Youngman flaps, which could not only be angled to slow the plane for landing or to reduce its turning circle in a dogfight, but which had extended aft to increase the wing area for cruising—an early form of variable geometry.

By the time the second Firefly prototype was delivered to Heston for flight testing in March 1942, the machine gun armament had been replaced by four wing-mounted 20mm Hispano cannon. Capable of a maximum speed of 316 miles per hour, the first production Firefly F Mark I was delivered to the Royal Navy on March 4, 1943.

After trials and training, the first Firefly unit, No. 1770 Squadron, was formed at Yeovilton on October 1, 1943, followed by No. 1771 Squadron on February 1, 1944. Assigned aboard the new aircraft carrier *Indefatigable*, 1770 Squadron blooded the Firefly in battle for the first time on July 17, 1944, when it participated in Operation Mascot, a raid against the German battleship *Tirpitz* in Altenfjord, Norway.

Fairey Firefly Mark Is of No. 1770 Naval Air Squadron aboard aircraft carrier *Indefatigable* unfold their wings for their first strike on Japanese oil refineries in Sumatra on January 4, 1945. (*Andrew Thomas*)

Earlier, during Operation Tungsten on April 3, British carrier planes had scored some 1,000-pound bomb hits that damaged *Tirpitz*'s guns and radar, but by mid-July her repairs were nearing completion, resurrecting her status as a threat in being to Allied convoys bound for the Soviet Union via the Arctic. As a participant in the follow-up raid, 1770 Squadron's duties were primarily to reconnoiter Altenfjord and strafe antiaircraft gun positions during the attack. A German smoke screen foiled the first strike, and natural fog contributed to the failure of the second. Two more carrier raids, on August 22 and 29, also failed to inflict serious damage to *Tirpitz*, but the information gathered by the Fireflies proved useful toward later air attacks, until Avro Lancasters dropping 12,000-pound Tallboy bombs finally succeeded in sinking the great ship in Tromsø Fjord on November 12.

Engaged in reconnaissance and anti-shipping sorties along the Norwegian coast, the Fireflies of No. 1770 Squadron,

joined in October by No. 1771 aboard HMS *Implacable*, seldom encountered aerial opposition in 1944. The two-seaters finally got their chance to engage enemy aircraft half a world away, however, when *Indefatigable* was transferred to the Indian Ocean. The vindictive eagerness of the Firefly crews to test their mettle against the Japanese was bolstered by confidence after a Firefly repeatedly won out in mock dogfights with a Grumman F6F Hellcat at the US Navy Test Center at Patuxent River. On January 1, 1945, carriers *Indomitable*, *Illustrious*, *Victorious*, and *Indefatigable*, commanded by Rear Adm. Sir Philip Vian, launched their aircraft for the first of a week-long series of strikes on the Pangkalan Brandan oil refinery on Sumatra, during which 1770 Squadron's Fireflies, led by Maj. Vernon B. G. "Teddy" Cheesman, contributed to the carnage with wing-mounted rockets. On January 4, the Japanese Army Air Force came up to fight, though the principal fighter encountered by the Fireflies was a slightly improved version of the same Nakajima

Ki.43, or "Oscar," that the British had first encountered back in 1941. Grumman Hellcats of Nos. 1839 and 1844 Squadrons based aboard HMS *Indomitable* downed three of the Oscars, while Vought Corsairs of *Victorious*'s No. 1834 Squadron accounted for five Ki.43s, a Mitsubishi Ki.21 bomber, and a Mitsubishi Ki.46 reconnaissance plane. During one of 1770 Squadron's encounters with the enemy, Sub-Lt. Dennis Levitt drew first blood for the Firefly by downing a Ki.43, while Sub-Lts. Vincent Redding and John Philip Stott shared in the destruction of a second. The British lost only one Grumman Avenger in the day's fighting.

The Pangkalan Brandan raid was only a dress rehearsal for a greater strike. Moving from Trincomalee to Sydney, Australia, the carrier force, renamed the British Pacific Fleet, launched Operation Meridians One and Two, a pair of massive strikes on the ex-Shell refineries at Pladjoe, near Palembang, on January 24 and 29. The Japanese defended the precious oil facilities with heavy antiaircraft fire, barrage balloons, Ki.43s, Nakajima Ki.44s, and Kawasaki Ki.45 twin-engine fighters. Roaring through flak and past the barrage balloons on the twenty-fourth, *Indefatigable*'s Fireflies again subjected the refinery to a deluge of rockets. As they left the targets and flak behind, Cheesman's Firefly pilots spotted a flight of Avengers under attack by enemy fighters and engaged the Japanese until Corsairs and Hellcats arrived to drive them off. During the fight, the Fireflies downed two more Ki.43s, Stott sharing in both victories.

The job was only half done, however, and on January 29 the carriers struck Palembang again, this time seeking out the Soengei Gerong refinery. None too astonishingly, the flak and aerial opposi-

tion were more intense than before. Four more Avengers went down, and the four fighters lost included 1770 Squadron's first Firefly and its crew, Sub-Lieutenant Levitt and Lt. John F. Webb. Stott paid the enemy back in kind by sharing in the destruction of two Ki.43s of the 26th *Sentai*.

When Vian's force returned to Fremantle, Australia, on February 4, it had lost sixteen aircraft in combat and twenty-five from other causes, while claiming the destruction of thirty-four Japanese aircraft in the air and thirty-eight on the ground. More important, it had reduced Pladjoe's production by half and put the Soengei Gerong refinery completely out of operation for six months.

So ended the Firefly's fighting debut, with almost six months elapsing between its first operation and its first air-to-air combat. Such a protracted blooding may have been appropriate, however. For the British Pacific Fleet, soon to be redesignated Task Force 57, the Okinawa campaign still lay ahead, during which Phil Stott added two Mitsubishi Ki.51 attack planes to his score on April 12, to become the only Firefly ace. After that, on July 10, 1945, Fireflies of No. 1771 Squadron from *Implacable* would make history as the first British warplanes to fly over Japan—and on July 24 Fireflies would be the first British aircraft to overfly Tokyo. And after all that, the end of World War II would be just the beginning for the remarkably versatile Fairey Firefly, which, through a series of progressive improvements, would fight again over Korea and during the 1956 Suez Crisis, and would serve the Fleet Air Arm into the mid-1960s.

Thirteen

The Axis Strikes Back

GERMAN, ITALIAN, AND JAPANESE FIGHTERS, 1941–44

For the first two years of World War II, Messerschmitt Me 109 was synonymous with German fighter—there simply was nothing else. Then, amid the cross-Channel aerial jousting between Royal Air Force and Luftwaffe that characterized the autumn of 1941, a new challenger entered the lists on the German side. Sharing the Messerschmitt's businesslike angularity, but replacing its sharp nose with the pugnacious, blunt snout of a radial-engine fighter, this new combatant was the Focke-Wulf Fw 190.

Stemming from a 1937 contract from the *Reichsluftfahrtministerium* (RLM) for a new single-seat fighter, the Fw 190 was designed by Focke-Wulf chief engineer Kurt Tank in two forms: one using the water-cooled inline Daimler-Benz DB 601 engine, and one using the BMW 139 air-cooled radial. The latter was selected for development in the summer of 1938, and the first prototype, the Fw 190V1, took off for the first time on June 1, 1939. With the fan-cooled 1,550-horsepower BMW 139 giving it a promising speed of 370 miles per hour, the Fw 190V1 originally used a large ducted spinner, but cooling problems led to a more conventional cowling and spinner arrangement. As the prototype was refined, the BMW 139 was replaced by the BMW 801, which, while heavier, had greater development potential. Although the engine gave trouble, the Fw 190 displayed excellent handling characteristics, while its wide undercarriage track made takeoff and landing far less tricky than it was for the Me 109. Eighteen preproduction Fw 190A-0s were ordered, the first seven of which had wingspans of 31 feet 2 inches, and the rest using wings 34 feet 5 1/2 inches in span, which improved the plane's already outstanding maneuverability. After preservice trials at Rechlin-Rogenthin, the larger wings were incorporated into the Fw 190A-1 *Würger* (Butcher Bird, or Shrike), which went into production in Bremen and Hamburg, the first hundred being completed in the late spring of 1941. Powered by a 1,600-horsepower BMW 801C engine, the Fw 190A-1 was armed with four wing-mounted 7.9mm MG17 machine guns.

In March 1941, thirty pilots were detached from II *Gruppe*, *Jagdgeschwader 26 "Schlageter"* and brought to Rechlin to test-fly the new fighter. Designated the

Erprobungsstaffel (Operational Test Squadron), the detachment was commanded by *Obltn.* Otto Behrens, a prewar Luftwaffe mechanic whose technical knowledge was to come in handy, as would his proficiency as a veteran fighter pilot.

One of *Erprobungsstaffel* 190's members, *Obltn.* Karl Borris, recalled his first impressions of the Focke-Wulf:

> From the first takeoff we were convinced of the robustness and the excellent flying qualities of the new aircraft. However, the BMW 801 engine, a new twin-row, 14-cylinder, air-cooled radial design, gave us nothing but misery. Whatever could possibly go wrong with it, did. We hardly dared to leave the immediate vicinity of the airfield with our six prototype machines. Oil lines ruptured. The heavily armored oil cooler ring in front of the engine broke often. The bottom cylinder of the rear row seized again and again, since the oil pump and the cooling surfaces were too small. Leaking fuel lines left the pilots in a dazed state from the fumes, unable to climb out of their airplanes unaided. The constant speed propeller often failed to work properly. . . . Professor Tank tried to meet our demands in the most direct manner, avoiding the bureaucracy. The RLM propeller specialists and the Rechlin test pilots could only shake their heads when we soared across the field, smoking and stinking. An RLM commission . . . wanted to scratch the Fw 190 from consideration for active service. We protested vehemently, because the airframe itself was truly outstanding.[1]

Behrens's strictly professional, constructive criticism of the new plane is often credited with saving the Fw 190 project from cancellation. Eventually, its flaws were sufficiently corrected for the type to be cleared for service in July 1941. Tank also responded to the Luftwaffe pilots' complaints that the Fw 190A-1's armament was inadequate, by replacing the inboard MG17s with two 20mm MG FF cannon and redesignating the modified fighter as the Fw 190A-2.

The *Erprobungsstaffel* moved to Le Bourget airfield outside Paris and began training pilots of II./JG 26 on the newly arrived production Fw 190A-2, starting with the *Gruppenführer*, *Hptmn.* Walter Adolph. *Oberleutnant* Walter Schneider's 6th *Staffel* was the first squadron fully equipped with the new type, followed by *Obltn.* Kurt Ebersberger's 4th and *Obltn.* Wolfgang Kosse's 5th *Staffeln.* By September 1 the entire II *Gruppe* was ready, *Erprobungsstaffel* 190 was dissolved, and the group moved up to frontline bases at Moorsele and Wevelghem in western Belgium.

Problems were still cropping up in the Fw 190A-2's BMW 801C-1 engine, but II/JG 26's training casualties only amounted to one pilot injured until August 29, when German flak downed one of the unfamiliar-looking machines near Dunkirk, killing its pilot, *Ltn.* Hans Schenk. The II *Gruppe* was cautious about committing its new fighter to the unforgiving environment of the Channel Front, but while eight of its Fw 190s were covering a large tanker on September 18, the ship came under attack off Blankenberge by three Bristol Blenheims of No. 88 Squadron RAF. The Germans laid into the bombers, one of which fell to *Obfw.* Wilhelm Roth of 4./JG 26 and a second to Adolph himself before the escorting Spitfire Mark VBs of No. 41 Squadron intervened, engaging the unfamiliar enemy in a lively dogfight. When the German flight reassembled and returned to Moorsele, one plane, Fw 190A-1 No. 110028, bearing the double chevron of

Newly decked out in unit markings, Focke-Wulf Fw 190A-1s undergo maintenance and refueling in Bremen prior to their delivery to the bases of II *Gruppe, Jagdgeschwader* 26 at Moorsele and Wevelghem in September 1941. (*Bundesarchiv*)

the *Gruppenkommandeur*, was absent. *Hauptmann* Adolph's body washed up on the beach at Knokke, Belgium, three weeks later.

The first combat fatality in an Fw 190, Adolph had been credited with one victory in Spain and twenty-four victories since 1939, and was a holder of the *Ritterkreuz*. He had apparently circled over his last victim one time too many, however, when he was jumped by Flying Officer Cyril Babbage, who, since the British had no idea of what his opponent was flying at the time, was credited with shooting down a "Curtiss Hawk."

On the day following Adolph's disappearance, *Obltn.* Joachim Müncheberg, then leader of 7th *Staffel* and victor over some fifty Allied aircraft, was promoted to *Hauptmann* and put in charge of II *Gruppe*. Two days later, on September 21, *Oberleutnant* Schneider's 6th *Staffel* demonstrated the Fw 190's potential with a vengeance when it engaged a squadron of Spitfires over Boulogne and shot down four of the vaunted British fighters without loss. These were but the first of thousands of Allied aircraft that would be destroyed or damaged by the *Würger* over the next four years.

The introduction of the new radial-engine German fighter took the RAF entirely unawares, and early descriptions of it were even discounted by British intelligence in spite of mounting losses throughout October 1941 to JG 26's Fw 190 pilots, whose confidence and aggressiveness were growing. On October 13, Circus 108A, an attack on the ship lift at Arques by six Blenheims of No. 139 Squadron, came under attack from below by JG 26's *Geschwaderkommandeur, Maj.* Adolf Galland, who sent one of the bombers down in flames. Galland's wingman, *Ltn.* Peter Göring—a nephew of *Reichsmarschall* Hermann Göring—was less fortunate, being struck by return fire from a Blenheim turret gunner and crashing to his death. Elsewhere, Me 109s of I./JG 26 and Fw 190s of 2./JG 26 picked off five Spitfires from the bombers' escort, suffering no losses. The returning British aircraft, however, carried clear gun-camera evidence of the Fw 190A's existence—at last, the RAF airmen were believed.

October also saw the conversion of *Maj.* Gerhard Schöpfel's III./JG 26 to the Fw 190, during which time the *Gruppe's* technical officer, *Obltn.* Rolf Schroedter, discovered that by simply rerouting the exhaust system, he eliminated the tendency of the BMW 801's lower rear cylinder to overheat. With more widespread modification, the Fw 190's Achilles' heel could be remedied.

Barring the early engine troubles, the Fw 190A-2 showed a marked superiority to the Spitfire V; but worse was to come for the RAF. In response to complaints about the engine and the slow-firing, low-velocity MG FF cannon, Focke-Wulf began producing the Fw 190A-3 in the autumn of 1941. Powered by the improved 1,700 BMW 801Dg radial, the Fw 190A-3 was not only armed with four faster-firing MG 151 cannon in the wings, but also with two 7.9mm MG 17s in the fuselage.

From that time on, the seesaw struggle for air superiority between Spitfire and *Würger* would be matched behind the lines as Supermarine and Focke-Wulf came up with progressively improved models of their superlative designs. Both underwent engine changes along the way, necessitating some redesign of airframes. In comparison to the Spitfire's transition from Rolls-Royce Merlin to Rolls-Royce Griffon, however, the Fw 190 underwent a radical alteration when it was adapted to the twelve-cylinder water-cooled 1,750-horsepower Junkers Jumo 213A with an MW 50 methanol/water fuel-injection device, to produce the Fw 190D series, a stretched-out variant called "*Langnasen-Dora*" (Long-Nose Dora) by its pilots. Entering production in August 1944, the Fw 190D gave Tank's fighter a new lease on life, with a speed of 426 miles per hour at 21,650 feet.

On its heels came the Ta-152H *Höhenjäger* (high-altitude fighter), with an even greater wingspan as well as greater length, and a Jumo 213E with GM-1 nitrous oxide injection that could briefly raise power to 2,050 horsepower. Given its prefix in Tank's honor, the Ta-152H could reach 472 miles per hour at 41,606 feet; but only forty-three were built, and only a handful saw use in JG 301 during the last month of the war.

While the Fw 190A served as a worthy complement to the Me 109F in the Luftwaffe's fighter arsenal, Italy's *Regia Aeronautica* was in desperate need of a new fighter just to restore parity with such British counterparts as the Hawker Hurricane and Supermarine Spitfire. The most numerous Italian fighter in 1939 had been the Fiat CR.42 biplane, essentially a refined World War I fighter. The Fiat G.50, Italy's first monoplane fighter, could barely outperform the CR.42, let alone its contemporary opposition.

Aeronautica Macchi's designer, Mario Castoldi, had already tried to redress those consequences of shortsightedness on the part of the *Regia Aeronautica*. His C.200 *Saetta* (Thunderbolt), which first flew on December 24, 1937, was a monoplane with retractable landing gear that strove to incorporate the aerodynamic refinements of Castoldi's Schneider Trophy racers, much as Reginald Mitchell did with his Spitfire. Unlike the Spitfire, however, the C.200 suffered from compromises. It had a humped upper fuselage to provide the pilot with a good field of vision, which was enhanced by the later omission of its enclosed canopy at the behest of conservative seat-of-the-pants pilots. Most telling, both from the standpoint of performance and aesthetics, was the installation of an 870-horsepower Fiat A74 RC.38 fourteen-cylinder double-row radial engine on the airframe. Looking as if it had been stuck on as an afterthought—which, for all intents and purposes, it was—the radial obscured the

Macchi C.202 *Folgores* of the 9o *Gruppo*, 4o *Stormo Caccia Terrestre*, at Ciampino airfield in September 28-29, 1941, during their move from Gorizia to Comiso, Sicily, for operations against Malta. (*Paolo Varriale*)

C.200's Schneider Trophy pedigree and added an inordinate amount of drag.

CR.42s were the only fighters committed to Italy's invasion of France on June 10, 1940, but on the following day, the C.200 joined battle over another target entirely: the British-held isle of Malta. As successive flights of Savoia-Marchetti SM.79 trimotor bombers of the 34o, 11o, and 41o *Stormi B.T.* left their Sicilian air bases for Malta, eighteen *Saettas*, drawn in equal part from the 79a and 88a *Squadriglie* of *Tenente Colonello* (Lieutenant Colonel) Armando Francois's 6o *Gruppo Caccia Terrestre Autonomo*, took off from Comiso, Sicily, to provide escort. As the eighth raid of the day neared Malta, the island's lone radar picked up the attacking formations, and the island's entire fighter force—three Gloster Sea Gladiator biplanes led by Flt. Lt. George Burges— rose to intercept them. The Italians were already bombing Valletta harbor and Hal Far airfield when the Gladiators split up to attack as many of the enemy as they could—with little damage inflicted by either side. It was the third Gladiator, N5520 flown by Flying Officer William J. Woods, that caught the attention of one of the escorting C.200s, flown by *Tenente* Giuseppe Pesola of the 79a *Squadriglia*.

"Timber" Woods had just completed his second attack on a five-plane bomber formation when he heard machine gun fire behind him, immediately went into a steep left-hand turn and then saw the enemy fighter diving at him.

"For quite three minutes I circled as tightly as possible and got the enemy in my sight," Woods reported afterward. "I got in a good burst, full deflection shot, and he went down in a steep dive with black smoke pouring from his tail. I could not follow him down, but he appeared to go into the sea." Pesola, who had fired 125 rounds at the Gladiator before having the tables turned on him, was credited to Woods as the first aerial victory to be scored in the long aerial siege of Malta— but in fact he brought his Macchi back to Comiso with little damage. For neither the first nor the last time in the war, Woods and other witnesses had mistaken the black exhaust smoke from a fighter diving away with its throttle suddenly opened for a burning adversary.[2]

The next encounter would be more conclusive. On June 23, three SM.79s made for Malta, escorted by five C.200s of the 88a *Squadriglia*. Burges, in Gladiator N5519, and Woods, in N5531, rose to intercept when Burges found one of the fighters diving on him. He evaded the *Saetta*'s fire and then engaged it in what he described as a tight dogfight out of World War I. At one point the Macchi overshot, allowing Burges, in his own words, to "belt him up the backside as he went past." After four or five such rounds, Burges got in a burst that set the C.200 on fire, and its pilot, *Sergente Maggiore* Lamberto Molinelli, bailed out over Sliema, where he was taken prisoner. Burges later paid him a visit in Intarfa Hospital, but found him less than friendly.[3]

Molinelli had reason for a sour mood because the C.200's first combats showed it to be almost as nimble as a biplane— but not quite nimble enough to make old school dogfighting a good idea. They had not highlighted any other merits in the plane.

The first confirmed victory for C.200s did not come until November 1, when Malta-based Short Sunderland N9020 of No. 228 Squadron, conducting a morning reconnaissance near Augusta, Sicily, was caught and shot down by *Tenente* Luigi Armanino and *Sergente Maggiore* Natalino Stabile of the 88*a Squadriglia*, with all nine crewmen perishing aboard. A second 228 Squadron Sunderland, L5806, was patrolling thirty-two miles from Malta at 1530 that afternoon when it came under attack by two more *Saettas* of the 88*a*, flown by *Tenenti* Pesola and Pio Tomaselli, along with Fiat CR.42s of the 75*a Squadriglia*, 23*o Gruppo*, 3*o Stormo*, flown by *Tenente* Ezio Monti and *Sgt.* Francesco Cuscuna. Although riddled by its attackers, the tough Sunderland managed to return to Kalafrana, where it was promptly taken upslip before it sank. The Italians reported their victim as only being "damaged," and rightly so, for it was repaired and flying again on November 22.

Over the next three years, *Saettas* would soldier on over Malta, North Africa, and the Soviet Union with sometimes creditable, but never spectacular, results. Although a sufficient improvement over the Fiat CR.42 and G.50 to have warranted production as a stopgap fighter, the *Saetta* was barely a match for the Hurricane and no match for the Spitfire. A closer examination of the C.200's airframe, however, revealed an essentially clean design with an excellent combination of stability and maneuverability. All it needed was a better engine.

With that in mind, Castoldi privately approached the Daimler-Benz A.G. and purchased a twelve-cylinder air-cooled DB 601Aa engine. He then commenced work on an aerodynamically refined adaptation of the C.200 airframe to accept the German engine, at the same time abandoning the C.201, another project to reengine the *Saetta*. The result of his efforts, which took to the air at Varese on August 10, 1940, restored the racy appearance of the Castoldi floatplanes to the basic C.200 design, as well as to its performance potential. So successful were its tests that the *Ministerio dell'Aeronautica* immediately ordered the new fighter into series production—not only at Macchi's Varese factory but also at Breda's plant at Sesto San Giovanni, near Milan. While more DB 601Aas were ordered to power the first production batch, Alfa Romeo acquired a license to manufacture the engine as the R.A.1000 R.C.41-I *Monsone* (Monsoon), which was rated at 1,040 horsepower at 2,400 revolutions per minute. The Macchi C.202 *Folgore* (Lightning), as the new fighter was designated, had a maximum speed of 372 miles per hour at 18,370 feet, featured self-sealing fuel tanks, a molded armor-plate pilot's seat, and an enclosed cockpit, although it lacked an armor-glass windscreen. Armament was initially the same as the C.200—two synchronized 12.7mm Breda-SAFAT machine guns—but ammunition capacity was increased from 370 to 400 rounds per gun. Later-production series *Folgores* added two 7.7mm Breda-SAFAT guns in the wings.

The first C.202s were delivered to the *4o Stormo C.T.* at Gorizia in July 1941. After accustoming themselves with the new fighter's characteristics, pilots of the wing's *9o Gruppo*, comprised of the *73a Squadriglia (Fotoricognitori)* and *96a* and *97a Squadriglie C.T.*, commenced operations against Malta from their base at Comiso on September 29, 1941. On the following afternoon, Italy's new lightning bolt struck for the first time when five Hurricane fighter-bombers of No. 185 Squadron, escorted by six other Hurricanes, attacked Comiso. Three C.202s of the *97a Squadriglia* scrambled up to intercept them, and in the running fight that followed, *Tenente* Iacopo Frigerio shot down Pilot Officer Donald W. Lintern, who was last seen bailing out near Gozo Island.

After returning to their base to refuel, five of the Hurricanes accompanied a Fairey Fulmar of the Kalafrana Rescue Flight in a search for Lintern. They never found him, but they did come under attack by the C.202s. *Tenente* Luigi Tessari and *Sgt.* Raffaello Novelli were jointly credited with downing an enemy fighter, which they reported to have fallen into the sea and blown up ten kilometers south of Cap Scaramia. Their victim was the Fulmar, but it ditched relatively intact, and its crew, Lt. D. E. C. Eyres and Sub-Lt. Bernard Furlong, were subsequently rescued by a Fairey Swordfish floatplane of their flight. One of the Hurricane pilots, Flt. Lt. Charles G. St. David Jeffries, claimed to have probably downed one of the unidentified enemy

fighters, while Pilot Officer Peter J. B. Veitch and Flt. Sgt. A. W. Jolly each claimed to have damaged one; Tessari returned with numerous holes in his fuselage.

The *9o Gruppo* carried the fight back to Malta on the morning of October 1, as *Capt.* Mario Pluda led seven C.202s to escort two C.200s on a reconnaissance mission. At 1150 hours, eight Hurricane Mark IIAs of No. 185 Squadron took off to intercept, but as they reached an altitude of 24,000 feet, thirty miles northeast of the embattled island, they were jumped by the *Folgori*. *Capitano* Carlo Ivaldi, *Tenente* Pietro Bonfatti, and *Sergente Maggiore* Enrico Dallari claimed two Hurricanes shot down and two probables in their first pass; but only one Hurricane was lost along with its pilot, Squadron Leader P. W. B. Mould—the same "Boy" Mould who, as a member of No. 1 Squadron, had scored the first confirmed Hurricane victory in France on October 30, 1939. Mould's total account stood at eight, plus one shared, when he became one of the C.202's earliest victims. The Italians did not get off scot-free, however. Sergeant Ernest G. Knight scored hits on Ivaldi's main fuel tank, and he only just made it to Sicily before the last of his fuel drained away, force landing on the beach near Pozzallo.

The *Folgore* quickly demonstrated its inherent mastery over the Hurricane, and by the end of 1941, at least one of the *9o Gruppo*'s pilots, Teresio Martinoli, had been credited with five out of an eventual personal total of twenty-two victories (one of them German while flying for the Allies in Italy's Co-Belligerent Air Force), including Peter Veitch, whom he shot down and killed off Malta on October 4. The C.202's numbers were too small to have a decisive impact over Malta in the late months of 1941, however. By the time it was available in significant quantities in 1942, Spitfire Mark Vs had arrived to engage the Italian fighters on roughly equal terms. Nevertheless, the C.202 gave a much-needed boost to the confidence of Italian fighter pilots and became the *Regia Aeronautica*'s fighter mainstay until Italy capitulated on September 8, 1943. A more potent variant with a license-produced version of the DB 605 engine and heavier armament, the C.205 *Veltro* (Greyhound), would continue to be a formidable fighter thereafter, in the hands of both Allied Co-Belligerent pilots and the diehard Fascisti of the *Aeronautica Nazionale Repubblicana*.

Macchi was not the only Axis aircraft manufacturer to benefit by installing the Me 109's power plant in one of its airframes. The Kawasaki Kogyo K.K. of Japan obtained similar results with a series of inline-engine fighter projects that culminated in the Ki.61 *Hien* (Swallow). As a result of its experience against the Soviets over Nomonhan in 1939, the Japanese Army Air Force (JAAF) began to reconsider its almost pathological obsession with maneuverability as the paramount asset in a fighter. One result was the beginning of a general trend among Japanese manufacturers to develop aircraft with higher speed and heavier armament. Like Macchi, Kawasaki, the Japanese aircraft builder that had had the most previous experience with water-cooled inline engines, looked into the idea of building two fighter types around the DB 601A. One, given the *kitai* (body) number Ki.60, was to be a cannon-armed specialized interceptor. The other, the Ki.61, was to be a lighter general-purpose fighter. Work began in February 1940, with chief engineer Takeo Doi and his deputy, Shin Owada, being instructed to place the higher priority on the Ki.60.

Armed with two wing-mounted Mauser MG 151 cannon and two 12.7mm machine guns in the fuselage, the Ki.60 had three wing spars and an unusually high wing loading by Japanese standards. The first of three prototypes was completed at Kagamigahara in March 1941; but during flight testing the controls proved to be excessively heavy, Kawasaki's chief test pilot declaring, "I have flown more maneuverable heavy bombers!" Overall performance was also disappointing, the top speed of 340 miles per hour being almost 9 percent below that called for in the type specification.[4]

The Ki.60 project was abandoned, but in December 1940 Doi and Owada began work on the Ki.61, employing a high aspect ratio wing with greater area than that of the Ki.60. The fuselage was longer and less deep than the Ki.60's, resulting in reduced drag. At the same time, preparations were made for license production of the DB 601A at Kawasaki's Akashi plant. Designated the Ha-40 (Army Type 2), the engine entered production in November, and the first Ki.61 emerged from Kawasaki's Kagamigahara factory in the second week of December. Early tests proved—to nobody's surprise—that the Ki.61 was no match for the Nakajima Ki.43 *Hayabusa* in a dogfight, but Doi and his design team were more than content with the acceptable handling qualities that it did display, as well as with its considerably greater speed of 367 miles per hour.

A strange twist of fate decreed that the Ki.61 would first fire its guns in anger during an early stage of its evaluation program. Lieutenant Ryuzaburo Umekawa had just departed Mita airfield in a prototype Ki.61 to perform firing trials at the Mito Army Flying School's target range on April 18, 1942, when he received a

A grounded Kawasaki Ki.61-I-Ko *Hien* serial No. 263, reputedly flown by Capt. Shogo Takeuchi, commander of the 2nd *Chutai*, 68th *Sentai*, found by the 1st Marine Division on Cape Gloucester (Tuluvu), on December 30, 1943. (*National Archives*)

radio message to intercept enemy bombers that had appeared over the mainland of Honshu. It was an unprecedented order, but the aircraft in question were North American B-25B Mitchell medium bombers led by Lt. Col. James H. Doolittle, which, unknown to the Japanese at the time, had been launched from the aircraft carrier *Hornet*. Umekawa spotted at least one of the raiders and after a long chase managed to get just close enough to fire one long-range, utterly ineffective burst before trouble with the guns and shortage of fuel compelled him to turn back for home.

In the summer of 1942, two prototype Ki.61s were put through comparative trials against an imported Me 109E-4, a captured Curtiss P-40E, a LaGG-3 acquired when a Soviet pilot defected to Manchuria, a preproduction Nakajima Ki.44 *Shoki*, and a production Ki.43. Aside from its inferior maneuverability in comparison to the *Hayabusa*'s, the Ki.61 proved to be superior to all of its counterparts in every respect. After further testing, the first full-production models of

the Ki.61 Type 3 *Hien* began to leave the assembly line in September 1942. Early Ki.61-I-ko models retained the two fuselage-mounted 12.7mm Ho-103 and two wing-mounted 7.7mm Type 89 machine guns used in the prototypes, whereas the Ki.61-I-otsu featured four 12.7mm weapons. In addition to a sturdier airframe and heavier armament, the Ki.61 was the first production JAAF fighter to feature self-sealing fuel tanks and armor protection for the pilot.

The first operational *koku sentais* (air regiments) to exchange their old Nakajima Ki.27s for Ki.61s were the 68th and 78th, in May 1943. On June 14, the 68th *Sentai* arrived at Wewak on the northeastern coast of New Guinea, while the 78th was deployed to Rabaul. The 68th flew its first combat patrol on June 16.

By then, the Japanese were on the defensive in the South Pacific, and New Guinea in particular was gaining a reputation among the Japanese similar to that of Stalingrad among the Germans—a miserable meat grinder of a place, from which few expected to come back alive. The heat and humidity, combined with the poor quality of fuel available, resulted in frequent overheating and mechanical failure in the Ki.61's relatively sophisticated engine. When the plane did work, it outperformed most of its early opposition—and certainly surprised the Allied pilots who first encountered so radical a departure from the light, nimble but fragile radial-engine fighters they had fought up to that time. American pilots mistook the Ki.61 for an Me 109 at first, while British Commonwealth airmen, many of whom were veterans of North Africa and the Mediterranean, thought the needle-nosed machine resembled a Macchi C.202. The latter opinion probably led to the adoption of "Tony" as the Ki.61's Allied code name.

The first clash occurred on July 18, 1943, when elements of the 78th *Sentai* ran into Lockheed P-38s of the 39th Fighter Squadron between Lae and Salamaua. First Lieutenant Gene Duncan claimed to have damaged a plane that "resembled a Me 109," while his squadron mates claimed to have shot down two, probably destroyed four, and damaged two. The 78th *Sentai* recorded no losses and in turn claimed its first victory, credited to First Lieutenant Fujishima.[5]

The 68th *Sentai* blooded its *Hiens* on July 20 during an attack on Allied positions at Benabena. Captain Shogo Takeuchi of the regiment's 2nd *Chutai* spotted a Consolidated B-24 and led his five-plane element to attack and shoot down the four-engine bomber. On the following day, B-25s attacked the Japanese base at Bogadjim, and when Ki.61s of both the 68th and 78th *Sentais* rose to intercept them, along with a unit of Ki.43s, they in turn came under attack by escorting P-38s of the 39th and 80th Fighter Squadrons. The Americans claimed no less than twenty-two Japanese fighters in the swirling dogfight that ensued between the Ramu Valley and Madang, but only four Japanese fighters were actually lost, two of which were Ki.61s of the 78th. In addition to the deaths of 1st Lts. Kunji Fujita and Kunishiko Suzuki, the 78th *Sentai*'s 2nd *Chutai* commander, 1st Lt. Tadashi Tomashima, claimed to have downed three P-38s before being wounded in the left arm, which resulted in his being evacuated to Japan in August. The 78th claimed a total of five American planes, but the Fifth Air Force had suffered no losses at all.

The *Hien* was a swallow destined to enjoy a short summer. By the autumn of 1943, new-generation Allied fighters, like the P-38 and the Republic P-47D Thunderbolt, were taking the place of P-39s and P-40s over New Guinea and generally dominating the Ki.61s. Arguably, the *Hien*'s most notable exponent was Shogo Takeuchi, who was reputedly the top-scoring JAAF ace over New Guinea and was credited with at least thirty Allied aircraft there, in addition to the three Hawker Hurricanes he had downed in a single fight over Singapore on January 31, 1942, while flying a Ki.43 in the 64th *Sentai*. On December 15, 1944, Takeuchi's unit took on P-47s over the Arawe peninsula, and after allegedly downing one Thunderbolt (not matched by any loss in American records) and driving another off the tail of his *sentai* commander, Maj. Kiyoshi Kimura, Takeuchi's own plane was hit. As he tried to land at Hansa airfield, his engine seized and the *Hien* crashed in the trees. Takeuchi was extricated from the wreck but died of his injuries three hours later.

By July 25, 1944, the 68th and 78th, as well as the Ki.43-equipped 77th and 248th *Sentais*, had been virtually annihilated and were officially disbanded, never to be re-formed. The Kawasaki Ki.61 would fight on, often heroically, in defense of the Japanese home islands against Boeing B-29s until the end of the war; but its period of aerial supremacy had been disappointingly brief.

There was, however, to be an unusual postscript to the *Hien* story. Responding to the threat posed by Boeing B-29 Superfortress bombers operating from the Marianas, in 1944, Kawasaki developed a high-altitude interceptor, the Ki.61-II-Kai, only to find itself producing more airframes than the Akashi factory could match in its uprated 1,500-horsepower Ha-140 engines. On October 26, 1944, the *Koku Hombu* suggested that Kawasaki adapt the Ki.61 airframe to the 1,500-horsepower Mitsubishi Ha-112-II fourteen-cylinder air-cooled radial engine instead.

As Soviet designer Semyon Lavochkin had done when ordered to change the inline-engine LaGG-3 into the radial engine La-5, Kawasaki's Takeo Doi initially balked at the proposal. With Japan's position becoming desperate, however, in December he and his staff took on the challenge, working day and night using an Fw 190A as their model to adapt the Ki.61 to a different thrust line and center of gravity. Working in their favor was the fact that there was only a ninety-nine pound difference between the two engines, largely compensated for by eliminating the Ha-140's radiator from below the fuselage. Aligning the exhaust pipes horizontally on either side of the fuselage, as Kurt Tank had done on his Fw 190A, augmented thrust to add six to nine miles per hour to the airspeed.

On January 19, 1945, Kawasaki's project was given added impetus when B-29s

Kawasaki Ki.100s of the 1st *Chutai*, 59th *Sentai* sit idle on a Japanese airfield after Japan's surrender. (*National Archives*)

bombed the Akashi plant, stalling engine production for 271 Ki.61-II-Kai airframes. On February 1, the first prototype of what was designated the Ki.100-I-Ko Model 5 fighter was ready for testing with Maj. Iori Sakai in the cockpit.

The tests astonished even Doi's team. Aside from being marginally slower than the Ki.61-II-Kai at high altitude, the radial-engine plane climbed faster, was more maneuverable, and had better takeoff and landing characteristics. Mock combats conducted at Yokota with a P-51C captured in China yielded similar results: although the Mustang could disengage at will, in a dogfight the Ki.100 outmaneuvered and out-dived it.

The "*Goshikisen*" (Type 5 Fighter), as the Ki.100 was unofficially called, had a top speed of 367 miles per hour and could climb as high as 35,008 feet, although its performance tended to fall off above 25,000 feet. Armament was standardized on two Ho-5 20mm cannon in the fuselage and two wing-mounted Type 103 12.7mm machine guns. While the engineless Ki.61s were being converted into Ki.100s, Doi and his team trimmed down the plane's fuselage and replaced the original canopy with a sliding-bubble type. With the completion of the first converted batch in June, the Japanese hoped to produce 200 of the improved Ki.100-I-

Otsu per month, but the American bombing campaign limited total output to 118.

Just as the first Ki.100s began arriving at the 18th *Sentai*, the new fighter underwent its baptism of fire on the night of March 9–10 as 279 B-29s made a low-level raid on the east side of Tokyo, using incendiary bombs to immolate almost sixteen square miles, 267,000 homes, and some 100,000 people. One Ki.100 joined the 18th's Ki.61s in what, unsurprisingly, proved to be a vain defense, although Ki.45 *Toryus* of the 53rd *Sentai* claimed ten of the Superforts.

First blood flowed in both directions on April 7 when 1st Lt. Mitsuo Oyake of the 18th *Sentai*, after repeatedly attacking a B-29 without effect, rammed it and bailed out safely. Its tail chopped off, the B-29 of the 73rd Bombardment Wing fell at Kugayama outside Tokyo, and was credited as Oyake's fourth bomber kill, for which he was awarded the *Bukosho*, an unprecedented medal issued by the Japanese to encourage just such deeds of desperate courage. Master Sergeant Yasuo Heima tried to emulate Oyake, reportedly making repeated attacks on the B-29s until their .50-caliber guns struck home. Crashing into a mountainside in Saitama Prefecture, the first Ki.100 to fall in combat, with Heima's remains, was not found until February 1972.

Like the Lavochkin La-5, whose genesis its own roughly resembled, the Ki.100 proved to be a remarkable achievement, not so much for its creation, for which the Fw 190 and the La-5 both set precedents, but for the three-month's time in which it occurred. On a par with Japan's best fighters, the *Goshikisen* achieved its share of successes in 1945; but with a grand total of 369 built, it could not stave off the inevitable.

An even more formidable product of Japan's change of heart regarding the future of lightweight general-purpose fighters would come from the builder of one of its quintessential embodiments, the Ki.43 *Hayabusa*. In November 1941—a month before the Ki.43's combat debut—the *Koku Hombu* asked Nakajima to design its successor. Anticipating an improved generation of Allied fighters, the *Koku Hombu* wanted a plane that would combine the Ki.43's maneuverability with the speed and climb of Nakajima's heavyweight Ki.44 *Shoki*. Powered by an army version of the navy's new Nakajima NK9A Homare eighteen-cylinder, twin-row direct fuel-injection radial engine, the new fighter was expected to have a maximum speed of 400 to 420 miles per hour. Like the Ki.61, it was also to incorporate heavy armament, armor protection, and self-sealing fuel tanks.

Early in 1942, Yasumi Koyama and the Nakajima design team began work on the new Ki.84 fighter. Their design was approved on May 27, and such was the priority placed on the project that the first prototype emerged from Nakajima's Ota plant in March 1943. The Ki.84 was much sturdier than the Ki.43: its windscreen had sixty-five millimeters of armored glass, and the pilot's seat had thirteen millimeters of head and back armor. Armament was comprised of two synchronized 12.7mm Ho-103 machine guns in the fuselage and two wing-mounted 20mm Ho-5 cannon. Racks under the wings could carry either two forty-four-gallon drop-tanks or up to 550 pounds of bombs. The first production example of the Army Type 4 Fighter Model I-ko *Hayate* (Gale) left the Ota plant in April 1944. The army called for 2,565 Ki.84-I-kos by the end of the year, but continuing development problems with the Ha-45 engines held up production until April 1944, when a rate of 100 engines per month was finally achieved.

Using a 1,860-horsepower Ha-45 Model 21 engine driving a four-bladed propeller, the Ki.84-I-ko had a maximum speed of 388 miles per hour at 21,325 feet, a normal cruising speed of 236 miles per hour, and an initial climb rate of 3,790 feet per minute. Service ceiling was 36,090 feet, and the 780-mile range at normal cruising speed could be increased to 1,410 miles with two drop-tanks. Intrinsically, the *Hayate* was the Allies' worst nightmare—a Japanese army fighter with all the virtues of the nimble Ki.43 but with none of its faults, such as inferior level and diving speed, weak armament, and vulnerability to battle damage.

The *Hayate*'s debut was calculated for maximum effect. On March 5, 1944, the 22nd *Sentai* was formed at Fussa, Yokota, with forty Ki.84-I-kos and a pilot cadre drawn from the *chutai* that had been evaluating the plane since October 1943. The new regiment's commander, Maj. Jyozo Iwahashi, was a veteran of the 1939 Nomonhan Incident and already had twenty victories to his credit. It was not until August 24, 1944, however, that the new fighter made its first appearance—at Hankow, China, at a time when Maj. Gen. Claire Chennault's Fourteenth Air Force had nearly half of its combat strength committed in Burma, and the rest was supporting Chinese and American efforts to halt a Japanese offensive toward the Yangtze River.

The 22nd had its first fight on August 29, when it engaged a large force of Curtiss P-40Ns of the Chinese-American Composite Wing (CACW), the 118th Tactical Reconnaissance Squadron of the 23rd Fighter Group, and the 51st Fighter Group, which were engaged in an attack

After the war a Nakajima Ki.84 *Hayate* of the 1st *Chutai*, 85th *Sentai* awaits American examination. Behind it is another Ki.84 of the 22nd *Sentai*, which first used the Ki.84 over China in August 1944. (*US Air Force*)

on the railroad yards at Yochow. As the Americans were returning from the target area, 1st Lt. Robert S. Peterson chased what he identified as a Zero off 1st Lt. James A. Bosserman's tail, while 1st Lt. Forrest F. Parham claimed to have shot down one "Hamp," probably downed a second, and damaged a third. Lieutenant James Focht claimed to have damaged an "Oscar," a "Tojo," and a "Hamp."

Clearly, the Americans had no idea of what aircraft the Japanese were flying, but whatever they were, they had proven to be difficult opponents. On the Japanese side, Iwahashi was credited with a P-40N (possibly Bosserman's, though he did, in fact, return safely to base) for the 22nd *Sentai*'s first victory.

Over the next five weeks, the 22nd *Sentai* ran roughshod over the best opposition the Allies could offer, and also attacked China-based B-29s of the Twentieth Air Force's XX Bomber Command as they flew missions to Japan. Iwahashi was leading a strafing attack at Xian airfield on September 21 when his plane was hit by ground fire. Apparently deciding that he could not make it back, the 22nd's commander dived his *Hayate* into the ground; according to some witnesses, he tried to crash into an enemy fighter parked on the field.

On that same day, twelve P-40Ns of the 75th Squadron set out on an offensive sweep against Ninsiang and Sinshih. One plane turned back because of engine trouble, but the rest dive-bombed two Japanese antiaircraft positions. Shortly afterward, Lieutenant Bosserman—ironically, the pilot who most likely had been overoptimistically claimed by Iwahashi on August 29—radioed that he was low on fuel and was heading home. Soon afterward, he reported that he had to bail out over Sinshih. He was never seen again.

The 22nd *Sentai*'s rampage over China abruptly ended in October 1944, when it was transferred to Leyte to counter the imminent American invasion of the Philippines. By September another China-based *sentai*, the 85th, had begun to supplement its Ki.44s with new Ki.84s, and its pilots would make the *Hayate*'s presence felt in the coming months.

Most notable among them was Capt. Yukiyoshi Wakamatsu, commander of the 2nd *Chutai* of the 85th *Sentai*, based at Canton. Although he had served in the JAAF since 1939, Wakamatsu did not come into his stride until July 24, 1943, when he shot down two P-40s of the 23rd Fighter Group's 74th Squadron over Kweilin while flying a Ki.44. He accounted for eleven more Allied aircraft while flying the *Shoki*. Just days after transitioning to the *Hayate*, he downed two North American P-51B Mustangs of the 23rd Group's 76th Squadron over Wuchow on October 4, 1944, raising his score to fifteen.

Although the Americans had been hearing about a new Japanese army fighter since early in 1944 and had given it the code name "Frank," the Ki.84 came as something of a surprise to them, and its resemblance to a cross between the Ki.43 and Ki.44 probably resulted in its being frequently misidentified in their combat reports as either an "Oscar" or a "Tojo." Captain David L. "Tex" Hill, a former ace of Chennault's original AVG who still fought on in the Fourteenth Air Force, described a bombing raid over Hong Kong on October 16 that must have involved Wakamatsu and his squadron of *Hayates*:

We arrived down in the Hong Kong area and the first thing I know I see three enemy aircraft up there that I couldn't identify. I knew they were Japs but I'd never seen the type before. We called them out and turned into them, and as we pulled up into them, why, they went straight up and I could see we were going to stall out and so I bent it over. These guys came right down on top of us and shot down the three guys who were with me, and they chased me all the way down to eight thousand feet, which was the altitude I needed to get back over some hills.

I swear I believe this Jap was trying to overrun me in a dive. His tracers were really going by my head. When I got back I talked to the Old Man, telling him about this new type Japanese fighter which was later identified as a Zeke 52 or Tojo, which was another term I think they used for them. I told him, "Well, I don't know, there's a new type here. I don't know if we're going to be able to beat these guys in the air." And Chennault thought awhile and very characteristically—he had an easy solution for everything—he said, "Well, Tex, don't worry about that. Get them on the ground, then you don't have to fight them in the air." So this was a strategy that was actually used. Some of the guys went up to the airfields around Tientsin where these airplanes were congregated and just cleaned them out right on the ground.[6]

Of the three Mustangs Tex Hill thought he saw shot down, two actually made it to friendly territory. The pilot of the third, 1st Lt. Robert Colbert, was wounded in the leg, but managed to bail out and was eventually smuggled back to Allied lines by the Chinese.

Chennault launched his biggest counterstrike of the year on December 18, coordinating his efforts for the first—and only—time with the Twentieth Air Force. While 84 of XX Command's B-29s dropped 511 tons of bombs on Hankow, the Fourteenth Air Force's B-24 and B-25 bombers added to the destruction, while fighters of the 23rd Group and the CACW flew interdiction strikes against known JAAF air bases.

The Japanese had anticipated the attack, and Wakamatsu's 2nd *Chutai* had moved to the satellite airfield outside Wuchang, to reinforce the 85th *Sentai's* 1st *Chutai* at Hankow. Reports of oncoming B-29s sent Wakamatsu scrambling up to intercept, but he had just taken off and retracted his landing gear when he was attacked by ten of his old enemies, Mustangs of the 23rd Fighter Group's 74th Squadron, and quickly shot down. He was probably the victim of Capts. Philip G. Chapman and John C. "Pappy" Herbst, both of whom claimed an "Oscar" over Wuchang. At the time of his death, Wakamatsu had at least eighteen enemy aircraft to his credit, half of them allegedly P-51s.

Hayates would frequently, albeit sporadically, give good accounts of themselves throughout the last year of the war, but they could not stave off Japan's defeat. They were invariably outnumbered, on top of which the attrition of war was rapidly thinning out the Japanese army's pool of trained, experienced pilots, whereas the number of proficient airmen on the American side was growing exponential-

ly. Moreover, the Ki.84's merits were often handicapped by the lax production standards that attended its rush into service, which resulted in chronic hydraulic malfunctions, brake failures, broken landing gear legs, and a general maintenance nightmare for the ground crews.

Only after Japan's surrender on September 2, 1945, did the Americans realize just how much deadlier an adversary their pilots could have faced. In 1946, a Ki.84-I-ko of the 11th *Sentai's* 2nd *Chutai* that had been captured intact at Clark Field, Luzon, was test-flown at Wright Field outside of Dayton, Ohio. Its evaluators concluded that the "*Hayate* was essentially a *good* fighter which compared favorably with the P-51H Mustang and the P-47N Thunderbolt. It could out-climb and out-maneuver both USAAF fighters, turning inside them with ease, but both P-51H and P-47N enjoyed higher diving speeds and marginally higher top speeds."[7] That the earliest model of Ki.84 had compared so well with the latest models of P-47 and P-51 said volumes about the soundness of its basic design. Of equal significance, however, was the fact that the captured aircraft was using one-hundred-octane fuel for probably the first time in its flying career—the *Hayates* that the American had met in combat used lower grade (eighty-octane) fuel, often mixed with dirt, water, and the odd tropical insect.

After being passed on to the Smithsonian Institution's National Air Museum in Washington, DC, and then to the Ontario Air Museum at Claremont, California, the last surviving Ki.84 was acquired in 1973 by Morinao Gokan, president of the Japanese Owner Pilots' Association, who brought the last *Hayate* home.

Fourteen

Red Resurgence

SOVIET FIGHTERS, 1941–44

The nonaggression pact between Adolf Hitler and Josef Stalin was rudely terminated on June 22, 1941, when German forces launched *Unternehmen* Barbarossa, the invasion of the Union of Soviet Socialist Republics. Equally rude was the awakening for the *Voyenno-Vozdushny Sili* (Soviet Air Forces). Most V-VS aircraft were caught and destroyed on the ground. The majority of those that were able to get into the sky were soon shot out of it by German fighters flown by airmen whose superiority was by that time based as much on experience as it was on training.

Numerically, the most important fighter in the V-VS arsenal at the time was still the Polikarpov I-16—revolutionary in 1935, but obsolescent in 1941. The second most numerous Soviet fighter, the I-153, was essentially a refined I-15 biplane with retractable landing gear, designed in 1937 after the I-16 because Nikolai N. Polikarpov, concerned that the higher wing loading of monoplanes reduced climb rate and agility, still thought the biplane might have a future. Powered by an 850-horsepower Shvetsov M-62 nine-cylinder radial engine, the I-153 had a fair

turn of speed for a biplane—280 miles per hour at 15,090 feet. Armament consisted of four fuselage-mounted 7.62mm ShKAS machine guns, which would later be supplemented by underwing racks for up to four fifty-five-pound bombs or six 82mm RS-82 rockets.

The new biplane entered service in the spring of 1939 and was soon committed to combat in the Khalkhin-Gol region at the border of Mongolia and Japanese-controlled Manchuria, where an undeclared war had been raging since May 11—and where Nakajima Ki.27s of the Japanese Army Air Force were outmaneuvering the I-16s and outrunning the older I-152 biplanes. Aside from the startling performance of the Japanese fighters, Soviet personnel assigned to Mongolia up to that time were generally substandard, since the region was regarded as a backwater.

So serious was the beating the V-VS had taken, however, that on May 28, People's Commissar for Defense Kliment Y. Voroshilov ordered a temporary suspension of combat sorties, and the decimated 70th IAP (*Istrebitelny Aviatsionny Polk*, or Fighter Aviation Regiment) was pulled

back to Bain Tumen for replacement aircraft and personnel. Among its new con-
signment of aircraft were twenty I-153s,
fresh off the assembly line and shipped by
road across the Soviet Union and assigned
to one of its squadrons. The regiment also
got an infusion of Spanish Civil War veterans, including its new commander, *Mayor*
Sergei I. Gritsevets, who had thirty victories over Spain to his credit, as well as two
Gold Stars of a Hero of the Soviet Union
(HSU).

The I-153's fighting debut came at 1100
hours on July 25, when Gritsevets took
nine of the planes into the sky, divided
into three-plane *zvenoi* (flights) led by
himself and two other pilots, Nikolai
Viktorov and Aleksandr Nikolayev. Within
seconds of noting a white marker in one
of the foothills of the Khamardaba, indicating the forward-most position of
friendly ground troops, Gritsevets's
deputy waggled his wings to attract his
attention and then pointed toward the
Uzur-Nur, a lake on the Japanese side of
the lines. There, about 5,000 feet above
the lake and climbing toward the Soviets,
was a gaggle of Ki.27s.

To the surprise of his pilots, Gritsevets
turned a few degrees away from the
Japanese and led them off at a leisurely
speed. It soon dawned on them, however,
that he was trying to draw the Japanese
into Soviet territory, at the same time
presenting an angle that would conceal
the I-153s' retracted landing gear from
their view so that they would think they
were dealing with I-152s—relatively easy
prey. As the Japanese eagerly closed in on
them, the Soviet pilots cocked their gun
triggers into the "fire" position, then,
once the Japanese had closed to 2,200
yards, Gritsevets made a circling gesture
over his head. Gunning their engines to
full boost, the Soviets broke formation
and turned on their pursuers.

Under camouflage netting, a Polikarpov I-153 of a
Soviet naval regiment is readied for takeoff with the
help of a truck-mounted Hucks starter in 1939.
Although a retrograde step, the I-153 biplane was sec-
ond only to the I-16 in the Soviet fighter inventory by
June 1941. (*National Archives*)

The wild dogfight that followed only
lasted five minutes, but it certainly left
the Soviet pilots satisfied, as all of them
returned to their base and reported seeing at least one Nakajima break up in the
air, a second spiral down trailing smoke,
and two others plunge down to the
steppe below before the rest retreated
over the lines. Their perception proved to
be somewhat exaggerated—the enemy
unit involved, the 24th *Sentai*, reported
the loss of only two planes and their
pilots, Sgt. Maj. Kiyoshige Tatsumi and
Cpl. Shunji Takagaki—but it was far more
conservative than the Japanese description of their run-in with the I-153
squadron that appeared in the Tokyo
newspaper *Yoshiura* a few days later, stating that "although flown by veritable dev-
ils, it had been bested by fighters of the
Imperial Japanese Army which had
accounted for no fewer than *eleven* of this
new warplane." In fact, only two of the
nine I-153s that returned had bullet holes
in their fabric skins to indicate that they
had been in combat.[1]

In spite of that promising start, the I-
153 soon lost its fleeting ascendancy over
the Ki.27. Allegedly some Soviet pilots
would fly with their landing gear down in

the hope of convincing Japanese pilots that they were flying I-152s rather than I-153s, retracting it as the enemy closed in on them—no mean feat, since the I-153's undercarriage had to be hand-cranked. More often, however, the I-153s took the worst of it against the Ki.27s, and they only proved effective if used in concert with a higher element of I-16s, just as the older I-15s had done over Spain. Its leading exponent, Gritsevets, added a total of twelve Japanese planes to his civil war tally, but his wealth of experience was to be denied his comrades later, when it would have been needed most. Appointed a regiment commander just before the Soviet invasion of Poland on September 17, 1939, he was taxiing for takeoff on the preceding day when another plane coming in for a landing, flown by fellow Spanish Civil War veteran *Mayor* Piotr I. Khara, suddenly stalled and crashed on top of Gritsevets's machine, killing him.

The overall success of the Khalkhin-Gol campaign did not blind the Soviet High Command, or Stavka, to the shortcomings displayed by its standard fighters, but by then Polikarpov was committed to production of the I-153; 3,437 were completed by the time manufacture finally ceased in early 1940. The embarrassing showing that the V-VS displayed during the 1939–40 Winter War against Finland reinforced the need for a major modernization of its arsenal, and, indeed, by that time Soviet designers were already developing a generation of airplanes to bring the V-VS back up to world standard. These newer types were only entering service, however, when the Germans struck in June 1941. In consequence, the first line of Soviet fighters, and the men who flew them, would have to buy time for production of the more advanced types to reach full tempo—time for which they paid in blood.

In January 1939, Stalin and the Commissariat of the People for the Aviation Industry issued a specification for a new general-purpose fighter to compete with the Messerschmitt Me 109. Ten design bureaus took part in the resulting competition, including those of Aleksandr S. Yakovlev, Semyon A. Lavochkin, and Nikolai Polikarpov. One of Polikarpov's proposals, more of a high-speed interceptor than a frontline fighter, was to be powered by the new 1,400-horsepower Mikulin AM-37 engine, which was expected to give it a normal maximum speed of 416 miles per hour and briefly boost it to 445 miles per hour by means of two turbosuperchargers.

The I-200 project was adopted for development in December 1939; but since Polikarpov was then engaged in developing the I-180—essentially a refined I-16—the Commissariat set up a special design department to proceed with the I-200, headed by Artyom Mikoyan, a talented engineer who was the younger brother of Anastas Mikoyan, the future deputy chairman of the Council of the People's Commissars of the USSR and People's Commissar for Foreign Affairs. Among Artyom Mikoyan's deputies was Mikhail Y. Gurevich.

When work on the AM-37 engine was abandoned, Mikoyan's team proceeded using the 1,200-horsepower AM-35A engine and produced a sleek fighter of mixed metal and wooden construction that made its first flight, with Arkady N. Yekatov at the controls, on April 5, 1940. Engine overheating problems were eventually overcome by redesigning the radiator, and after overflying Moscow in the May Day parade, the I-200 was put into production on May 25 as the MiG-1. An improved version, the MiG-3, was flown by Yekatov in February 1941. Its maximum speed was 397 miles per hour at its normal service ceiling of 25,500 feet, but during a test flight on March 13, 1941, the supercharger impeller suffered damage and the prototype went down out of control, killing Yekatov. In spite of that and subsequent crashes, the MiG-3 was also hastily put in production, resulting in operational aircraft of such crude manufacture that performance almost invariably fell below the intrinsic specification, and the armament was not properly harmonized. The MiG-3 also suffered from increased weight, most of which shifted its center of gravity aft, making it a difficult plane to fly and land.

By mid-1941, 1,289 MiG-3s had been delivered to V-VS units. Piotr M. Stefanovsky, a MiG test pilot sent to convert the 4th and 55th IAPs of the 20th SAD (*Smeshannaya Aviadiviziya*, or Combined Air Division), later recalled:

After their disastrous debut as frontline fighters, MiG-3s of the 6th *Istrebitelny Aviakorpus, Protivovozdushnaya Oborona*, are put to more productive use in defense of Moscow in December 1941. (*National Archives*)

> The division had two complete sets of fighter equipment comprising the outdated I-16 and I-153 and the brand-new MiG-3, but by the time I arrived they wouldn't dare even to make a test flight with the latter type, so I had to push it. Surprisingly, the pilots weren't at all enthusiastic about the new aircraft, and nobody volunteered to master it. Well, I had to show off right away with the MiG-3. So I took off and squeezed everything I could out of the aircraft, and a bit more. Once I landed it appeared that their suspicious attitude toward the fighter had gone. . . . The ensuing conversion training had to be completed in some haste as time was pressing, and we flew from dawn to dusk.[2]

One of the more talented pilots who managed to master the MiG-3's idiosyncrasies was *Starshy Leitenant* Aleksandr Ivanovich Pokryshkin of the 55th IAP who, after flying I-16s, was instantly captivated by its sleek lines. "I liked it at once," he later said. "It could be compared to a frisky, fiery horse—in experienced hands it was to run like an arrow, but if you lost control you finished up beneath its hooves."[3]

Another 55th IAP member, *Mladshy Leitenant* (Junior Lieutenant) Valentin I. Figichev, had barely completed his training at Pyrlitsa airfield, near the Romanian border, before he became the first pilot to employ the MiG-3 in the interceptor role. While leading a patrol near the Romanian border on June 9, he reported sighting a Junkers Ju 88 with German markings in Soviet air space, and tried to force its pilot to land. The intruder—more likely a Dornier Do 215B of the 3rd *Staffel* of *Aufklärungsgruppe* (*Fernaufklärung*) *des Oberfehlshabers der Luftwaffe*, which conducted high-altitude photoreconnaissance missions over the Black Sea and the southern USSR from Pipera airfield near Bucharest—turned back, and Figichev and his flight pursued it several kilometers into Romania before abandoning the chase. This caused a diplomatic incident with Romania that would have led to Figichev's arrest had events thirteen days later not caused it to be forgotten. Figichev went on to be credited with eight German and one Romanian plane in the next month's fighting over Moldavia and the southern Ukraine. The 4th IAP, stationed at Kishinyov, claimed to have brought down three Romanian aircraft that had intruded into Soviet air space prior to the German invasion. In most cases, however, Stalin ordered his fighters to refrain from attacking Axis intruders, lest the Germans use any such incident as an excuse for war.

That all became meaningless at 0300 hours on June 22 when the Luftwaffe launched a comprehensive strike that destroyed about 2,000 Soviet aircraft, mostly on the ground, in the first twenty-four hours. At that time, 917 MiG-3s were distributed among units in the Baltic, Western, Kiev, and Odessa military districts, with an additional 300 defending Moscow and Leningrad. Two days later, only 234 MiGs were operational, the highest proportion of losses to any fighter type in the V-VS in that time period. Worst hit was the 9th SAD, which lost virtually all of its 37 MiG-1s and about 200 MiG-3s—out of total divisional losses of 347 out of 409 aircraft—by June 25. On the twenty-sixth the 9th SAD's commander, *General-Mayor* (Major General) Sergei A. Chernykh, a five-victory Spanish Civil War ace and Hero of the Soviet Union, was court-martialed and executed by firing squad.

Ironically, that sort of comprehensive disaster was avoided by the older I-16s of the 67th IAP based at Bolgarijka (Bulgarica) airfield in southern Bessarabia, which had prudently dispersed some of its *zvenoi* to temporary "ambush fields" closer to the Romanian border. From these they were able to react and intercept enemy strikes from the outset, starting with a Romanian Bristol Blenheim of *Escadrila* 1 *Recunoastere* on a reconnaissance sortie destroyed at 0400 hours on June 22, along with its four-man crew, by *Mladshy*

Leitenant Nikolai Yermak—the first Axis loss during Barbarossa.

Shortly afterward, two successive groups of nine I-16s each from the 67th intercepted two waves of Savoia-Marchetti SM.79B bombers of Romanian *Grupul* 1 *Bombardement*, escorted by IAR-80s of *Grupul* 8 *Vânātoare*, as they attacked airfields at Bolgrad and Bolgarijka. According to the Soviet account, *Kapitan* Ivan Artamonov led his flight against the fighters, which soon disengaged and retired, leaving *Kapitan* Feodor Chechulin's flight free to disperse and drive off the bombers, all Soviet fighters returning intact. The Romanians, in contrast, claimed four victories for the IARs—including one by *Sublocotenent Aviator* (Pilot Officer) Ioan Miháilescu for his first of an eventual five victories—and two by SM.79 gunners, as well as four Soviet fighters destroyed on the airfield, for the loss of two bombers. IAR-80, No. 56, was driven down near Brãila in Romanian territory, and *Sublocotenent* (Junior Lieutenant) Gheorghe Postelnicu returned battle-damaged IAR No. 18 to base with wounds to his head and neck.

This encounter marked the combat debut of an indigenously designed Romanian fighter with something of an international pedigree. After building foreign aircraft under license since 1925, the Industria Aeronauticã Românã (IAR) at Brasov embarked on a fighter design of its own in October 1937. Designed by Professors Ion Grosu and Ion Cosereanu and engineers Gheorghe Zotta and Gheorghe Vollner, the prototype IAR-80 was completed in December 1938, but did not undergo its first test flight until April 12, 1939, with *Capitan* Dumitru Popescu in the cockpit. The plane's tail surfaces were clearly influenced by the PZL P-11f that

Romanian-built IAR-80 No. 7 of *Grupul 8 Vânătoare* escorts a Heinkel He 111 of *Grupul 5 Bombardement* on a mission into southern Russia in the first days of Operation Barbarossa, June 1941. (*Dénes Bernád*)

IAR had been building under license, while its 870-horsepower IAR K14-IIIC32 radial engine was a license-built version of the French Gnome-Rhône 14K-II Mistral Major. In between them, the sleek fuselage and duralumin-skinned wing were entirely original. With a speed of 317 miles per hour at 13,000 feet, a ceiling of 39,000 feet, and a climb rate of 16,400 feet in six minutes, the IAR-80 was judged somewhat inferior to the Spitfire Mark I and Me 109E, but superior to anything produced in the Romanian arsenal, and competitive with everything else in the world at the time. The installation of a 930-horsepower K14-IIIC36 engine necessitated a longer fuselage with the cockpit set even farther back to compensate for the altered center of gravity. The adverse effect this had on the already poor pilot's forward visibility was also offset by installing a raised bubble canopy.

Ultimately, a total of 170 IAR-80s and 176 more potent IAR-81 variants were built and served with considerable success over the Eastern Front, as well as in defense of the Ploesti oil fields against American bombing attacks. The appearance of improved Soviet fighters and the North American P-51 Mustang, combined with the growing availability of Me 109Gs, led to the IAR fighters being phased out of first-line service by the end of July 1944; but a two-seat training version, the IAR-80DC, was not retired from the Romanian air force until 1952.

Other Soviet units besides the 67th IAP managed to get some planes into the air. Situated only twelve kilometers from the Polish border, the 129th IAP at the advanced base at Tarnovo had fifty-seven MiG-3s, fifty-two I-153s, and only forty combat-ready pilots. Many of the MiGs were not airworthy, and so few of the pilots had accustomed themselves to its flying characteristics that, given the choice, most of them flew the old biplanes. The unit's senior political instructor, Anatoly M. Sokolov, was still engaged in converting pilots from I-153s to MiG-3s when the Germans struck. Commissars already had an unsavory reputation for being more politically reliable than competent, but when the chips were down that morning, Sokolov proved to be an exception. Joining the twelve MiG-3s and eighteen I-153s that managed to get airborne at dawn and intercept a dozen Me 109s at 0405, Sokolov shot down one of them. Another interception by the

129th caused eighteen Heinkel He 111s to drop their bombs short of Tarnovo before retiring. Inspired more by Sokolov's example than they ever could have been by his words, his comrades claimed another five German aircraft for the loss of only one of their own in aerial combat. At 1000, however, a series of strikes by small formations of Ju 88s and Me 109s succeeded in destroying twenty-seven MiG-3s, eleven I-153s, and eight training aircraft on the ground, as well as rendering Tarnovo airfield inoperable.

The badly mauled 129th IAP withdrew to Balbasovo four days later, but it would go on to greater things. Sokolov would be credited with eight enemy planes and received an HSU before being shot down and killed by enemy fighters on January 25, 1942. Later equipped with Lavochkin LaGG-3s, La-5s, and finally La-7s, the 129th produced a number of outstanding aces—the highest scoring of whom, Vitaly I. Popkov, was credited with forty-one and awarded two HSUs—and on December 6, 1941, was given an honorary *Gvardiya* (Guards) redesignation as the 5th GvIAP. By the end of the war the 5th GvIAP had been credited with 737 victories, the highest score of any fighter regiment in the V-VS, as well as the destruction of 1,832 military vehicles and 283 artillery emplacements.

The 124th IAP had its forces divided between two airfields on June 22. One, Wysokie-Mazowieckie, was a small field forty miles from the border, hosting seventy MiG-3s and twenty-nine I-16s. There was little taxiing room, handicapping the MiGs in particular, since unlike the I-16s they needed at least 540 yards of field to take off. Consequently, when the base underwent six successive fighter-bomber attacks by Me 110Es of II *Gruppe*, *Schnellkampfgeschwader* 210, the field and its aircraft were wiped out. A few MiGs

managed to take off and 124th's deputy commander, *Kapitan* Nikolai Kruglov, was credited with bringing down a Dornier Do 215 at 0415 hours. Fifteen minutes later *Mladshy Leitenant* Dmitry V. Kokorev fired a few rounds at what he identified as a Do 215 before his gun jammed. Kokorev then ran his MiG into the enemy plane's empennage, chewing it away with his propeller until it fell, then managed to bring his plane home. According to German records, II./SKG 210 lost two Me 110s, and one, downed near Zambrow, may have been the first—but by no means the last—victim of Kokorev's desperate tactic; as many as eighteen other such aerial rammings, called *tarans*, were recorded on that first day alone.

The 55th IAP, based at Beltsy, got some MiGs into the air to intercept twenty He 111s of III *Gruppe*, *Kampfgeschwader* 27, escorted by eighteen Me 109s, claiming two Heinkels and a Messerschmitt for three MiGs damaged. The Germans acknowledged a wounded *Ltn.* Gerhard Krems force landing his He 111H-2 near Sileni with 30 percent damage, and two other Heinkels returning with wounded crewmen. Seven more claims were made by the Soviet regiment in the course of the day.

The next day saw the 55th IAP carry out a dubious interception, as its MiGs attacked six light bombers with single radial engines, one of which force landed and three others of which returned to base—in Soviet territory, since the MiGs' prey had in fact been Sukhoi Su-2s of the 211th *Blizhnebombardirovochny Aviatsionny Polk*, which suffered the loss of one navigator killed and another wounded. One of the claimants was "Sasha" Pokryshkin. Soon after this embarrassing combat debut, however, he redeemed himself by shooting down an Me 109, followed by two Henschel Hs 126s in one sortie. More important than the aerial victories were the lessons Pokryshkin was learning, which he would later convert into something the V-VS pilots needed even more than new fighters: an effective tactical fighting doctrine. Pokryshkin proved to be a great teacher and leader, as well as being the second-ranking Allied ace of World War II with a final tally of fifty-nine victories. He would also be one of only two Soviet pilots to be awarded the HSU three times, as well as the US Army's Distinguished Service Medal.

On the morning of June 22, V-VS commander Gen. Pavel F. Zhigarev ordered ninety-nine new fighters to be rushed to the front, but the chaos of the German advance made that impossible. There were no new fighters to oppose the Luftwaffe by June 24, but two hundred fighters arrived the next day, and after that a new regiment reached the front almost daily. Prominent among the aircraft were MiG-3s; but they never fully measured up to expectations and were eventually relegated to units of the PVO (*Protivovozdushnaya Oborona*, or Defense Group), protecting Moscow and other cities, and serving in the high-speed reconnaissance role. Overall, the design team of Mikoyan and Gurevich was not a particularly successful one during World War II, although the postwar jet age would make their names world famous.

Ironically, the MiG's high-altitude performance gave it no advantage as the kind of "frontal fighter" that the V-VS needed to establish air superiority over the battlefield. Fortunately for the Soviets, there was another type arriving at the front that would—the Yak-1.

Conceived in May 1939 and completed eight months later, Aleksandr Yakovlev's fighter began as the I-26, a low-wing monoplane of mixed construction with a steel-truss forward fuselage structure skinned in duralumin, a fabric-covered after fuselage, wooden wings, and duralumin ailerons and flaps. Powered by a Klimov M-105P twelve-cylinder inline water-cooled engine, the I-26 first flew on January 13, 1940, reaching the promising speed of 360 miles per hour before its flight had to be curtailed owing to a dangerous rise in oil temperature. The first prototype crashed on April 27, 1940, but it did not deter Stavka's decision to authorize development and manufacture. By the autumn of 1940, the production version of the I-26 was entering service as the Yak-1.

Although praised for its handling characteristics and overall performance—which was certainly an improvement over that of the I-16—the Yak-1 had gained considerable weight in the course of its development. Consequently, the 1,250-horsepower M-105PF engine used in the production machine gave it a maximum speed of 348 miles per hour, which was barely competitive with an Me 109E, let alone with the new Me 109F and Fw 190A. Armament consisted of one 20mm ShVAK cannon firing through the propeller hub and two synchronized 7.62mm ShKAS machine guns.

Preproduction Yak-1s were delivered for operational trials to the 11th IAP at Kubinka, just east of Moscow, and by mid-May 1941 the unit had sixty-two production machines. It was something of a training/conversion center for subsequent Yak-1 outfits, such as the 20th, 45th, 123rd, 158th, and 91st IAPs, all of which had some on hand prior to Barbarossa. Most of the aircraft were kept around Moscow, but 125 Yak-1s had been sent to the Western Military District when hostilities commenced. Only thirty-six pilots of the 20th IAP, based at Sambora in the Kiev Special District, had mastered the new fighters when war began. In other units only regimental and squadron commanders were flying Yaks.

One such pilot was *Mayor* Boris N. Surin, commander of the 123rd IAP, based at Strigivo. Primarily equipped with sixty-one I-153s, the 123rd received twenty Yak-1s, but they were not assembled until June 19. Responding to Luftwaffe attacks on the Fourth Army headquarters at Kobrin, Surin was apparently flying one of the new Yaks when he was credited with an Me 109 at 0500 hours, though his likely opponents, the 11th *Staffel* of *Jagdgeschwader* 51, suffered no losses at the time. He accounted for another two German aircraft in a subsequent sortie, but probably reverted to an I-153 in the course of the day. Other members of the regiment were credited with anything from twenty to thirty enemy planes that day, including an Me 109, two Ju 88s, and two He 111s by *Lt.* Ivan N. Kalabushkin; but the 123rd lost nine I-153s and eight pilots killed, including Surin while leading his fourth mission of the day, all apparently to JG 51 Me 109s.

Such successes as the Soviet fighters achieved, exaggerated or not, were miniscule in the face of the overall disaster that attended the German invasion, however. Within ten hours of the start of Barbarossa, the Luftwaffe had attacked sixty-six airfields and destroyed more than eight hundred aircraft on the ground, as well as shooting down almost four hundred more. Although the handful of skilled pilots trained to fly the Yak-1s acquitted themselves comparatively well, their aircraft suffered from the maintenance problems attendant on a new and hastily manufactured design. At Leningrad, for example, it was discovered

A lineup of early model Yakovlev Yak-1s prepare to deploy to the front. Although initially underpowered, the Yak fighter series would develop into an instrument of Soviet victory. (*National Archives*)

that there were twice as many faulty Yak-1s on hand as there were operational ones, compelling the 158th to give its aircraft over to the 123rd IAP and retire to the rear to pick up more fighters. As the airframe and engine defects were remedied, however, the basic soundness of the Yak-1's design became apparent—even to its opponents, as expressed in a Luftwaffe assessment in 1942:

> The Yak-1 fighter is presumably the best Soviet fighter. It had a better speed and rate of climb compared to the MiG-3 and came close to the performance of the Bf 109F, but was inferior to the latter in speed. It was appreciably more difficult to hit the Yak-1 from behind than the MiG-3. It retained a good rate of climb up to 6,000 m. (19,685 ft.), but its maneuverability fell off at that height. For this reason the pilots dived from high altitudes to lower ones where they accepted combat.[4]

While little more was done with the MiG-3, the Yak-1 was only the first of a progressively improving series of fighter. There were, in fact, two divergent lines of development based on the original Yak-1 configuration. *Tyazhely* (or heavy fighters), designed for maximum range, armament, and protection, included the Yak-7 and superb Yak-9, which would be one of the most prolific fighters in history and respected even by German pilots as one of their most formidable adversaries at the end of the war. A line of *legky* (lightweight) interceptors based on the Yak-1 airframe began with the Yak-1M and culminated in the little Yak-3, equally feared by the Germans for its scintillating performance and indisputably the most successful lightweight fighter of the war.

A third contender in the 1939 competition—in the form of a fighter of allwooden construction, to alleviate possible shortages of aluminum and other strategic alloys—had originally been proposed in 1938 by Vladimir P. Gorbunov. "Even if only one small grove of trees is left in Russia," he grandly declared, "even then we shall be able to build fighters." His idea was accepted, and in May 1938 a design team was placed under his supervision, including Semyon A. Lavochkin and Mikhail I. Gudkov. The fighter they labored to develop was initially designated the I-22.[5]

The principal element in the I-22's construction was delta-drevsina, a composite of birch strips pressed together in crossgrained fashion and impregnated with phenols derived from birch tars. The resultant plywood was strong, fire-resistant, and could be molded into shape and polished to a drag-cheating finish. Working under a daunting time constraint, designers managed to have their first prototype, the I-22, ready for its first test flight on March 30, 1940. Powered by a 1,050-horsepower Klimov M-105P engine, the I-22 suffered its share of initial problems and at least one crash; but development continued, and an improved prototype, the I-301, underwent state acceptance trials on June 14, 1940, during which it achieved a speed of 376 miles

per hour at 16,404 feet—faster than the Yakovlev I-26, even though the wooden fighter was heavier. Although it displayed several disadvantages, including stiff ailerons and elevators, on July 29, 1940, the I-301 was cleared for mass production as the LaGG-3. Armament consisted of two Berezin UBS 12.7mm machine guns in the upper fuselage decking and a third firing through the propeller hub.

Tests of the production LaGG-3 conducted near Moscow in May 1941 showed that its performance had fallen rather short of the prototype's, with a maximum speed of 357 miles per hour and a rate of climb (2,412 feet per minute) which was inferior to that of the I-16 it was meant to replace. The weight of additional military items such as seat armor, extra fuel tanks, and a radio downgraded that performance even more. Elevator control response was found to be so poor that horn balance weights were added to the control surfaces in later production series.

Only 322 LaGG-3s had been built when the German invasion occurred, and the first unit to which they were assigned, the 19th IAP of the Leningrad PVO, only got a few, as did the 157th and 24th IAPs of the Moscow PVO. Most were shipped to the Far East as a precaution against a renewed Japanese threat. Only when Stalin was certain that the Japanese would respect the neutrality pact they had signed with the USSR on April 13, 1941, did he authorize the LaGG-3s to be transferred to the west.

Initially withheld, by and large, from the western borders, the vast majority of LaGG-3s were spared the decimation meted out to their stablemates, although the 166th IAP sadly logged four destroyed and another ten damaged in its first day of combat on June 22. July saw the LaGGs being rushed out to replace the fighters that had been lost in the initial German onslaught. I-16 and Yak-1 pilots alike expressed one complaint or another about them, but at least 143 fighter regiments eventually had LaGG-3s on strength.

One unit that got the LaGG-3 in the later summer of 1941, the 5th IAP of the Baltic Fleet (*Krasny Baltisky Flot*, or KBF), which served around Leningrad and Tallinn, showed what the plane could do when flown by a pilot who mastered its quirks. Such pilots included *Leitenanti* Ivan I. Tsapov (fifteen victories), Semyon I. L'vov (six, plus shares in twenty-two others), and Dmitry M. Tatarenko (sixteen), the last mentioned acquiring a reputation for striking his opponents on the first pass, which earned him the nickname of "*Rubaka*" (broadsword). *Starshy Leitenant* Georgy D. Kostylev scored his first four victories in I-16s before getting a LaGG-3 in August, eventually being credited with a total of nine solo and thirty-two shared victories. Yet another member of the 5th, *Kapitan* Igor A. Kaberov, survived the war with eleven personal and eighteen shared victories, and stands out as one of the few pilots to praise the generally maligned LaGG-3. The 5th IAP-KBF was credited with 105 enemy planes in 5,899 sorties by January 18, 1942, when the unit was given *Gvardiya* status as the 3rd GvIAP-KBF. With a final total of 500 aerial victories plus 17 planes, 71 armored vehicles, and 750 other vehicles destroyed in ground attacks, the 3rd GvIAP-KBF was the highest-scoring Soviet naval regiment of the war.

The LaGG-3 underwent sixty-six production series by the time the 6,528th and last machine arrived at a V-VS unit in the summer of 1942, with improvements and refinements introduced in at least seven of them. Lavochkin often went to the front to listen to pilots' complaints

and then do what he could to remedy them within the constraints of a desperate production schedule that would tolerate no delays. He also brought with him Konstantin A. Gruzdev and Aleksei Grinchik, two highly skilled test pilots who had both enjoyed early successes in the LaGG-3, and who tried to show the demoralized frontline pilots the means of wringing out the best performance from their machines. Gruzdev, who scored nineteen victories before being killed in a flying accident on February 2, 1943, recommended fifteen degrees of flap to tighten the fighter's turn in a dogfight. Grinchik's most notable experience in the LaGG-3 was hardly typical—hit in the engine by Me 109s, he was gliding down to a forced landing when one of his attackers miscalculated and passed right in front of Grinchik's gunsight, at which point he opened fire and saw the German go down.

Most LaGG-3 pilots were not that skilled, however, and few of the biplane pilots were accustomed to a monoplane with such high wing loading mated to an engine of such inadequate power. It had a habit of going into a sudden, vicious spin amid a steep banking turn, and during the landing approach it would suddenly nose up and then stall. The landing gear was weak, and faulty hydraulics were known to cause it to retract while on the ground, or extend while in flight. The inadequate view from the cockpit was further impaired by the insufficient clarity of the nitrocellulose compound used on the canopy of early versions, compelling most pilots to fly with it open or to discard it entirely, at the sacrifice of more than nine miles per hour to the added drag. At least one disparaging song was written about the LaGG-3, also known as the "mortician's mate," and some pilots quipped that the wooden fighter's acronym really stood for *lakirovanny garantirovanny grob* (varnished guaranteed coffin). If the plane had any indisputable virtue, it lay in the delta-drevsina, which could stand up to a lot of punishment, and which made a bullet-riddled LaGG-3 a lot less prone to go down in flames than a stricken Yak-1.

While the Yak-1's fundamentally sound airframe lent itself to progressive improvements that culminated in the superb Yak-9 and Yak-3, it took a more radical step to turn the LaGG-3 into something more than a deathtrap: the replacement of its inline engine with Arkady Shvetsov's M-82 radial. Ironically, other Soviet designers had experimented with the radial on their existing airframes, such as Mikhail Gudkov's Gu-82, Mikoyan's MiG-9, and Yakovlev's Yak-7 M-82, while Lavochkin hesitated. By early 1942, only the Sukhoi Su-2 short-range bomber was using the M-82 when Lavochkin and Shvetsov were called in to a conference of the People's Commissariat of the Aircraft Industry in Moscow. In essence, Lavochkin was told that reports on his LaGG-3 were so unsatisfactory that if something significant were not done soon, production of the fighter would have to be canceled. And since hundreds of unwanted M-82s were piling up at Shvetsov's Plant No. 19 in Perm, Lavochkin was strongly urged to try fitting the radial in his plane.

Lavochkin protested. Modifying the LaGG-3 airframe to take an air-cooled radial that was eighteen inches greater in cross section and 551 pounds heavier than the inline M-105P would be complicated by a shift in the center of gravity. The M-82's propeller shaft could not accommodate a 20mm cannon. He also feared that production would cease before he and his design team would effect such complex alterations. There was already a precedent for such a fighter, however. As early as March 1941, Gudkov had lifted an M-82 directly from an Su-2 and had worked out a way of mounting it on a LaGG-3 airframe, and on October 12 the Commissariat announced a willingness to put his Gu-82 into production at the Gorky plant instead of the LaGG-3. Gudkov's attention was then diverted by a project to mount a 37mm cannon to fire through the LaGG-3's propeller hub, and the more resolute Aleksandr Yakovlev secured a contract to produce his new Yak-7B fighter at Gorky.

News that LaGG-3 production at Plant No. 31 in Tbilisi was to be halted in April gave Lavochkin some added incentive to intensify his efforts. The LaGG-3's fuselage midsection was widened and the engine mount reworked. Two variable cooling flaps on the fuselage sides and altered cooling-air baffles provided uniform cooling. Two 20mm ShVAK cannon were mounted above the engine. The machine was completed in February 1942, and Lavochkin anxiously awaited the results of its first evaluation. "The aircraft is good, pleasant to control and responsive, but the cylinder heads became hot," reported test pilot G. A. Mishchenko. "Measures should be taken to correct this." He also reported that level speed was 10 percent greater than that of the LaGG-3. Encouraged, Lavochkin and his team did further work on the prototype, which got its first official evaluation from May 9 to 14, 1942. Cooling and controllability problems were encountered, but with a speed of 372.8 miles per hour at its service ceiling of 21,000 feet, a climb rate of 16,400 feet in six minutes, and maneuverability that was superior to foreign as well as indigenous designs, the M-82-powered LaGG was good enough to completely reverse Lavochkin's shaky fortunes. Since Gorbunov had left the design team by then, the new fighter was designated the LaG-5 and ordered into production, the first example rolling out of Gorky's Plant No. 21 on June 20. Gudkov also parted company with Lavochkin soon afterward, and from September 1942 the radial engine fighters were officially referred to simply as La-5s.[6]

An early Lavochkin La-5 of the 302nd *Istrebitelnaya Aviatsionnaya Diviziya* on the Belorussian Front in early 1943. Among the pilots assigned to this division was Ivan Kozhedub, who did not have any success with the new "work-in-progress" fighter until July. (*Library of Congress*)

In August 1942 the first operational LaG-5s replaced the I-16s and LaGG-3s of the all but decimated 49th IAP on the Northwestern Front. In the course of flying their first 180 sorties, LaG-5 pilots of the 49th claimed sixteen German aircraft in the course of seventeen combats. The regiment lost ten planes, however, and five of its pilots were killed in action.

Soon after the 49th IAP, fifty-seven LaG-5s were assigned to four regiments of Col. Stefan P. Danilov's 287th *Istrebitelnaya Aviatsionnaya Diviziya* (IAD), attached to the First Air Army near the embattled city of Stalingrad. The Lavochkins flew their first combat missions on August 20, but they displayed the unmistakable signs of hasty production, only two-thirds of them being combat capable. One plane crashed during take-off, while two others collided while taxiing, due to two aspects of the Shvetsov radial engine: a greater degree of propeller torque that took getting used to, combined with poor visibility from the cockpit. Again, impaired visibility compelled the pilots to fly with their canopies open, and cooling problems and lack of confidence in the retractable tail wheel resulted in flying with the cowling side flaps fully opened and the tail wheel down, all contributing to an 18.6 to 24.8 miles per hour reduction in speed. In the

first three days of fighting, the LaG-5 pilots claimed eight German fighters and three bombers but lost seven of their own planes—including three to Soviet antiaircraft gunners who mistook them for German Fw 190As.

Among the first standout pilots was twenty-three-year-old *Lt.* Evgenny P. Dranishchenko, who joined the 287th IAD's 437th IAP on August 20 and scored his first victory, over a Ju 88, just three days later. He was credited with two Ju 88s of II./KG 76 on September 8, and by the thirteenth he scored his fifth victory in the course of ten combats. Dranishchenko's total stood at twenty-one individual and seven shared victories in 120 missions and fifty combats when he was killed in action on August 20, 1943, exactly one year since his arrival at the front.[7]

The LaG-5's debut yielded mixed results at best. Pilots of the 287th IAD's 27th IAP concluded that their planes were inferior to the Me 109F-4—and, even more so, to the newer Me 109G-2 in speed and vertical maneuverability. "We have to engage only in defensive combat actions," they reported. "The enemy is superior in altitude and, therefore, has a more favorable position from which to attack."[8] Concentrating on German bombers for a time, the LaG-5 pilots claimed fifty-seven of them within a month, but continued to suffer heavy losses whenever they encountered enemy fighters.

Again, Semyon Lavochkin was eager to read and respond to the criticisms leveled at his fighters. For a start, he removed two of the five fuel tanks that had been intended to extend the plane's range, but whose added weight adversely affected performance. Aerodynamic improvements, lightening of the airframe, and the introduction of the new supercharged M-82F engine resulted in a better fighter, which entered production in January 1943 as the La-5F (for *forsirovanny*, or "boosted"). In addition, the ninth production La-5 batch, produced in November 1942, had control surfaces of reduced area, redesigned trim tabs, and larger flaps, which improved both controllability and maneuverability. The after part of the dorsal fuselage was also lowered, and a new teardrop-shaped canopy of armored glass was installed, greatly improving visibility from the cockpit. With the subsequent introduction of the fuel-injected M-82FN engine, which boosted takeoff power from 1,700 to 1,850 horsepower in the La-5FN during the Battle of Kursk in July 1943, the curious transition of Lavochkin's wooden "grand pianos" from "mortician's mates" to instruments of ultimate Soviet victory was nearly complete. As the tide of war turned in favor of the Red Army, production standards would improve. Lavochkin continued to refine his now-proven design, culminating in late 1943 with the La-7, one of the cleanest radial-engine fighters of its time.

Amid the heady successes that attended Operation Barbarossa, it may have been difficult for Luftwaffe pilots to imagine the V-VS recovering at all from the initial blow dealt it, let alone do so sufficiently to replace the obsolescent or flawed new fighters that they had first encountered. It would have been harder for even the Soviet airmen to imagine that the unpromising LaGG-3, or even the less than world-beating LaG-5, were steps on the way to one of the great fighters of World War II. Nevertheless, the La-5FN and La-7, which were at their best at low altitudes, did much to clear the skies over the battlefield for the Red Army's resurgent ground forces. It might be added that the leading Allied ace of the war, Ivan Nikitovich Kozhedub, scored all sixty-two of his victories—including one Messerschmitt Me 262 jet—exclusively in Lavochkin fighters, from the LaG-5 to the La-7. His last wartime La-7, displaying his three HSUs, can still be seen at the Air Force Museum at Monino.

Fifteen

Aerial Supremacy Over the Islands

Vought F4U Corsair and
Grumman F6F Hellcat, 1943

The Japanese pilots were momentarily puzzled as they peered down at the remarkably mixed bag of American planes that had come to bomb their ships in the Buin-Shortland area of the Solomon Islands on February 14, 1943. Nine of the intruders were four-engine bombers—Consolidated PB4Y-1s, naval versions of the B-24 Liberator, from bomber squadron VB-101—and ten of their escorts were Lockheed P-38F Lightnings of the 339th Fighter Squadron, Thirteenth Air Force. Twelve other American fighters, however, were of a type the Japanese had never seen before: large, single-engine monoplanes with inverted gull wings. Nevertheless, at that moment aces Isamu Miyazaki, Bunkichi Nakajima, and twenty-five other Mitsubishi A6M2 Zero pilots of the 252nd *Kokutai* had more important matters with which to concern themselves than recognition. Registering the strange new fighters in their minds as either Bell P-39 Airacobras or Curtiss P-40 Warhawks, the Zero pilots plunged into the American fighter formations, while fifteen Nakajima A6M2-N and Mitsubishi F1M2 floatplane fighters of the 802nd *Kokutai* went after the bombers.

The Japanese had been expecting trouble that day. On the morning of February 13, the Americans had launched a similar raid from Henderson Field on the newly secured island of Guadalcanal. It had come in two waves, the first of which consisted of nine PB4Y-1s of patrol squadron VP-51, which struck at Buin Harbor at 1130 hours and failed to hit any ships, the Japanese not even noticing the eleven new bent-wing fighters that had accompanied them. Half an hour later six B-24Ds of the 424th Squadron, 307th Bomb Group, arrived, escorted by four P-38s of the 339th Fighter Squadron and seven P-40Fs of the 44th Fighter Squadron. The Army B-24s scored one 1,000-pound bomb hit on a cargo vessel, but lost two of their number to the ship's antiaircraft fire. Afterward, in a fifty-minute running fight with Zero fighters, another B-24 was set on fire but managed to reach Choiseul before ditching off the north coast of the island, while two P-40s and a P-38 were shot down, and three other P-38s, their fuel exhausted, had to ditch off the Russell Islands. Six Zeros were credited to the US Army Air Force fighters, including two Zeros and a "probable" to 2nd Lt. Robert P. Rist before his

P-38 was shot down. In reality, the Japanese lost two Zeros and their pilots, Leading Seaman Hifumi Yamamoto of the 204th *Kokutai* and Petty Officer 2nd Class Takano Kotaro of the 252nd.

The new fighters whose presence had gone unnoticed by the Japanese on February 13, and which accompanied the renewed effort on the fourteenth, were Vought F4U-1 Corsairs. Designed as carrier fighters, they were nevertheless making their debut with a land-based US Marine fighter squadron, VMF-124, because the Navy didn't want them.

Chance Vought built his first Navy fighter—a conversion from his VE-7 trainer—shortly after World War I ended. Although he produced a number of successful naval aircraft in the subsequent two decades, his attempts to produce an outstanding naval fighter had been in vain until February 1938, when his design bureau, headed by Chief Engineer Rex Beisel, set out to fulfill a Navy specification for a high-speed, high-altitude fighter. Built around a 1,800-horsepower Pratt & Whitney XR-2800-2 Double Wasp eighteen-cylinder air-cooled radial engine, the XF4U-1 Corsair featured an all-aluminum fuselage that was spot-welded to reduce surface drag. Another drag-reducing measure was to assemble the wing at as close to a ninety-degree angle to the fuselage as possible. Since the thirteen-foot Hamilton Standard propeller used on the XF4U-1 made the optimum midwing attachment impractical, Beisel's design team attached the wing at an angle to the lower part of the fuselage and gave it an inverted gull configuration, thus allowing for a shorter, sturdier undercarriage. Armament originally consisted of two synchronized .30-caliber machine guns in the fuselage and two .50-caliber guns in the wings, but that was later revised to six wing-mounted .50-caliber weapons.

First test-flown by Lyman Bullard on May 29, 1940, the XF4U-1 immediately demonstrated its potential. The first single-seat fighter to exceed 400 miles per hour in level flight—reaching a record 405 miles per hour in October 1940—the XF4U-1 also had an outstanding roll rate. The Navy was impressed and ordered the Corsair into production, but the new fighter soon showed problems as well as promise. The narrow canopy was confining, and forward visibility was reduced even further when the cockpit of the production F4U-1 was situated almost three feet farther aft to accommodate a 237-gallon self-sealing fuel tank in the fuselage, the original wing tanks having been replaced by the six .50-caliber machine guns. On top of that, the port wing had a tendency to drop at low landing speeds. Vought later remedied the visibility problem somewhat by devising a more rounded canopy with reduced metal framing for the F4U-1A and F4U-1D models, and the eventual installation of a small triangular wedge on the leading edge of the starboard wing, just outboard of the guns, alleviated the wing dropping problem. When the first F4U-1s were delivered to Navy squadrons VF-12 and VF-17 in October 1942, however, their landing characteristics were found to be too dangerous for carrier operations. VF-12 would eventually go into combat with Grumman F6F-3 Hellcats. VF-17, originally slated to serve aboard the new carrier *Bunker Hill*, was instead assigned to land bases in the Solomons, where its Corsairs established an outstanding record. Other production Corsairs were assigned to the Marines, starting with VMF-124, commanded by Maj. William E. Gise, and based at "Fighter 2" airfield at Kukum on Guadalcanal.

The Marines were accustomed to receiving Navy castoffs, but they soon found the Corsair to be a serendipitous case of the Navy's loss being their gain. Powered by a 2,000-horsepower Pratt & Whitney R-2800-8 engine and capable of 425 miles per hour, the production F4U-1 was the hottest plane they had flown yet. And with a range of 1,015 miles—twice that of the F4F-4 Wildcat—the Corsair gave the Marines a fighter capable of escorting bombers as well as defending their island air bases. When VMF-124's twenty-four planes were shipped to Guadalcanal aboard the escort carrier *Kitty Hawk* in January 1943, however, its twenty-nine pilots had only had an average of twenty hours' flight training, including one high-altitude flight, one gunnery exercise, and one night flight, before being rushed to the front. One hour after the first seventeen Corsairs arrived at Henderson Field on February 12, they flew their first mission, escorting a Consolidated PBY-5A Catalina flying boat to Kolombangara, where it picked up two downed Marine Wildcat pilots—Lt. Jefferson DeBlanc (who would subsequently receive the Medal of Honor for outstanding valor in the action in which he had been brought down) and Tech Sgt. James A. Feliton—as well as an Army P-38 pilot who had ditched off the south coast of New Georgia. Although they came within fifty miles of a Japanese air base, the twelve F4U-1 pilots encountered no enemy planes on that sortie, nor did they during their escort mission on the following day.

On February 14, the Americans prepared for their next strike with a photoreconnaissance flight by a Lockheed F-5A that had the misfortune to be intercepted over Kahili airfield, Buin, and was shot down by four 252nd *Kokutai* Zeros, killing 2nd Lt. Ardall A. Nord of the 17th Squadron, 4th Photoreconnaissance Group. The mission proceeded as planned; but Japanese coast watchers reported the oncoming Americans, and by the time they reached Buin the enemy was ready and waiting.

Ten P-38s, arranged in two three-plane sections and a four-plane flight, were providing top cover for the bombers when the 252nd *Kokutai*'s A6M2s tore into them. Captain J. A. Geyer, leader of the four-plane flight, claimed two Zeros shot down, while 1st Lt. William M. Griffith sent another crashing into the sea; but two other P-38s of that flight were lost along with their pilots, 2nd Lts. Donald G. White and Joseph Finkenstein. Two other P-38s from the three-plane sections were also lost, though the pilot of one, 2nd Lt. John A. Mulvey Jr., ditched near the Russell Islands and was subsequently rescued. Second Lieutenant Wellman H. Huey was less fortunate—bailing out over Kahili airfield, he was captured and subsequently died sometime after being transferred to Rabaul. Two P-40s were also shot down by the Zeros.

Below the dogfight, the bombers scored two hits on one cargo ship, *Hitachi Maru*, killing four crewmen and compelling it to run aground in the shallows off Moila Point on southern Bougainville before settling. The bombers also claimed near misses on two other vessels, but came under attack by the floatplanes and Zeros that broke through the American fighter scrimmage. One of the A6M2-Ns scored a hit in a PB4Y's cockpit, and it went down near Shortland Island. A second bomber, already damaged by antiaircraft fire, was pursued by the aggressive floatplanes until it was forced to ditch in the sea twelve miles from New Georgia.

The Marines tried to defend the bombers, Capt. Joseph Quilty and 1st Lt. James English being credited with down-

A Vought F4U-1 assigned to Marine fighter squadron VMF-124 takes off on the unit's second operational sortie from Fighter One airfield on Guadalcanal on February 13, 1943. The Corsair's first encounter with Mitsubishi A6M Zeros the next day was far from an unqualified triumph. (*National Archives*)

ing an A6M2 and an F1M2 floatplane, but most of the Corsair pilots discovered that they had a lot to learn about fighting Zeros. First Lieutenant Howard J. Finn left his formation to pursue a lone Zero, only to find several others on his tail. He ended up seeking shelter under a PB4Y, the gunners of which claimed one of the Zeros and sent Army intelligence officers to Henderson the next day, asking Finn to confirm their kill. "Some big hero," Finn later remarked sarcastically of his performance in this first fight, although he would be credited with the destruction of six enemy planes in the months to come.[1]

At least he lived to learn from his experience. Second Lieutenant Gordon Lee Lyon Jr. died in a head-on collision with a Zero. One of VMF-124's element leaders, 1st Lt. Lloyd B. Pearson, described what became of one of his men, 2nd Lt. Harold R. Stewart:

Stewart's plane was racked with machine-gun fire diagonally across the main fuel cell. When Stewart rejoined me after the melee, I could see the gaso-

line spraying out of the numerous bullet holes. He appeared to be okay. After approximately ten minutes with us his fuel gave out, he waved goodbye and nosed down to the water from about 20,000 feet. The Zeros then followed him down, shooting at him all the while. He made a successful water landing and I thought I could see a yellow spot (his life raft) beside the cockpit. However, the Zeros continued to strafe him. We never heard from Stewart again.[2]

Upon their return to Henderson Field and Kukum, the Americans claimed a total of fifteen enemy planes, including three by VMF-124 and no less than nine by VB-115's gunners. In actuality, one F1M2 had come down damaged and three Zeros had been shot down, but only one Japanese pilot, Petty Officer 2nd Class Yoshio Yoshida of the 252nd *Kokutai*, was killed—probably in the midair collision with Lyon.

The raid went down in American records as the "St. Valentine's Day Massacre." Given the losses it had suffered on February 13 and 14, VB-101 canceled further daylight bombing missions and flew night raids for some time thereafter. As for VMF-124, it had given the F4U a less than glorious baptism of fire. One of its chastened pilots, 2nd Lt. Kenneth Ambrose Walsh, later admitted: "Being the first unit to go out in the Corsair, we didn't know exactly how to employ it, so we had to establish a doctrine." When Walsh asked one of Guadalcanal's veteran Marine Wildcat aces about how to deal with the Zero, his only answer was, "You've gotta go after them." "Well," said Walsh, "we knew it would take more than that!"[3]

Ultimately, Walsh and his comrades learned that having the advantage of altitude was the key to success. He also discovered that the Corsair could outfight the Zero at high speeds, although it was suicide to be caught by one in a slow climb or to try to outmaneuver it at low speeds. As with most other American fighters, the Corsair could take much more punishment than the Zero, which the F4U's six .50-caliber guns had little difficulty in setting afire.

Ken Walsh's next opportunity for a rematch with the Japanese came on April 1, 1943, when Adm. Isoroku Yamamoto launched Operation I, an all-out aerial counteroffensive against Allied forces in the Solomons. As fifty-eight Zeros from the 204th and 253rd *Kokutais* swept down on Guadalcanal, twenty-eight Wildcats, eight Corsairs, and six Lightnings scrambled up from Henderson Field to intercept them. At about noon, F4U-1s of VMF-124 and P-38Gs of the 12th Fighter Squadron got into a fight with A6M3s of the 204th *Kokutai* between the Russell Islands and Baroku. One of the P-38s was lost, but the 12th Squadron's commander, Maj. Paul S. Bechtel, claimed a Zero. Of the Marines, 1st Lt. Dean Raymond sent

a Zero down in flames, while Walsh accounted for two Zeros and an Aichi D3A1 dive bomber. The 204th *Kokutai* lost two pilots that day—Lt. j.g. Shigeto Kawahara and Petty Officer 2nd Class Eichi Sugiyama. In addition to the results of that fight, fifteen other Zeros were claimed by Navy and Marine pilots in the course of three hours of combat for the loss of five aircraft. Besides the 204th's losses, the 253rd *Kokutai* lost five pilots on April 1.

VMF-124 had gotten the measure of its adversary by May 13, when it got into another battle royal with fifty-four Zeros over the Russells, during which Walsh claimed three to become the first Corsair ace. Warrant Officer Hayato Noda and Petty Officer 2nd Class Yuhi Kariya of the 204th *Kokutai* and Petty Officer 2nd Class Shogo Sasaki of the 582nd were killed in the fight. However, VMF-124 lost Major Gise, probably killed by one of the 204th *Kokutai*'s best pilots, Warrant Officer Ryoji Ohara, who downed a Corsair in his first diving attack. Ohara then became separated from his flight and was attacked by two Corsairs, which chased him halfway to New Georgia before he suddenly turned on his pursuers and put three cannon shells through the F4U-1 of 1st Lt. William Cannon. Although his Zero sustained thirty-eight hits, Ohara managed to reach Kolombangara, where he force landed and was subsequently credited with downing Cannon's Corsair as well as Gise's. Once again, the Japanese had overestimated the damage they had done—Cannon made it back to Henderson Field.

By the time Walsh's first tour ended on September 7, he had been awarded the Medal of Honor and had twenty victories to his credit—of which sixteen were Zeros—to which he would add one more Zero while commanding VMF-222 in

June 1945. VMF-124 downed a total of sixty-eight enemy planes during its first deployment, but paid for its growing font of experience with the loss of thirty aircraft and eleven pilots, including its commander, Gise. Walsh himself was shot down three times and crash-landed two other F4Us as a result of combat, as well as bringing back about a dozen shot-up aircraft. Nevertheless, he noted that by the time more F4U squadrons arrived in the Solomons, "I had a lot more to tell them about than just 'you gotta go after them.'"[4]

Britain's Fleet Air Arm was also receiving Corsairs, and its pilots were as appalled as their American colleagues at how such an excellent fighter could be such a horrendous carrier plane. In spite of numerous shipboard crashes, one of which resulted in the death of a squadron commander aboard the carrier *Illustrious*, the British managed to devise a landing technique, coming in out of a turn so that the pilot could see ahead until the last few seconds, which made it possible to operate their Corsairs from carriers. Their plight was somewhat alleviated later in 1943 when the F4U-1A, with an improved blown Plexiglas canopy, joined the FAA as the Corsair Mark II. On April 3, 1944, Corsair IIs of Nos. 1834 and 1836 Squadrons from HMS *Victorious* became the first of their breed to participate in a full-scale carrier strike, Operation Tungsten, when they escorted bombers against the German battleship *Tirpitz*. US Navy and Marine squadrons began operating Corsairs from carriers later that year, and the F4U eventually replaced the F6F Hellcat as the standard Navy fighter. Some eleven thousand Corsairs were built—more than any other US Navy fighter—and they would continue to serve with distinction in the Korean and Indochina wars and in the

1956 Suez Crisis. The US Navy did not retire its last F4Us until August 1957, the French navy used them until 1964, and Corsairs saw their last aerial combats during the brief war between El Salvador and Honduras in 1969. From its inauspicious combat debut and its initial rejection as a carrier plane, the Vought Corsair ultimately emerged as one of the most successful combat planes ever built.

By the fall of 1943, the Japanese were fighting a losing battle over the Solomons against Allied aircraft that were improving in quality and increasing in quantity. In addition to the P-38 Lightning and the F4U Corsair, that new generation included a formidable carrier fighter—the Grumman F6F Hellcat.

Codesigned by Leroy R. Grumman, William T. Schwendler, and Richard Hutton as a larger, more powerful replacement for the stalwart F4F Wildcat, the F6F embodied progressive improvements to increase survivability, being even sturdier than the Wildcat and devoting 212 pounds of armor to pilot protection. As was the case with the US Army Air Forces' P-38 and the Marines' F4U, the main question facing the F6F was how it would fare against its principal Japanese opponent, the Mitsubishi A6M Zero, or Zeke, as the Allies had officially codenamed it.

Contrary to popular myth, the Hellcat was not designed on the basis of data gathered from a downed A6M2 that had been recovered intact at Akutan Island, near Dutch Harbor, Alaska, in June 1942—the XF6F-1 prototype first flew on the twenty-sixth of that very month. Test-flying the Zero, however, provided the F6F pilots with invaluable information on its performance, from which they could formulate tactics to cancel the

Japanese fighter's strengths—primarily its maneuverability—and take full advantage of its weaknesses, which included slower level and diving speeds and a complete lack of protection for its pilot and fuel tank.

Grumman engineer Leon Swirbul returned from Hawaii after interviewing Lt. Cmdr. John S. Thach and other Wildcat pilots about their experiences at the Battle of Midway, with a general request for more power. Grumman responded by replacing the prototype's 1,700-horsepower Wright R-2600 radial engine with Pratt & Whitney's new 2,000-horsepower R-2800 Double Wasp with turbosupercharger to produce the XF6F-2. A two-stage supercharger on the XF6F-3 earned that variant on the production contract in October. About 1,000 had been built before Grumman, eschewing the sexual connotations of the name "Tomcat," settled on its ultimate, mildly profane sobriquet.

Aside from being radial-engine naval fighters, the F6F and A6M were diametrically opposed in conception. The Zero had sacrificed all except armament in order to accommodate its 925-horsepower Nakajima Sakae 12 nine-cylinder radial engine within the sleekest, lightest airframe possible, whereas the Hellcat was a brutish heavyweight that relied upon its eighteen-cylinder R-2800-10 Double Wasp to haul its loaded weight of more than six tons at a maximum speed of 375 miles per hour, and had the largest wing area of any single-engine American fighter (334 square feet) to endow it with surprising agility.

Production of the Hellcat coincided with a general buildup in US Navy carrier strength. The nucleus of the new fleet was the twenty-seven-thousand-ton *Essex*-class fleet carrier, larger and more heavily armed successors to the *Yorktown* class,

of which only one, *Enterprise*, remained (very much) operational after the costly battles of 1942. Supplementing the *Essexes* were the eleven-thousand-ton *Independence*-class light carriers, built on the hulls of *Cleveland*-class light cruisers. Crowding the decks of those carriers like vengeful birds of prey were veteran Grumman TBF-1 Avenger torpedo bombers and Douglas SDB-5 dive bombers, as well as newer Curtiss SB2C-1 dive bombers and F6F-3 Hellcat fighters, all manned by aircrews that had been intensely trained by their combat-experienced forebears.

As his fleet's strength grew, the American commander in chief in the Pacific (CINCPAC), Adm. Chester W. Nimitz, formulated a plan to advance on Japan by a more direct route than was being carried out through the South Pacific. Except for some early retaliatory raids by US Navy carriers between February and April 1942, the great naval confrontation at Midway Island in June, and a diversionary attack by US Marine Raiders in the Gilbert Islands in August, the Central Pacific had seen relatively little activity, while the main contest for initiative in the Pacific went on in New Guinea and the Solomons. Now, however, Nimitz planned to seize strategic Japanese-held island groups in the Central Pacific, one by one, while at the same time eliminating the main Japanese air and naval base in that region, Truk Atoll in the Carolines. First, however, he wished to blood his new task forces with some minor raids, to give their personnel experience and confidence.

The first such raid took place on August 31, 1943, when Task Force 15.5, built around carriers *Yorktown*, *Essex*, and *Independence*, and commanded by Rear Adm. Charles A. Pownall, launched its aircraft against Marcus Island, 1,568 miles from Midway and less than 1,000 miles from Tokyo. In the course of six strikes totaling 275 sorties, the Americans destroyed several Mitsubishi G4M2 bombers (code-named Betty by the Allies) on the ground for the loss of three Hellcats and one Avenger to antiaircraft fire.

The first aerial victory for the Hellcat was scored on the following day when Lt. j.g. Richard L. Loesch and Ensign A. W. Nyquist of VF-6, attached to the light carrier *Princeton*, teamed up to shoot down a snooping Kawanishi H8K2 four-engine flying boat—code-named Emily by the Allies—near Howland Island. Two days later, Lt. j.g. Thaddeus T. Coleman of VF-6 was flying a patrol about fifty miles south-southwest of Baker Island when he spotted another Emily at 1303 hours. A thirty-mile chase ensued, ending at 1325 when Coleman finally sent his quarry crashing into the sea. Coleman's feat was repeated on September 8, while *Princeton* was prowling the Gilbert Islands and encountered another H8K. On that occasion, Lt. Harold N. Funk, executive officer of VF-23, teamed up with Lt. j.g. Leslie H. Kerr Jr. to dispatch the Emily.

Later, Admiral Pownall led another force, built around carriers *Lexington*, *Princeton*, and *BelleauWood*, on a raid in the Gilberts. Between September 18 and 19, seven strikes were made on Tarawa and Makin. Again, four American aircraft were lost; but half of the eighteen Japanese planes present on Tarawa Atoll were destroyed on the ground, another four planes were destroyed on the ground at Makin, and Hellcats of *Princeton*'s VF-23 shot down a G4M2. The Japanese "Water Defense Section" in the Gilberts—three picket boats and minecraft—was wiped out, and a transport ship was sunk in the lagoon. More important were a set of low-oblique photos taken by *Lexington*'s planes of the lagoon side of Betio Island,

which would prove to be useful toward planning for the assault on Tarawa two months later.

Aside from snooping bombers and flying boats, the American carrier air groups had met little aerial opposition thus far, leaving the Hellcat pilots still anxious to see how they would fare against the Zero. Unknown to them, however, that question had already been answered, back in the Solomons, where a unit of land-based Hellcats had already come to grips with the notorious Mitsubishis.

Commissioned at Espiritu Santo on August 15, 1943, VF-33, under the command of Lt. Cmdr. Hawley B. Russell, began its first tour of duty from Henderson Field on August 30. The unit had its first conclusive encounter with Zeros while escorting bombers over Morgusaia Island on September 6, one of the enemy fighters being credited to Ensign James A. Warren. Japanese records do not record a pilot lost as a result of that fight, nor for a subsequent scrap between the Shortlands and Choiseul, in which Lt. j.g. James J. Kinsella claimed a probable victory over an A6M.

VF-33 had its first real air battle on September 14 as part of a mixed force of about two hundred Thirteenth Air Force and Navy bombers, with an escort of Navy F6Fs, Marine F4Us, and P-38s of the Thirteenth Air Force, which was opposed by 258 Zeros. The statistics that emerged from the day's series of confusing melees were inevitably inflated on both sides, the Americans claiming twenty-nine Japanese aircraft (eight by VF-33), while the Japanese 201st *Kokutai* alone claimed sixty, including six probables.

One of the earliest American claims for the day came from a Marine F4U-1A pilot of VMF-222, Maj. Donald H. Sapp, who

A Grumman F6F-3 of VF-33, the first Hellcat squadron to take the measure of the A6M3 Zero, comes to grief in a bomb crater on Barakoma airfield, Vella Lavella, in late 1943. (*US Navy*)

downed two Zeros fifteen miles southeast of Ballale at 0916 hours. The Navy's contribution consisted of seventy-two SBDs and TBFs, escorted by sixteen F6Fs, which left Munda to bomb Ballale. Their arrival stirred up a hornet's nest of A6M3 Zeros at 1315, and ten of VF-33's Hellcats became embroiled in a running dogfight between Ballale and Kahili. Ensign Jack O. Watson downed one of the Japanese, Ensign Frank E. Schneider accounted for another, and Lt. Carlos K. Hildebrandt was credited with three.

"Ken" Hildebrandt's first opponent was attacking one of the retiring Dauntlesses when he got on the Zero's tail. "I poured lead into him and he rolled over on his back smoking, at 200 ft.," Hildebrandt reported afterward. "Tracers went by me then, so I pulled up sharply and collected 7.7mm slugs through the cockpit enclosure. They went into my jungle pack and my back. The Zero turned away as I

turned into [Ensign Jack] Fruin who had another one following him. Firing from 100 yards, I continued through his pull-out and roll. He went in when his port wing was shot off.

"Then I was jumped at 100 ft. by a Zero," Hildebrandt continued. "Using the hand lever to dump my flaps, I saw the Jap go by and pull up in a turn. I just held the trigger down until he blew up. Suddenly the sky was empty. Fruin was nowhere to be seen and I headed home."[5]

Actual Japanese losses for the day totaled five Zeros in the air and another nine aircraft destroyed on the ground. Three of the Zeros were from the 201st *Kokutai*, which recorded the deaths of Petty Officers 2nd Class Eiji Nishida and Hiroshi Mure. The 204th *Kokutai* lost the other two Zeros along with their pilots, Petty Officers 2nd Class Makato Terao and Tokuji Yoshizaki. The top Japanese scorer was Chief Petty Officer Takeo

Okumura of the 201st *Kokutai,* who flew three sorties in the course of the day and was credited with one SBD and a share in the destruction of one B-24, an F4U, two P-40s, and five of the new Hellcats—a one-day Pacific War record for a Japanese pilot. In fact, all of VF-33's planes returned to Munda, although two pilots were wounded and their F6Fs so badly damaged that they had to be written off. In addition to those, Okumura was probably credited with Hildebrandt and Fruin, whose stricken planes may have given the impression of being in more trouble than they actually were—by no means the last time that a Japanese pilot underestimated the rugged Hellcat's ability to bring its pilot home.

On the following day, September 15, Schneider downed another Zero over Ballale for the fourth of an eventual total of seven victories. By then, a second Hellcat squadron had joined the fray—VF-38, which had been formed in June and subsequently joined VF-33 at Henderson Field. At 1115 on September 15, Lt. Oscar Chenoweth Jr. downed a Zero five miles northwest of Ballale, and between 1415 and 1500 hours on the following day, Lt. j.g. Leland B. Cornell of VF-38 was credited with two Zeros and one "probable" north of Ballale.

By the time VF-33 was withdrawn from its first tour on September 21, it had claimed a total of twenty-one Japanese aircraft for the loss of two pilots killed and two injured. James Warren was with the squadron when it resumed operations from Segi Point, and later from Ondonga on the island of New Georgia, but before he could add to his score, he was shot down over New Guinea while flying his fortieth mission on December 23. On the following day, he was taken to a prison camp run by the notorious *Kempei Tai* (military intelligence), where he and

sixty-one other Allied soldiers subsequently died—either by execution, torture, mistreatment, or neglect.

Two days after VF-33 was withdrawn, a carrier unit, VF-12 from *Saratoga,* was stationed at Henderson Field. On September 25, Lt. John Magda, a VF-12 pilot who had seen previous combat during the Battle of Midway, shot down a Zero over Barakoma on the island of Bougainville. Technically, Magda was the first pilot from a shipboard Hellcat unit to be credited with a Zero, although he was not operating from a carrier at the time—VF-12 was reassigned to *Saratoga*'s decks on September 29.

While VF-33 was ending its first tour in the Solomons, the US Navy had assembled its largest carrier force to date: Task Force 14, comprised of carriers *Essex, Yorktown, Lexington, Cowpens, Independence,* and *Belleau Wood,* under Rear Adm. Alfred E. Montgomery. Their next target was Wake Island.

As a result of the increase in American naval activity in the Central Pacific, Wake's air defenses had recently been bolstered by a detachment of Zeros from the 252nd *Kokutai*—the same unit that had drawn first blood in combat with the F4U Corsair back in February. After establishing its headquarters at Roi in the Marshall Islands, the 252nd distributed its sixty fighters among air bases at Wake, Nauru, and Taroa on Maloelap Atoll. Although the Japanese had introduced a more powerful and aerodynamically refined version of the Zero, the A6M3, the veteran pilots of the 252nd *Kokutai* still preferred to use the older A6M2, desperately clinging to its lighter weight and consequent better maneuverability as their only hope against the newer Allied fighters that had taken their measure over the Solomons.

In addition to the 252nd *Kokutai*'s twenty-six-plane fighter detachment led by Lt. Motonari Suho, a veteran credited with eleven victories in China and another four over the Solomons, the 22nd *Koku Sentai* (Air Flotilla) had placed a contingent of twenty-five medium bombers of the 755th *Kokutai* on Wake, comprised mainly of G4M2s, but including a few older Mitsubishi G3M3s, known to the Allies as "Nell."

The Americans launched their first strike before dawn on October 5. "It was not only black dark," exclaimed one pilot later. "There was also a hell of a storm. Every time I thought of a couple of hundred planes rendezvousing in that mess my teeth chattered. I had to kick myself in the pants to get going—and I was fighter skipper!"[6]

The early launch resulted in the first operational losses when several aircraft crashed while trying to form up after takeoff, but it failed to completely surprise Wake's Japanese defenders, who had been warned either by a picket boat or radar. By the time the American planes arrived, they were greeted by heavy antiaircraft fire, while Japanese troops manned their trenches as though expecting an invasion. The 252nd *Kokutai* had twenty-three of its fighters up by 0540, and the carrier-based Hellcat pilots finally got their first opportunity to duel their Zero-flying counterparts.

The first combat of the day involved Ensign Robert Duncan of *Yorktown*'s VF-5, who was flying as wingman to Lt. Cmdr. Melvin C. Hoffman when he encountered a Zero at 0547, sent a burst of fire into its cockpit and saw it go down. He then intercepted another Zero in the act of attacking a Hellcat just ahead of him.

Circumstantial evidence suggests that Duncan's opponent was Warrant Officer Toshiyuki Sueda, who had been credited with one of the first Zero victories, as well as an additional five over China, and three during the Pacific War when he took off on the first flight to intercept the Americans. Informed of who his second victim probably was in 2000, Duncan made some retrospective comments that reveal something about the difference the Hellcat's improved performance made:

> After shooting down the first Zero at Wake, I spotted Lt. (j.g.) Hugh A. Kelley coming up on the tail of a Zero who wasn't aware of him. Nor was Kelley aware of Sueda coming up on him. I turned toward Sueda. I knew that the distance was too far for me to hit him. I felt I had to do something to draw him off Kelley. I hiked up the nose of my plane, much as one would do when trying to water the lawn, and found the hose too short to reach the outer perimeter. My radio was working only intermittently, mostly with a lot of squelch rather than clear. I lobbed a burst at Sueda and he took an immediate defense of doing a wingover and coming at me. I pulled around and in on him where he couldn't get into position to lead me enough to do much damage. He did put a few holes into the fuselage about halfway between the cockpit and the tail. As he recovered I made a hard turn and came up on his tail. As I did he pulled up into the start of a loop. As he did the thought came to me that "this guy must be right out of flight school, he's going to get himself killed trying a stunt like that."

Sueda was anything but green, of course, but realizing that, Duncan attributed his fateful tactic to his unfamiliarity with the new Hellcat as compared to its predecessor, the Wildcat:

The initial climb speed of a Wildcat was 2,190 feet per minute, while the Hellcat's was 3,650 feet per minute. The Wildcat's maximum speed was 318 miles per hour at 19,400 feet; the Hellcat's was 376 miles per hour at 24,400 feet. The Wildcat's service ceiling was 33,700 feet and the Hellcat's was 38,100 feet. The Hellcat, like its predecessor, could outdive the Zero and could out-turn the Zero at the higher speeds, though as both aircraft's speeds began to fall off, the Zero's turn radius began to become more favorable to the Japanese plane. When that began to occur, it was time for the Hellcat to dive out.

It is now my belief that Sueda had either faced the Wildcat previous to the Wake raid, or that he had the tactics he used at Wake drilled into him religiously by Zero pilots who had used those tactics successfully. When he started up into the loop, we both had a lot of speed to work with. I had picked up close to maximum speed in trying to catch up to him, as he was getting ready to open up on Kelley. I feel certain that he mistook my Hellcat for a Wildcat and his plan was to arrive at a speed near stalling at the top of the loop, where a Wildcat would fall off into a spin. At that point, he would bring his plane around and dive down onto his opponent, who lost control of his craft. The only problem was that a Hellcat wasn't going to stall; instead, it would follow his Zero on around and when he looked back to see the Hellcat spinning out of control, he would find it right on his tail.

In our actual fight I didn't follow him completely to the apogee or zenith. I caught him just before he got to that point. I strongly believe that the way I have outlined it is how it occurred. Sueda was too experienced a pilot to have taken the action he did had he known the

specifics of the Hellcat. Probably, had he been successful enough to get through Wake alive and had time to learn more about the Hellcat, he might very well have come through World War II alive. Lastly, Kelley shot down the Zero ahead of him. He had watched me shoot down my first Zero earlier, but wasn't aware that Sueda had almost had him for breakfast.[7]

"Boogie" Hoffman accounted for a Zero of his own, while VF-5's squadron leader, Lt. Cmdr. Edward M. Owen, downed another Zero about half an hour later. Yet another Zeke kill in the first dawn encounter was claimed by Lt. George C. Bullard, a veteran of the desperate carrier battles of 1942 who was now leading a twelve-plane detachment of VF-6 from the light carrier *Cowpens*.

Lieutenant Commander Philip Torrey's VF-9 from the carrier *Essex* encountered enemy fighters at about 0600. First blood for that unit was drawn by Lt. j.g. Hamilton McWhorter III, in a display of tactics which led to his being known as "One Slug" for the rest of his naval career. As he reported it:

> Suddenly, I was aware of Zekes all around me. There must have been twenty. I didn't see any of my gang around so I headed to join up with three planes from another carrier. Just then it happened: there he was, sitting in my sights. So I let go. Just one burst. That was all. And I had my first Zeke. It was almost funny. . . . You just sat back, pressed the button and he blew up and wasn't there anymore.[8]

McWhorter was credited with one Zero destroyed and one probably shot down. Another VF-9 pilot, Lt. j.g. Mayo A. Hadden Jr., was a little less fortunate, shooting down the rearmost of two

Zeros—though he'd aimed at the leader—and then spending too much time watching his victim explode upon crashing into the sea until he found himself under attack by the flight leader he'd missed. After bore-sighting the Hellcat with 7.7mm rounds, the Zero pilot added five 20mm cannon shells, riddling both wings, the right elevator and stabilizer, blowing away ten inches of rudder, shattering the oil tank and wounding Haddon's leg with shell fragments. Unable to outturn his assailant, Haddon remembered to use the F6F's superior roll rate by whipping around in a 180-degree turn and then escaping in a high-speed dive. Although he was undoubtedly counted in the Zero pilot's tally, Haddon made it back to *Essex* and even circled while less badly damaged planes landed before coming in with no flaps and no hydraulics to catch a wire—with just a gallon and a half of fuel left. Among the first of numerous pilots who could attribute their survival to the Hellcat's phenomenal durability, "Mike" Haddon would go on to score seven more victories, placing him among twenty aces produced by VF-9.

Lexington's VF-16, commanded by Lt. Cmdr. Paul D. Buie, ran into its first aerial opposition at 0610. Buie himself shared in the destruction of one Zero, while Lt. j.g. Alfred L. Frendberg destroyed one and damaged another. Lieutenant William E. Burkhalter got another Zero, while Ensign John W. Bartol destroyed one in flames after a head-on gun duel.

By the end of the early morning raids, the first and second waves of American fighters were claiming to have encountered thirty-three Zeros and to have downed twenty-seven of them, while Task Force 14 analysts stated that none of the American aircraft lost in that first strike

had been downed in air-to-air combat. Such claims were exaggerated, but the truth stood on its own merits, to mark a most auspicious debut for the carrier-based Hellcats. Of the twenty-three Zeros it sent up, the 252nd *Kokutai* lost fifteen, and three of the eight pilots who returned were wounded. Strafing and bombing eliminated eight more Zeros on the ground, as well as nineteen of the 755th *Kokutai's* twenty-five G4M2s and G3M3s, virtually eliminating it as a threat to the carriers. The jubilant Americans predicted "no further air opposition from Wake-based planes," but that boast would prove to be somewhat premature.[9]

Independence launched her first combat air patrol (CAP) of the day at 0615, while Rear Adm. Ernest G. Small's Southern Bombardment Group—comprised of heavy cruisers *Minneapolis*, *New Orleans*, and *San Francisco*, and three destroyers—left the task force to punish the Japanese further with their guns. A second flight of eight F6Fs of VF-6 left *Independence* at 0915 with orders to cover Small's group until relieved, then strafe targets of opportunity on Wake before returning.

The cruisers were about to take up station off the southern coast of the island at 1145 when *Minneapolis's* radar detected several bogeys (unidentified aircraft) twenty-nine miles to the west, climbing as if they had just taken off from Wake. The cruiser's fighter director officer, Lt. j.g. Nelson H. Layman, relayed the information to the VF-6 leader, Lt. Cmdr. Edward H. O'Hare, and he led his Hellcats to investigate.

"Butch" O'Hare was already a US Navy legend, becoming its first ace in dramatic fashion by hurtling into a formation of G4M1s that were on their way to attack his carrier, *Lexington*, on February 20, 1942, and shooting down five of them in as many minutes. O'Hare was awarded the Medal of Honor and toured the United States to raise public morale before returning to duty training—and inspiring—a new generation of Navy fighter pilots. Ill at ease with resting on his laurels, he was glad to be back in the combat zone as commander of VF-6, and eager to resume the job he had enlisted to do—as a fighter pilot.

As Layman continued to direct the eight Hellcats toward their quarry, O'Hare's keen eyes spotted the bogeys at 1205—three brownish-green Zekes in a loose V formation, heading back toward Wake. The hapless trio had been sent up at 1130 and consisted of three survivors of the morning fracas—all wingmen and petty officers second class—Yasuo Matsumoto leading Magoichi Kosaka and Kazuo Tobita.

As O'Hare led his wingman, Ensign Henry T. Landry, into position for a quick hit-and-run ambush of the Zekes, he was closely followed by Lt. j.g. Alexander Vraciu, whose radio was malfunctioning, and Vraciu's wingman, Ensign Allie W. Callan Jr. Lieutenant Sy E. Mendenhall, leading three other VF-6 Hellcats also watched O'Hare intently, since his radio was also not working. Making a diving turn to catch his target from above and to the right, O'Hare sent a stream of .50-caliber bullets into the cowling and cockpit of Kosaka's Zero. Kosaka, either wounded or killed in that first pass, slumped forward, and his smoking plane nosed downward. O'Hare then turned slightly to nail the lead Zero, but Matsumoto had been alerted by the sudden sight of tracers whizzing through his formation and took violent evasive action. O'Hare tried to follow, but his F6F-3 was going fifty miles per hour faster than the A6M2. Unable to stay with Matsumoto's turn, O'Hare wisely climbed to position himself for another

pass at the Zero, while his inexperienced wingman, Hank Landry, unwisely continued to pursue Kosaka's descending Zero.

Meanwhile, Vraciu had attacked Tobita's Zero from above and to the right. The Zero's engine smoked, then burst into flame, and Vraciu had to pull up hard to avoid a midair collision. Suppressing his elation, Vraciu climbed to rejoin O'Hare but lost track of Willie Callan, who, like Landry, was probably attacking Kosaka's already doomed plane. Mendenhall, following Callan, also got on the tail of a burning Zero—Vraciu's victim—and noticed another one on fire when Matsumoto's Zero suddenly flashed by in a wingover and disappeared into a cloud. Mendenhall then briefly flew alongside Kosaka's Zero before seeing it go into a shallow dive with flames streaming from the engine.

Thinking his firing pass had accounted for the Zero, Callan later said that he was "ready to shoot anything that moved," and went after a fighter he saw below him. He put one bullet hole through its rudder before he realized to his horror that it was Mendenhall's Hellcat. Once they had reestablished themselves as mutual "friendlies," Mendenhall and Callan climbed to find the remaining Zero, but in the meantime Matsumoto had discovered Landry's F6F and dived on it. Landry reacted with an evasive roll, causing Matsumoto to overshoot and recover in front of him. Landry then opened fire and saw the Zero roll over on its back and go into what the American thought to be a fatal dive. Moments later, Landry again found tracers coming at him from above, some of which struck his rear fuselage behind the cockpit. As he turned and pulled up to engage his antagonist, Landry was astonished to discover it to be an equally shocked Butch O'Hare.[10]

Callan had rejoined Vraciu when the latter noticed a Zeke racing toward Wake. Both men roared down and chased the enemy fighter all the way down to the runway. Matsumoto managed to land, and as soon as his plane had sufficiently slowed down, he taxied off the runway onto the sand and hastily exited the cockpit—bare moments before Vraciu set his Zero ablaze. Vraciu then noticed a Betty on the ground, and he and Callan came around to destroy that, too, before heading north back to *Independence*. O'Hare and Landry, also diving in pursuit of Matsumoto, contented themselves with shooting up ground installations and riddling another Betty on the field. They then returned to their station over the cruisers, with O'Hare reporting to Layman: "Tally ho, shot down two, other one not sure."[11]

The day was not yet over for O'Hare. At 1220, *New Orleans*'s radar picked up a large bogey ninety-five miles to the southeast, and Layman dispatched O'Hare and Landry to intercept it. About twenty miles south of Wake, Butch recognized the now-familiar profile of a Betty and attacked it head on. Only one of his guns was working, but it scored hits on an engine and the wing root. Landry let the G4M pass, then turned to make a high-side pass at one of its engines—only to end up directly behind the bomber, trading shots with its rear gunner, who was armed with a 20mm cannon. O'Hare, meanwhile, had climbed back into an attacking position and made a second pass, finishing off the bomber. Then, while the cruisers began bombarding Wake, he and Landry, their fuel and ammunition nearly exhausted, returned to *Independence*. O'Hare's victim was one of two G4M2s of the 755th *Kokutai*, piloted by Warrant Officer Godai Kawano and Chief Petty Officer Nobukichi Wakizara,

which had left Maloelap that morning to search for the American task force. Neither plane made it to Wake, the other bomber apparently falling victim to Lts. John R. Behr and James H. McConnell, and Lts. j.g. James A. Bryce and Donald C. Stanley of *Independence*-based squadron VF-22 at 1354.

During the debriefing, O'Hare determined that he and Callan had mistaken Landry's and Mendenhall's F6Fs for Japanese aircraft because of the recent painting over of the upper right and lower left insignias on the wings of all of VF-6's planes, leaving nearly circular blotches of shiny paint that looked like red *hinomarus* when the sun glinted off them at the right angle. O'Hare also asked to share his Betty with Landry despite his wingman's protests that he had contributed nothing to its destruction. Ultimately, O'Hare was given sole credit for the victory—his seventh—and was awarded the Distinguished Flying Cross for his activities that day. For Alex Vraciu, Tobita's Zero was the first of an eventual nineteen victories, including six on June 19, 1944, which would make him the US Navy's fourth-ranking ace.

Another, more serious, case of mistaken identity may have resulted in the loss of a Curtiss SOC Seagull floatplane from *New Orleans*, which was shot down at 1312 by two fighters. Its crew, Lt. Alford M. Robertson and Aviation Radioman 1st Class George W. McCarthy, bailed out, and although strafed and wounded as they parachuted seaward, both airmen were rescued by the destroyer *Schroeder*—which was fortunate, indeed, when a postwar examination of Japanese records revealed that the crippled 252nd *Kokutai* had no Zeros in the air at that time. At 1330, Lt. Clement M. Craig of *Independence*'s VF-22 was leading a sweep northeast of Wake when he reported encountering a "Dave"—the Allied designation for the Nakajima E8N1, an obsolescent reconnaissance float biplane. Thus, the first of an eventual wartime total of 11 3/4 victories for Craig was probably over an American plane.

With Wake under attack, the 22nd *Koku Sentai* dispatched reinforcements from Maloelap, comprised of seven G4M2s of the 755th *Kokutai*, led by Lt. Cmdr. Kaoru Ishihara, and an escort of seven Zeros of the 252nd *Kokutai*, led by Lt. Yuzo Tsukamoto. After a grueling six-hundred-mile flight, the Japanese were about forty miles south of their goal when they were attacked at 1515 hours by four F6Fs from VF-6's *Belleau Wood*-based contingent led by Lt. j.g. Harvey G. Odenbrett. Tearing through the bomber formation, the Hellcat pilots claimed one of them but lost Ensign Edward L. Philippe.

Meanwhile, other American fighters were up dealing with what little air activity remained around Wake. Ensign Cyrus J. Chambers, joined by two colleagues of VF-6 from *Cowpens*, accounted for a G3M at 1456 hours. At 1505, O'Hare led three F6Fs from *Independence* and, upon learning of the Japanese reinforcements coming from the Marshalls, tried to join the running fight. After much searching, however, he only made a positive sighting at 1556—three Zeros fleeing westward at eighteen thousand feet, too far away to overtake.

Judging it impossible to get through to Wake, Ishihara ordered his bombers to turn back for Maloelap. He wanted the Zeros to land on Wake, but he could not communicate with them because their pilots had discarded their radios to lighten their planes. The retiring flight was about a hundred miles away from the island at 1620 when it was attacked by eight more Hellcats from *Belleau Wood*.

The Americans claimed three Bettys and three Zekes; in fact, they downed two of each. The two slain Zero pilots included Chief Petty Officer Bunkichi Nakajima, a veteran of numerous actions in the Solomons who was credited with sixteen American aircraft. The remaining Japanese bombers returned to the Marshalls in various shot-up states that evening, one being forced to ditch offshore, as did one of the Zeros. Tsukamoto managed to lead four Zeros to Wake through the cordon of American fighters, landing at 1730 and taxiing carefully down a runway pockmarked with bomb craters. Another Zero managed to make it back alone; its lucky pilot, Chief Petty Officer Isamu Miyazaki, would survive the war with thirteen victories to his credit.

The carriers struck at Wake again at 0640 in the morning of October 6, during which Lt. j.g. Eugene A. Valencia of *Essex*'s VF-9 damaged a Zero. It was not confirmed as shot down, but Valencia would subsequently be credited with twenty-three, making him the third-ranking US Navy ace. The Americans flew two more follow-up strikes on Wake, adding greatly to the destruction inflicted on the previous day. The only unusual event occurred when *Independence*'s catapult malfunctioned, throwing Cy Mendenhall's Hellcat into the drink in front of the carrier. Mendenhall managed to get out of the sinking plane and was rescued by the destroyer *Schroeder*.

Wake also saw the first successful use by the US Navy of submarines to rescue downed airmen. Prior to the raid, Admiral Pownall had asked Vice Adm. Charles A. Lockwood, Commander Submarines Pacific Fleet (ComSubPac), for the use of some of his vessels for lifeguard duty. As a result, *Snook* had stood off Marcus on September 1, and *Steelhead* had lain off Tarawa on September 20; but American losses had been few during both raids, and neither submarine got a chance to perform its assigned task.

At dawn on October 5, *Skate*, under Cmdr. Eugene B. McKinney, surfaced off Wake to support Task Force 14's strike. She was attacked several times by Japanese aircraft and shore batteries, and one of her officers, Lt. j.g. Willis E. Maxon III, was mortally wounded, but she rescued six downed airmen and proved the value of submarine "lifeguarding," which became standard doctrine for carrier strikes beyond the range of rescue planes. The effect her activities had on the morale of the carrier aircrews was summed up in a radio message that McKinney subsequently received from *Lexington*'s Capt. Felix B. Stump: "Anything in *Lexington* is yours for the asking. If it is too big to carry away, we will cut it up in small parts!"[12]

As Task Force 14 retired to the northwest on the evening of October 6, Admiral Montgomery felt satisfied with its results. Over the past two days, his force had made six strikes totaling 758 sorties, which had landed 340 tons of bombs on Wake—three times as much as had been dropped on Marcus—as well as 520 tons of shells. A force of PB4Y bombers also contributed to the carnage. The overall result was extensive damage to the island's fuel, water, and ammunition storage facilities, and the destruction of some sixty to seventy buildings. A gasoline-loaded tanker was also blown up in the lagoon.

As to Japanese planes, twenty-two (out of sixty-five claimed) were destroyed. Among the 252nd *Kokutai*'s losses were two aces: Sueda—who was posthumously promoted to the commissioned rank of ensign—and Nakajima. Other Japanese fatalities included Chief Petty Officers

Yukuo Miyauchi, Hisashi Hide, and Kazuo Tobita; Petty Officers 1st Class Saburo Fujiuma, Tamotsu Okabayashi, and Soyo Shibata; and Petty Officers 2nd Class Katsunobu Shiba, Yoshio Shiode, Kiyoshi Takei, and Kazuyoshi Tokuhara. The Japanese, in turn, claimed fourteen American aircraft shot down. In fact, Task Force 14 lost ten Hellcats and two Avengers, but the majority of them had fallen to antiaircraft fire or accidents, rather than in aerial combat.

After the Americans departed, Suho, and Wake's other surviving Zero pilots, flew back to Roi aboard a bomber. With only twelve flyable aircraft left on Wake, the Japanese had to send fresh reinforcements from the Marshalls. They would need all they could muster, for the Wake raid was only a harbinger of what was to come. On November 23, 1943, the Americans would return to the Central Pacific—not to raid its far-flung island bases, but to begin the series of invasions that would eventually carry them to the very heart of the Japanese empire.

In the vanguard of that advance would be the Grumman Hellcat, which, having achieved an ascendancy over the Zero in its first combats, would continue to wrest and maintain air superiority for the US Navy, often in spectacular fashion, during all the island campaigns to follow. By the time Japan formally surrendered on September 2, 1945, Hellcats had been credited with 5,200 Japanese aircraft and had produced 307 aces, the most for any single American fighter type.

Sixteen

Heavy Hitters
Cannon-Armed Fighters, 1940–44

The French first made the cannon a practical aerial weapon in 1917, when their Spad 12.Ca1 successfully fired a 37mm gun through the hollow propeller shaft of its geared Hispano-Suiza engine. Although its effectiveness was limited by the fact that its weapon had to be reloaded by hand after each shot, the basic arrangement which the Spad 12 pioneered was later used more successfully in fighters that fired rapid-fire cannon through their propeller shafts, such as the Messerschmitt Me 109 and Bell P-39. Cannon were also mounted on twin-engine fighters such as the Messerschmitt Me 110 and the Lockheed P-38, and advances in cantilever structure allowed them to be mounted in the wings of fighters such as the Supermarine Spitfire Mark IIb and Mitsubishi A6M2 Zero.

As such armament became practicable, the urge inevitably arose to see how much punch could be packed into a single airplane. France certainly pursued that goal, and by May 10, 1940, it had five *groupes de chasse* equipped with a new cannon-armed fighter, the Bloch MB.152, while another three groups were in the process of reequipping with it.

Originally conceived by Maurice Roussel in 1935 and developed as a private venture until finally accepted by the *Armée de l'Air*, the MB.152 was powered by a 1,080-horsepower Gnome-Rhône 14N49 radial engine, which gave it a maximum speed of 310 miles per hour at 18,045 feet. The plane was powerfully armed with two 7.5mm machine guns and two 20mm HS 404 cannon mounted in the wings, and they would take their toll on the bombers that swarmed over the border on that first day.

May 10 saw a Heinkel He 111 claimed over Calais by *Sgt.-Chef* Jean Honorat of GC II/8 in concert with an MS.406 flown by *Sgt.* Albert Durand of GC III/1, and another by *Sgt.* Amaury Montfort of GC II/1. It was GC I/8 that did the most, however. As Heinkels of *Kampfgeschwader* 55 swept over the Lorraine region to bomb Nancy, Toul, and Epinal, *Sous-Lt.* Robert Hollon and *Sgt.-Chef* Henri Liautard destroyed an He 111 of the II Gruppe (II./KG 55) near Château-Salins. Leading the *2ème Escadrille* into action, *Capitaine* Adrien Astier downed a Do 17 near Mars-la-Tour, but was himself shot down and killed soon afterward. *Sergent*

Left, Bloch MB.152 No. 249 of 3ème *Escadrille, Groupe de Chasse* II/8 at Marignane after the armistice of June 18, 1940. Note shark marking under the radio antenna. (Collection Bernard Régnier via Chrstophe Cony) Right, Back from a sortie in which *Sgt.* René Darlay shot down a Messerschmitt Me 110C over Pont-à-Mousson, Bloch MB.152 544 No. 10 of 1ème *Escadrille, Groupe de Chasse* I/8 has its own damage attended to at Vélaine-en-Haye on May 10, 1940. (*A. Trégouët via Christophe Cony*)

Darlay was credited with an escorting Me 110, which was probably the one lost over Rethel, southwest of Charleville, by 4th *Staffel, Zerstörergeschwader* 26 (2./ZG 26), along with crewmen *Fw.* Hannes Reimann and *Obergefreiter* (Leading Aircraftman) Heinrich Röwe.

In spite of the loss of its squadron commander, the *2ème Escadrille* of GC I/8 was up again the next day to intercept another wave of bombers, one of which, an He 111 of I./KG 53, fell victim to *Sous-Lts.* Robert Thollon and Augustin Flandi, aided by two MS.406s of GC II/7. Again, however, any cause the Bloch pilots had for celebrating was dampened by the loss of Flandi, who was killed in the course of the action.

On the Flanders front on May 14, GC II/8 downed two more Heinkels: one over Escaut, shared between *Adj.-Chef* Henri Mir and *Sergent-Chef* Goubert, and the other at Ostende, credited to *Adj.* Joseph Clerc. The more intense activity, however, shifted to Lorraine, where the French were realizing that Gen. Heinz Guderian was launching a major offensive across the Meuse River, a move that would ultimately split the Allied defenses. GC II/1's MB.152s were up in force,

with *Adj.-Chef* Ernest Richardin destroying a Junkers Ju 87B over Sedan, and *Sous-Lt.* Victor Belland and *Sgt.-Chef* Michel Rouquerbe accounting for another. *Capitaine* Albert Coiral and *Lt.* Philippe Maurin caught a Henschel Hs 126B of 1st *Staffel (Heeres), Aufklärungsgruppe* 14 (1.(H)/14) scouting for the 2nd Panzer Division and shot it down over Vrigne-au-Bois. Its crew bailed out and parachuted safely in German-held territory.

Over Belgium that same day, GC II/1 had its first encounter with German fighters, during which *Sgt.* Jean Robert claimed two Me 110s over Dinant. *Sous-Lieutenant* Victor Belland, *Sgt.-Chef* Pierre Patoor, and *Sgt.* René Brisou teamed up to down another, and *Lt.* Pierre Matras got a fourth, against which claims *Zerstörergeschwader* 26 lost two Me 110s with their crews, and a third was damaged. Brisou also claimed an Me 109. *Capitaine* Coiral, however, did not return from the sortie.

Taking off at 1215 hours on May 14, GC I/8's MB.152s were escorting Amiot 143s of *Groupe de Bombardment* I/34 over Sedan at about 1300 when they too were accosted by Me 109Es. *Sous-Lieutenant* Pierre Gouachon-Noireault claimed one of the Messerschmitts before his oil tank was shot up, and he was forced to belly-land alongside the road between Tahure and Suippes. *Sergent* H. Choulet's plane also suffered damage and he force landed at Bétheniville.

The Me 109Es involved in both actions were from *Jagdgeschwader* 53, while the Me 110s were from 1/ZG 52 which lost two planes. The confused nature of the overall fighting might be deduced from the fact that, between 1130 and 1235 hours that day, "Bloch 151s" "south of Sedan" were credited to *Hptmn.* Rolf Pingel, *Ltn.* Alfred Franke, *Ltn.* Rudolf Schmid, and *Uffz.* Willi Ghesla of 2./JG 53. Another Bloch, credited to *Ltn.* Julius Haase of 3./JG 53, probably accounts for the death of *Capitaine* Coiral.

Subsequent encounters with enemy fighters proved the MB.152's overall inferiority to the Me 109E, as *Sgt.* Edouard Courteville of GC II/9, who survived the war after sharing in the destruction of an Hs 126 of 5.(H)/13 on June 10, later recalled:

> From our first melees with the Messerschmitt Bf 109Es we were aware that our Bloch 152s needed two or three hundred more horsepower. Our opponent was faster and could outclimb us, and we had good cause to be grateful for the Bloch 152's small turning radius and ability to accept heavy punishment. Many a Bloch fighter staggered back to its airfield full of holes but with its pilot unscathed. We frequently complained to the Gnome-Rhône technical staff attached to our *Groupe de Chasse* that their 14N49 engine, admittedly reliable, was giving us insufficient power, but their answer was invariably the same: 'Just wait until the 14R engine comes along and then you will see something!'[1]

Unfortunately for Courteville and his squadron mates, they could no longer wait.

On the other side of the Channel, Britain had also been striving to up-gun its fighters with the 20mm Hispano cannon. Indeed, the Hawker Hurricane Mark IIC doubled down on the formula with no less than four wing-mounted weapons, but their weight handicapped the performance of a fighter that was already becoming outdated, though Hurricane IICs did yeoman service in the night intruder and ground-attack roles. Meanwhile, the Royal Air Force had been working apace on more specialized—and, it was hoped, more advanced—cannon-armed fighters.

The first such airplane had, in fact, sprung from Specification F.37/35, issued by the Air Ministry for a fighter to mount four 20mm cannon in 1935, just a year after F.5/34 had requested a fighter to be armed with eight .303-inch machine guns. One of the companies that strove to meet the requirement was Westland, seeking a chance to produce its first fighter for the RAF. William Edward Willoughby Petter directed the design team that settled on a twin-engine format with all guns in the nose, using then-new 860-horsepower Rolls-Royce Peregrine I V-12 engines. To reduce drag, Petter decided to place the radiators in the wing, inboard of the two streamlined engine nacelles. The slim monocoque fuselage was skinned aft of the cockpit with magnesium sheet, which had a better strength-to-weight ratio than aluminum, and the plane's one-piece rearward-sliding glazed cockpit canopy was also ahead of its time. Fowler flaps extended under the fuselage and even included the rear fairings of the engine nacelles. A twin-tailed arrangement was originally envisioned for the plane, but wind tunnel testing resulted in a single fin and rudder with a distinctive high-mounted tail plane.

Completed in the late summer of 1938, the Westland P.9, soon christened the Whirlwind, had its first flight on October 11. So radical a plane mated to so new an engine inevitably suffered from development problems, but in January 1939 an order for two hundred Whirlwinds was placed, with the first machine being delivered in June 1940. The first planes were earmarked for a night-fighter unit, No. 25 Squadron, but then the Air Ministry decided that that unit's experience was better suited for the Bristol Beaufighter. Instead, the new fighters were delivered to No. 263 Squadron, which, after flying Gloster Gladiators in Norway, had lost all its planes when the carrier *Glorious* was sunk on June 8, and was in the process of reorganizing when its intended replacement aircraft were switched from Hurricanes to Whirlwinds. In July the squadron commenced training at North Weald and Drem—much of the time being spent wrestling with numerous difficulties with both their Hispano Mark I cannon and the Peregrine engines—then they moved to Exeter and resumed combat operations.

The new cannon-armed fighter did not show much hint of fulfilling its promise until January 12, 1941, when two Whirlwinds, operating from Saint Eval in Cornwall, attacked a Junkers Ju 88A southwest of the Scilly Isles. After silencing its rear gun, Pilot Officer David Stein and Sgt. Dennis W. Mason saw the bomber go into a spiral dive into the clouds but were unable to witness its demise, and it entered 263 Squadron's records as "damaged." On the following day, the Whirlwinds encountered a

Westland Whirlwind P6969 HE-V of No. 263 Squadron was flown by Pilot Officer K. A. G. Graham on February 8, 1942, when he destroyed an Arado Ar 196 floatplane—at the cost of his own life. (*Imperial War Museum*)

Heinkel He 111 and pursued it until they ran low on fuel.

The next opportunity came at 0900 hours on February 8, when two Whirlwinds encountered an Arado Ar 196A of *Bordfliegerstaffel* 5./196 about twelve miles south of Start Point. The first Whirlwind pilot, Sgt. Clifford P. Rudland, came down on the floatplane's tail, but then thought he saw British roundels on the fuselage and broke off his attack, actually flying alongside the other plane until it vanished in a cloud. The floatplane turned up again and was attacked by the second Whirlwind pilot, Flying Officer Joseph S. Hughes, who closed to two hundred yards but failed to achieve a hit. Meanwhile Flt. Lt. David A. C. Crooks, a Canadian in RAF service, and Pilot Officer K. A. G. Graham had taken off from Saint Eval at 0906 and began patrolling south of Dedman Point, only to lose contact with one another in the clouds. Crooks saw Graham pass below him, heading west, but by the time he turned, Graham had disappeared again. Crooks then saw an enemy plane emerge from the cloud, descending to the northeast until it hit the water. Overflying the crash scene, Crooks noticed two floats and a piece of wing marked with a black cross. British Coast Guard observers subsequently reported seeing two aircraft, the first of which was in flames, plummet into the sea three miles offshore at 0950. Graham was credited with the Arado, but 263 Squadron's first confirmed victory had come at a grim price: Graham, who had been flying the fourth production Whirlwind (P6969 HE-V), was also killed, either in a head-on gun duel—the Ar 196 packed two wing-mounted 20mm MG FF cannon of its own—or by a collision in the cloud.

Whirlwinds damaged Ju 88As on March 1 and 5, but on the eleventh Pilot Officer Herbert H. Kitchener was wounded in a fight with a Ju 88 south of the Lizard, after which he crash-landed, seriously injuring himself and burning up his plane. The Luftwaffe struck back at Saint Eval at 2055 hours on March 12, when an He 111 dropped four bombs on the field and succeeded in damaging seven Whirlwinds and seven Hurricanes—three of the latter being write-offs. Ill fortune continued to attend 263 Squadron on March 14, as Pilot Officer Patrick Glynn Thornton-Brown crashed Whirlwind P6973 upon landing. On April 1, Squadron Leader Arthur Hay Donaldson and Flight Lieutenant Crooks attacked a Do 215 five miles north of Predannack, but as Crooks moved in to finish off the fleeing bomber its gunner scored a telling hit on his Whirlwind, P6989, which crashed in flames near Helston. Although damaged, the Dornier escaped.

Soon afterward, No. 263 Squadron's pilots were told that in order to take the most advantage of their armament, their Whirlwinds would be used primarily to make pinpoint attacks on German airfields on the other side of the Channel—a somewhat ironic reprisal for what the Luftwaffe had already done to them. The first such mission was launched on June 14 against two Me 109 bases at Maupertus and Querqueville on the Cherbourg peninsula, but poor visibility prevented this from being any more auspicious a start than that of the Whirlwind as a day fighter. Further attacks, called Warheads, were more successful in July and August, and in the latter month Whirlwinds escorted a daylight bombing mission to Cologne.

In late September, members of No. 263 Squadron were withdrawn to form the nucleus of a second Whirlwind unit, No. 137 Squadron, at Colerne. The two squadrons spent the rest of their fighting careers in the low-level strike role, particularly against German shipping in the Channel. Their destructive efforts were aided by the later provision of underwing racks for two 250- or 500-pound bombs, in which form their planes were designated Whirlwind Mark IAs or, unofficially, "Whirlibombers."

The Whirlwind, although liked by its pilots and which was a reasonably good fighter at low altitude, fared worse at altitudes above 15,000 feet, where its speed peaked at 360 miles per hour. In spite of its generous flaps, the Whirlwind's landing speed remained a "hot" 80 miles per hour, handicapping its ability to operate from grass airstrips. Its main problem, however, was the RAF's pressing need for Rolls-Royce Merlin engines, production of which was being interfered with by the parallel production of Peregrines. Moreover, by the time the Whirlwind finally became operational, cannon-armed fighters such as the Hurricane IIC and Typhoon Mark IB were available to do their job with just one engine. In consequence, only 112 Whirlwinds were built before the original 400-plane order was canceled. In June 1943, No. 137 Squadron was reequipped with Hurricane

Mark IVs, and No. 263 relinquished its last Whirlwinds for Typhoons in November.

The Typhoon's genesis began before the end of 1935, when the Air Ministry's Specification F.37/35 was followed by another, F.18/37, calling for a single-engine fighter armed with four cannon. Hawker's Sydney Camm, who had already proposed such armament, needed no prodding to get the project under way, though the principal obstacle to making such an arrangement work on what was essentially an enlarged Hurricane was in finding a sufficiently powerful engine. Prototypes were built to use several new engines that promised 2,000 horsepower, with two water-cooled power plants being selected for final consideration: the twenty-four-cylinder Rolls-Royce Vulture—essentially two Peregrines arranged in X form around a common crankshaft—powering the Hawker Tornado; and the twenty-four-cylinder flat-H sleeve-valve Napier Sabre installed in the Typhoon.

The Tornado first took to the air on October 6, 1939, but exhibited a high rise in drag at speeds above 400 miles per hour. This was traced to uneven airflow around its ventral radiator, which was repositioned directly under the engine cowling. The first Typhoon, powered by a Sabre I, made its first flight on February 25, 1940. By then the exigencies of war gave the Air Ministry impetus to order 1,000 of the two new fighters, but problems with both the Sabre and Vulture engines held up development. By early 1941, persistent failures of the Vulture's connecting rod bolts led to its abandonment along with the Tornado, while work proceeded apace on both the Sabre engine and the Typhoon. Since Hawker was still committed to producing the

proven and still much-needed Hurricane, Gloster Aircraft was selected to build the Typhoon Mark IA, armed with twelve .303-inch machine guns in the wings, and the Mark IB, with four 20mm Hispano cannon.

After six difficult months of further development, in September 1941 the first Typhoons began to leave Gloster's assembly line and were delivered to the Air Fighting Development Unit and No. 56 Squadron at Duxford. Operational evaluation revealed the new fighter to be still far from ready for action, however. The view from the cockpit—entry to which in early Typhoons was made through a car door on the left side, as on the Bell P-39 Airacobra—was poor, carbon monoxide seeped in through the engine firewall, and the cannon feed mechanism was unreliable. Hawker had projected a maximum speed of 464 miles per hour for the new machine, but the early aircraft barely achieved 400—though they did have the distinction of being the first production fighters in the RAF to achieve that speed—and at its best the Typhoon never exceeded 412 miles per hour. On top of all that, the thick wing section and high wing loading of the Typhoon gave it a poor rate of climb, and performance dropped off at high altitude. Tail-plane flutter, and structural weakness in the fuselage just forward of the tail, also plagued the early Typhoons. There was a growing wave of opinion within the RAF to abandon the whole Typhoon program, but Hawker labored on to rectify the shortcomings of its "ugly-looking beast," while RAF Fighter Command put No. 56 Squadron's planes, along with those delivered to No. 266 and 609 Squadrons, to work patrolling the Channel against hit-and-run *Jagdbomber*, or *Jabo*, raids by Me 109Gs and Focke-Wulf Fw 190A-4/U4s of *Jagdgeschwader* 2 and 26.

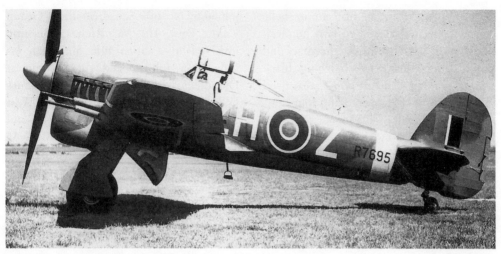

Hawker Typhoon Mark IB R7695 ZH-Z of No. 266 Squadron at Duxford came to a tragic end when it broke up and crashed at Glanvilles-Wooton, Dorset, on October 24, 1942. (*Andrew Thomas*)

The Typhoon's operational career began on May 28, when one of No. 266 Squadron's planes scrambled up after an unidentified intruder and intercepted the "bogey," which turned out to be a Spitfire. The first squadron-sized operation by the Typhoons commenced at 1515 hours on June 20, 1942, as Wing Cmdr. Denys E. Gillam led Typhoons of Nos. 56 and 266 Squadrons from Duxford, accompanied by the station commander, Group Capt. John Grandy. Climbing to twenty thousand feet, the Typhoons crossed the French coast at Gravelines at 1538 and turned south, sweeping the coast from Calais to Boulogne in hopes of encountering German fighters. Some unidentified planes were seen in the distance south of Boulogne, but no contact was made, and all fighters had returned to Duxford by 1645, the only problem being encountered by Sgt. Norham J. Lucas of 266 Squadron, who had to land almost immediately due to hydraulic trouble. Eight Fw 190s were encountered during the next offensive sweep on June 23, but the Germans chose not to engage the new fighters.

Spitfire pilots seemed to have had no such compunctions, as two Canadians downed what they identified as Fw 190s in the Channel, only to learn that they were Typhoons. As No. 56 Squadron's Typhoons were escorting Douglas Bostons on a bombing raid to Abbéville on July 30, a Norwegian squadron member, Flying Officer Erik Haabjoern, dropped out with engine trouble, but on his way home he was shot down by Spitfires, fortunately surviving to be rescued from his dunk in the Channel.[2]

Patrols continued to be fruitless until August 9, when No. 266 Squadron's Pilot Officers Ian M. Munro, in a cannon-armed Typhoon IB, and Norham Lucas, in a machine gun-armed Mark IA, engaged in a sea patrol, spotted an aircraft approaching at 2045 hours, and as it passed under them they identified it as a Ju 88. The Typhoon pilots turned to engage, Munro lining himself astern of the German and Lucas approaching from the port side. The German gunners opened up with inaccurate fire, while their pilot started weaving, which only brought the plane dead ahead of Lucas.

Munro opened fire, saw his cannon shells strike the water, closed to two hundred yards and fired another three-second burst. Lucas, his plane belching black smoke as he gave it full boost, also fired short bursts at six hundred and four hundred yards, then loosed a longer one as he closed to two hundred yards. At that point, flames appeared inboard of both of the Ju 88's engine nacelles. A German crewman tried to bail out but was caught in the gunfire that was still being directed at his plane, and slumped back into the cabin. The Ju 88 bounced on the water, dropped a wing and then went straight into the sea fifty miles off Cromer, Norfolk. Lucas and Munro were credited with shares in the first confirmed Typhoon victory.

At 2015 hours on August 13, Flt. Lt. A. C. Johnston, Sgt. Gerald G. Osborne, and Pilot Officers John D. Miller and W. J. A. Wilson of 266 Squadron were flying another sea patrol fifty miles northeast of Southwold when they encountered another twin-engine enemy plane. A running fight ensued, during which Johnston, Osborne, and Wilson all got shots into the fleeing machine, but Johnston was judged to have gotten in the fatal shot that set its engines on fire. At least one crewman was seen to bail out, and Johnston orbited the crash site to pinpoint it for a Supermarine Walrus flying boat before returning to Duxford. Johnston was credited with a Ju 88, but his late adversary had more likely been a new Messerschmitt Me 210, VN-AV of *Erprobungsgruppe* 210, whose pilot, *Ltn.* Heinrich Manger, was killed, but whose observer, *Uffz.* E. Rudolf, was rescued and taken prisoner.

Thus far the Typhoons had been engaged in what Denys Gillam described as "sweeps . . . on the fringe of things," but on August 19 his Duxford Wing, by then joined by No. 609 Squadron under Group Capt. Paul H. M. Richey, became involved in a bona fide air battle as Operation Jubilee and the landing at Dieppe got under way. At 1400 hours Gillam led all three squadrons from West Malling, with No. 609 in the lead and Nos. 56 and 266 flying "top cover." As they arrived sixteen thousand feet over Le Tréport, Dieppe, the wing received word that enemy bombers were heading for the Channel, and shortly afterward No. 266, led by Squadron Leader Charles L. Green, spotted three Do 217Es of *Kampfgeschwader* 2 with some Fw 190s in attendance. Green ordered Flt. Lt. Roland H. L. Dawson and Pilot Officers Wilfred S. Smithyman and Ian Munro to peel off after the Dorniers. Minutes later, Smithyman was heard on the radio reporting to have seen one Dornier crash, which Roland Dawson subsequently claimed as his victory. Munro dived on another bomber, fired a hundred-round burst as he closed the range from three hundred to fifty feet and then reported seeing his opponent descend in a steep dive with smoke issuing from the port engine and fuselage, although it was credited to him only as a "probable." Diving to the deck and pulling out over the coast, Munro saw another Typhoon, and upon rejoining it he found it was Dawson's.[3]

Meanwhile, another flight of 266 Squadron Typhoons, led by Flight Lieutenant Johnston, went after ten Fw 190s, eight of which broke to port and two of which kept diving straight ahead. Johnston went after the latter two, caught up with one and fired a series of short bursts between six hundred and four hundred yards, after which he and Pilot Officer Miller saw white smoke issue from the Focke-Wulf and its dive steepen to near vertical. At that point, Johnston's speed was 480 miles per hour, and he thought it prudent to pull out. Due to low ground haze, his Fw 190 had to be recorded as "probably destroyed."

Dawson and Munro were about halfway across the Channel and hastening to rejoin their now-retiring sections when they saw a squadron of Spitfires approaching fast on the starboard quarter. Suspecting trouble, the Typhoon pilots boosted their engines to get clear and weaved as the Spitfires closed in, but Dawson and Munro turned head on at the ones approaching them. One of the Spitfires opened fire, and Munro saw pieces fly off Dawson's plane, which half-rolled up to a hundred feet and then plunged into the sea.

The remaining Typhoon pilots returned to West Malling, all complaining of the Spitfires that had attacked them, though Dawson's was the only case in which one actually fired. They later learned that the Spits were from No. 331 (Norwegian) Squadron, led by Maj. Helge O. Mehre, who explained—as was becoming all too usual—that his pilots had mistaken their planes for Fw 190s. Both Gillam and Munro claimed to have damaged Fw 190s in the course of the patrol, but Smithyman never returned, and it was presumed that he had been shot down in the first fight with the Do 217s.

Gillam led another sweep at 1700 hours, but when his gun mounting came open he turned back—followed by No. 609 Squadron, which was not on the correct radio frequency. Squadron Leader Charles Green of 266 Squadron took charge of the remaining Typhoons and led the sweep from the coast just north of the Somme River to Cap Gris Nez. Cloud cover resulted in no enemy aircraft and no flak, and the British landed at Duxford without further incident. A final sweep of the day, flown from Fernes to Dunkirk, came under extremely accurate flak, but

no aircraft were damaged, and while two Fw 190s dived on the formation, they did not press home an attack.

So ended the Typhoon's participation over Dieppe, with only one enemy bomber confirmed as destroyed for the loss of two planes and pilots—one a victim of "friendly fire." As a precaution against further such incidents, the Typhoons were given a succession of distinctive identification markings, starting with yellow bands under the wings, followed by a white nose and black underwing bands, and finally, in 1943, black and white underwing bands similar to those that would be universally employed during the Normandy landings in June 1944.[4]

Although the Typhoon went on to a highly successful career as a low-level fighter-bomber, earning particular notoriety for the havoc it wrought on German armor and road transport during the Battle of Falaise in August 1944, it never really excelled as a fighter. Even with the bugs ironed out of its engine, the Typhoon never overcame its tendency to fly erratically at high speeds, a problem that was eventually traced to compressibility—local airflow exceeding the speed of sound.

As early as 1940, Camm had been working on a solution in the form of a laminar-flow wing of elliptical configuration similar to that of the Spitfire. The new wing was installed on the Typhoon II, but its thinner cross section—with a root thickness five inches less than that of the original Typhoon—necessitated the transfer of fuel tanks from the wings to the fuselage, which consequently had to be lengthened. A dorsal tail fin was added, along with so many changes that the final result was given a new name entirely. As first test-flown on September 2, 1942, the prototype Hawker Tempest Mark V was powered by the same 2,180-horsepower Napier Sabre II engine as the Typhoon, but sustained a higher speed of 430 miles per hour at 20,300 feet. Other variants with other engines were under development, but the RAF's demand for a fighter with the Tempest V's performance to match Luftwaffe counterparts, such as the Me 109G, led to the decision to put the proven version into production without further delay. First flown on June 21, 1943, the production Tempest Mark V reached a speed of 434 miles per hour at 22,800 feet and climbed to 20,000 feet in six minutes and thirty-six seconds. Armament consisted of four short-barreled 20mm Hispano V cannon, as well as eight underwing rockets or 2,000 pounds of bombs.

The first Tempests were slated to replace the Typhoons of No. 486 (New Zealand) Squadron in January 1944, but as that unit was engaged in attacks on German flying bomb sites being established along the French coast, it had no time to train in the new fighters. Instead, the first machines went to No. 3 Squadron at Manston in February, while No. 486 got its first Tempest in March. Both units, along with Typhoon-equipped No. 56 Squadron, made up No. 150 Wing under Wing Cmdr. Roland Prosper Beamont.

A veteran fighter and test pilot, "Bea" Beamont still believed in leading by example—as attested by his scoring the Hawker Tempest's first aerial victory. On June 8 Beamont, flying Tempest V JN751 bearing his personal initials "R-B," was leading some of his wing on a sweep north of Rouen when they encountered Me 109Gs, and he shot down the first of them. Moments later, Flt. Lt. A. R. Moore and Pilot Officer George A.

Whitman of No. 3 Squadron accounted for two more Messerschmitts.

Soon after that fair start, however, the Tempest squadrons were withdrawn from cross-Channel operations to defend Britain against a new threat. On June 13, the first of thousands of pulse-jet-powered guided glider bombs—officially given the deceptive designation of FZG-76 (*Flakzielgerät*, or Antiaircraft Target Device), but more widely referred to as *Vergeltungswaffen* (Vengeance Weapons) or V-1s by the Germans; and "buzz bombs," "doodlebugs," or "divers" by the British—hurtled skyward from ten launching ramps in France, to fall with often devastating effect on London and other cities.

When it came to intercepting these prototypical cruise missiles, the Tempest proved to be a textbook case of the right plane entering service at the right time. It was fast enough to overtake a V-1, and its guns could blow one up at sufficient distance to avoid the sometimes fatal damage that the exploding robot bomb could otherwise inflict on its pursuer. Because of its controllability at high speed, however, Tempest pilots also had the option of coming up alongside the V-1 and gradually raising a wingtip under one of the robot bomb's, letting the resulting air pressure lift the robot's wing until its gyrocompass guidance system was upset, and the "diver" would fall out of control and crash in an unpopulated area.

One RAF response to the V-1 menace was to form a fighter flight within its Fighter Interception Unit, comprised of Tempest Vs and Merlin-engine Mustang IIIs, to patrol the southern approaches against the robot weapons. Among the flight's personnel was Flt. Lt. Joseph Berry, who earlier had flown Bristol Beaufighter Mark VIFs over the Mediterranean, downing a Ju 88 on September 9, 1943. Now flying a

Hawker Tempest Mark V JN751, bearing the initials of Wing Cmdr. Roland P. Beamont, leader of 150 Wing, was used by him to shoot down a Messerschmitt Me 109G west of Rouen on June 8, 1944, as well as most of the thirty-one V-1 flying bombs credited to him. (*Andrew Thomas*)

Tempest, he opened his V-1 account with two "divers" on June 28, followed by one the next day and three on the night of June 30. Berry scored regularly throughout the month, including a record seven V-1s on the night of July 23. His sourest moment occurred on July 27, when he pursued a V-1 over West Malling airfield before finally hitting the buzz bomb and suffering damage to his Tempest in the explosion—and then learning to his chagrin that he would have to share credit for the victory with a Mosquito which had also been involved in the chase, in spite of testimony from his FIU mates that the Mosquito had fired at 1,000 yards and "missed hopelessly." Of the eighty-six and a half divers destroyed by the FIU, Berry accounted for fifty-two and a half before being placed in command of No. 501 Squadron on August 16, 1944. He had added seven more to his tally by the end of September, making him the highest scoring of the "diver aces." Beamont also excelled at hunting doodlebugs, adding a total of thirty-two of them to the six enemy planes and shared credit in two others on his final wartime record. Flight Lieutenant George A. "Lefty" Whitman, a Pennsylvanian who had joined the Royal Canadian Air Force in order to make his way into No. 3 Squadron RAF, added six and one-quarter V-1s to the Me 109 he'd downed on June 8.

Once the V-1 threat was under control, the Tempests returned to the offensive, serving with Nos. 2, 33, 56, 80, 222, 274, and 486 Squadrons. Operating from bases in the Netherlands and Belgium, Tempests of the Second Tactical Air Force added twenty manned jets, in the form of Me 262s, to their laurels before hostilities ceased. Although the Tempest never attained the fame of its stablemate, the Spitfire (but then, what British fighter did?), it was nevertheless the fastest and most powerfully armed single-engine fighter to see RAF service during World War II, and went on to a worthy postwar career as well.

In June 1942, while the Hawker Typhoon was flying its first missions over the Channel, another cannon-armed fighter was having an even less auspicious debut over China. Armed with a 20mm cannon and two 12.7mm machine guns, the

Japanese Army Air Force's first Kawasaki Ki.45 *Toryu* twin-engine fighters were about to test their mettle against more nimble single-seat fighters, flown by some of the deadliest pilots in Asia at that time—the "Flying Tigers" of the American Volunteer Group.

Built to a requirement issued to Mitsubishi, Nakajima, and Kawasaki by the *Koku Hombu* in March 1937, the Ki.45 was one of several twin-engine multiseat strategic fighters that came into vogue in the mid-1930s, other examples including the Bell YFM-1 Airacuda, the Messerschmitt Bf 110, the PZL P.38 *Wilk*, the Fokker G.I, and the Potez 63. The idea of a modern long-range general-purpose fighter had obvious appeal to the Japanese army, but its requirements were so broad and poorly defined that Nakajima designer Hideo Itokawa withdrew from the project, while Mitsubishi declined even to tender a proposal. Kawasaki, on the other hand, was eager to try its hand at a modern airplane design, and in spite of the firm's inexperience its chief project engineer, Isamu Imashi, came up with the Ki.38, a semimonocoque cantilever monoplane with semiretractable landing gear and powered by two Kawasaki Ha.9-II-Ko twelve-cylinder water-cooled engines. Just as the Ki.38 mock-up was nearing completion, however, the *Koku Hombu* ordered further work to be stopped, pending the resolution of differences between the army and the Air Technical Research Institute in regard to speed, endurance, and maneuverability requirements. A compromise was reached two months later, and Kawasaki was ordered to revise its design to use 820-horsepower Nakajima Ha.20-Otsu nine-cylinder radial engines, with an armament of one 20mm Ho.3 cannon and two 7.7mm Type 89 machine guns.

By then, Takeo Doi had succeeded Imashi as chief project engineer, and work proceeded on the revised and redesignated Ki.45, the design for which was completed in October 1938, and of which the first of three prototypes rolled out of Kawasaki's Gifu plant in January 1939. Problems immediately surfaced with the unreliable engine, combined with excessive drag, which several redesigned engine cowlings seemed unable to remedy. The hand-operated chain-and-sprocket retraction system for the landing gear also was unpopular, though it was eventually supplemented with an electrical system. The Ki.45's maximum speed of 298 miles per hour fell short of the 335-miles-per-hour requirement, and mock combats showed it to be unable to compete with Kawasaki's old Ki.10 biplane fighter, let alone with the Nakajima Ki.27. The plane's flexibly mounted rear 7.7mm machine gun was judged to be virtually useless at high speeds, and in general the Ki.45 was pronounced "incapable of performing the missions considered necessary for a two-seat fighter."[5]

Further testing was curtailed while the *Koku Hombu* reviewed the feasibility of the entire strategic fighter concept. Then, Kawasaki was ordered to try again—by adapting its airframe to the Nakajima Ha.25, a fourteen-cylinder twin-row radial that produced 1,050 horsepower, while possessing a slightly smaller diameter than the Ha.20. After being redesigned to handle greater stress, having its center of gravity altered and being given deeper nacelles to accommodate both the engines and its fully retractable, electrically operated undercarriage, the fourth preseries Ki.45 underwent its first test in July 1940. During the first flight, the plane was barely controllable and suffered damage in a forced landing, but after repairs and further modifications it showed much improvement, including a speed of 323 miles per hour. Doi and his design team continued to refine their design until it bore only a basic conceptual resemblance to the first prototype, completing the Ki.45 *Kai* in May 1941. The new fighter's overall lines were straighter and less curvaceous than the original, but during flight testing in September–October 1941, it showed viceless flying characteristics that contrasted radically from those of the earlier prototype.

The Kawasaki design team's persistence was finally rewarded with a production contract in November 1941, the first batch of which were to be powered by Ha.25 engines, later to be supplanted by more reliable 1,080-horsepower Mitsubishi Ha.102 fourteen-cylinder radials. In its first production form, the Ki.45 *Kai-Ko*, dubbed *Toryu* (Dragon Slayer) by the army, retained the 20mm Ho.3 cannon firing through a ventral fuselage tunnel offset to starboard; but nose armament was increased with twin 12.7mm Ho.103 Type 1 machine guns, while the rear observer was furnished with a 7.92mm Type 98 machine gun (based on the German MG 15). In addition to its heavier punch, the Ki.45 was one of the first Japanese fighters to have pilot armor and fuel tank protection. Although it was still criticized for its inability to maneuver as well as Japan's single-seat fighters—a virtually impossible task for any plane of its configuration—the Ki.45 was nevertheless one of the most agile twin-engine aircraft of World War II.

Given the many setbacks in the course of its development, there is a certain consistency to the Ki.45's baptism of fire. The first five production planes were sent to Canton, China, and attached to the 21st *Hikotai*, a composite unit of bombers and fighters also known as Nagano Force. The flight was led by Sgt. Jiro Ieiri, and the new fighters' first mission was to supplement Nagano Force's Ki.27s in escorting Kawasaki Ki.48 medium bombers in a series of bombing attacks on Liuzhou, Kweilin, and on other airfields in East China between May and June 1942. For four weeks the raids went unchallenged, and Ieiri decided to put his *Toryus* to use in the ground-attack role, loading the forward guns with explosive shells and mounting 110-pound bombs under the wings.

The tedium of the *Toryu* crews' routine was about to change, however. On the evening of June 11, eleven Curtiss P-40Cs and new P-40Es of the 1st Squadron of Col. Claire L. Chennault's American Volunteer Group arrived at the hitherto unoccupied airfield at Kweilin, which the Japanese had bombed earlier that day. In anticipation of another visit, Chennault called reveille at 0300 the next morning, and the appearance of Japanese reconnaissance planes at 0500 gave his men all the warning they needed. By 0558, when five Ki.48s of the Nanjing-based 90th *Sentai*, eight Ki.27s of the 54th *Sentai*, and the five Ki.45s of Nagano Force arrived over Kweilin, all three flights of Flying Tigers were up and circling at altitudes of 15,000, 18,000, and 21,000 feet, ready to pounce on their would-be assailants.

After dropping their bombs from an altitude of 16,600 feet, the Ki.48s turned for home, but in the process their formation strayed from those of the escorts, at which point George T. Burgard, leading the uppermost flight of Flying Tigers, put his P-40C into a dive and attacked the bombers. He scored some hits on one of them before being attacked by one of the Ki.27s and was soon engaged with three of the fighters. Allen Wright's P-40C was hit by one of the Ki.48 gunners, but he went on to attack three of what he thought were "light bombers," until driven away from them by another Ki.27. Burgard spotted Wright's plane going down with smoke streaming from its engine, and descended 2,000 feet to drive the Ki.27 off his tail. He then escorted Wright to the airfield, where the damaged P-40 bellied in and flipped over, spraining Wright's back.

Climbing to rejoin the fight, Burgard was joined by Lt. Romney Masters, a US Army Air Forces pilot who had recently joined the AVG in anticipation of its imminent transition from the Chinese Air Force to the USAAF. The two pilots encountered a Ki.27 and a twin-engine plane, the former of which Burgard took on, while Masters engaged the other. Burgard's quarry eluded him, so he had a go at the twin-engine machine. "The bomber was light, fast, and exceptionally maneuverable," Burgard reported afterward. "He dove sharply for the ground and made a sharp turn."

At that moment, the Ki.27 returned and put a hole in Burgard's left aileron, then dived away with Burgard in pursuit, firing long bursts until he saw it skid and fall off on a wing. "I pulled up," he wrote, "and saw he had run into the side of a sharp peak and blew up."

Burgard could not see Masters, and his .50-caliber machine guns had stopped, but he still wanted to deal with the twin-engine "light bomber," which he saw "hedgehopping through the sharp hills." After half a dozen unsuccessful attempts to get on the evasive machine's tail, Burgard caught it in a valley, "but he turned almost 90 degrees and began making 360-degree vertical turns around a sharp cone. . . . We were never more than 150 feet from the ground." During one pass, however, Burgard's machine guns started working again, and he managed to set the enemy plane's left engine on fire.

"I throttled back," he wrote, "and continued to shoot until he dragged his wing on a small knob and mushed in."[6]

Unknown to the AVG pilots at the time, Wright had been attacking three of the new Ki.45s when the Ki.27 jumped him, while Burgard had accounted for their leader. Sergeant Ieiri was killed, but his gunner, Cpl. Kei Honda, survived to be taken prisoner.

Camille J. Rosbert, flying a new P-40E in the lowest AVG formation, attacked another "bomber," only to see it loop up and over him and speed off in the opposite direction. Using his dive speed to regain altitude, he gave chase and fired a burst that tore off part of the enemy plane's wing, after which he said, "the Jap spun towards the sharp mountains below."

Robert H. Neale was attacking a Ki.48 when he found one of the new *Toryus* on his tail. "Thought it was a Me 110," he wrote. "Had a hell of a time getting away."[7]

Charles R. Bond Jr. attacked another Ki.45, but its gunner put a bullet through the coolant system of his P-40E. His hydraulic system and his guns also failed, as smoke curled from behind his instrument panel. He then noticed two Ki.27s on his tail, one of which followed his descending plane to 1,000 feet before turning and climbing away, probably convinced that he was done for. Bond's propeller stopped moments later, and he glided in and made a crash-landing in a rice paddy. Although he gashed his forehead on his gunsight, Bond was otherwise unhurt, and with the aid of local farmers and a Catholic missionary, Herbert Elliott, he eventually made his way back to his squadron.

The Flying Tigers claimed a total of nine enemy aircraft in the fight, which were credited to George Burgard, Joe Rosbert, John J. Dean, William E.

Bartling, and John R. Rossi. Their celebrations that night were enhanced when Reverend Elliott telephoned to report that "old hard-luck Bond" was all right, and when the Chinese brought in Ieiri's gunner, Corporal Honda. A chicken farmer before joining the army, Honda had been a bomber gunner for two years before joining Nagano Force in February 1942. His reference to his aircraft as a "Model 45" marked the first time that Chennault heard a JAAF plane referred to by its *kitai* number, and his description of its nose cannon, in which every second shell was an explosive round, was sobering news to the Americans, who had hitherto considered their superior firepower to be a consistent advantage. "No more head on runs for me," Neale remarked. After Burgard was introduced to him as the man who had brought him down, Honda posed for a photograph with his captors in front of one of their shark-mouthed fighters before being returned to the Chinese.[8]

In Tokyo, a newspaper reported that "Nine American aircraft of the Curtiss-Hawker P-40 type were downed by the Japanese Air Unit in a fierce combat" on June 12 for the loss of only two Japanese planes.[9] The JAAF aircrews in China knew better—that the actual casualties in the day's fighting were roughly the opposite. One Ki.48 had crashed near Kweilin—apparently shot down by Bartling—and two others were so shot up that they were written off upon their return. Only one Ki.27 was lost (though Burgard, Bartling, and Rossi were credited with one each); but two of the new *Toryus* had been destroyed near Kweilin by Burgard and Rosbert, while a third, also credited to Rosbert, had crashed on the way back, and the fuselages of the remaining two Ki.45s showed wrinkles which suggested that they could not stand

up to the stress of high-G maneuvering. The Japanese had brought down two P-40s, but both of their pilots, Wright and Bond, had survived. After analyzing the raid's results, the JAAF summarily canceled further operations against Kweilin.

The *Koku Hombu* also subjected the Ki.45 to yet another reappraisal (whether or not it noted the irony of the heavily armed fighter's first victory being scored by a rear gunner's relatively puny weapon is unknown). Although the *Toryu* had not exactly disgraced itself in its first combat, its performance as an escort fighter had fallen short of expectations, much as the Me 110 had done during the Battle of Britain. When the Ki.45 *Kai-ko* entered service in quantity over China in November 1942, it was with a *chutai* of the 16th *Sentai*, replacing that squadron's Mitsubishi Ki.30 light bombers. At about the same time, the newly activated 21st *Sentai* in Hanoi, Indochina, replaced its initial complement of Ki.27s with Nakajima Ki.43s and Ki.45s, and in April 1943 the 13th *Sentai* in New Guinea began to receive Ki.45s to supplement its Ki.43s. In all of those units, and in the 5th *Sentai*, which joined the 13th in New Guinea in July 1943, the Ki.45s served primarily in the fighter-bomber role, attacking ground targets and small Allied vessels.

Late in 1943 the Ki.45 *Kai-Otsu* was introduced, with the 20mm cannon moved to the nose in place of the twin 12.7mm machine guns, and its place in the ventral tunnel taken up by a 37mm handloaded Type 98 cannon. In early 1944, an increase in nocturnal bombing raids by Consolidated B-24 Liberators led Ki.45 units in the South Pacific area to turn their planes into improvised night fighters by removing the fuel tank in the upper fuselage aft of the cockpit and installing twin oblique 12.7mm Ho.103

Strafed and disabled on Cape Gloucester (Tuluvu) airfield, Papua New Guinea, this Kawasaki Ki.45-Kai of the 13th *Sentai* was photographed after US Marines occupied the area on December 30, 1943. (*National Archives*)

machine guns or 20mm Ho.5 cannon in its place, similar to the *Schräge Muzik* arrangement employed in Luftwaffe night fighters. Although the modified Ki.45s carried no radar, the 5th and 13th *Sentais* claimed some successes over the East Indies, New Guinea, and Rabaul, prompting Kawasaki to standardize the oblique 20mm Ho.5 mount in a final production variant, the Ki.45 *Kai-Hei*, which also replaced the Type 98 cannon with a semi-automatic recoil-operated Ho.203 cannon, which had a slightly lower muzzle velocity than the handloaded weapon, but which could use a twenty-five-shell magazine with a rate of fire of 130 rounds per minute. One Ki.45 *Kai-Hei* was trial-fitted with air intercept radar, but it was never standardized in production aircraft.

In April 1944, the first Ki.45 *Kai-Hei* fighters emerged from Kawasaki's Akashi plant and were delivered to the 4th *Sentai*, based at Kozuki, and the newly formed 53rd *Sentai* at Matsudo. By then, the new-model *Toryu* had a new dragon to slay—the Boeing B-29 Superfortress. On the night of June 15, 1944, sixty-eight B-29s of the Twentieth Air Force's 58th

Bombardment Wing (Very Heavy) left their base at Chengtu, China, to bomb the Imperial Iron and Steel Works at Yawata in northern Honshu—the first Allied bombing raid against mainland Japan since the Doolittle raid of April 18, 1942. Among the Japanese fighters that rose to challenge the bombers were eight Ki.45s of the 4th *Sentai*, which claimed seven of the B-29s. First Lieutenant Isamu Kashiide claimed two, plus one whose fate could not be positively determined, while 2nd Lt. Sadamitsu Kimura was credited with two destroyed and three damaged, for which he received a personal citation from the commanding general of the Western Military District and was awarded a ceremonial sword.

The Americans had indeed lost seven bombers, but their reports told a different story as to the causes. One B-29 from the 468th Bomb Group crashed shortly after takeoff with no casualties, and a second, from the 444th Group, subsequently crashed near Kiangyu, killing its entire crew. Four planes aborted their mission early and nine others, which failed to reach the objective, struck at alternate targets of opportunity. The forty-seven bombers that reached Yawata were hampered by cloud cover and haze over the target, preventing the 221 tons of bombs they dropped from causing much damage. A B-29 of the 368th Group's 792nd Squadron was shot down and six other planes damaged by antiaircraft fire. The Americans reported sixteen attacks by Japanese fighters, only three of which were by twin-engine aircraft and none of which scored significant damage on them. Two 368th Group B-29s crashed into mountains during the return flight over China with the loss of both crews and *Newsweek* correspondent Robert Schenkel, while another plane from the 444th was forced to land at Heihsiang, where Japanese fighters and bombers destroyed it on the field the following morning. A total of fifty-five men were killed in the Yawata mission, but none of those losses were directly attributed to enemy aircraft.

Unaware of the Americans' assessment, the *Koku Hombu* was satisfied that the *Toryu* had found its operational niche at long last. Even though the Ki.45 lacked the radar and high-altitude performance to deal consistently with the B-29s, its 37mm shells could cause terrific damage whenever they were able to strike home. In the months that followed, Ki.45 crews of the 4th and 53rd *Sentais*, joined by the 5th, which had been recalled from New Guinea, distinguished themselves in defense of the Japanese homeland.

While some Western nations were still wrestling with the feasibility of mounting cannon on single-engine aircraft, Japan entered the Pacific War with two 20mm cannon—albeit low-velocity weapons—in the wings of its principal air superiority fighter, the Mitsubishi A6M2 Zero.

That in itself was a significant increase in armament, though the fighter which carried it had made extraordinary concessions in structure and protection in order to compensate for its weight. Even while the A6M was under development, however, discussions took place between its chief designer, Jiro Horikoshi, and the Bureau of Aeronautics for the Imperial Navy regarding a radical departure from the established Japanese formula—a powerful, heavily armed interceptor built with an emphasis on speed and climb rate rather than maneuverability. The need to complete the A6M held up further work on the project; but in September 1939 the navy issued a specification for an interceptor capable of 373 miles per hour at 19,685 feet, the ability to reach that altitude in 5.5 minutes, and provision for two 7.7mm machine guns, two 20mm cannon, and an armored pilot's seat.

Given a choice of the 1,185-horsepower V-12 Aichi Ha-60 Atsuta—based on the Daimler-Benz DB 601A—and the 1,440-horsepower Mitsubishi Ha-32 Kasei Model 13 fourteen-cylinder twin-row radial engine, Horikoshi opted for the latter as he, Yoshitoshi Sone, and Kiro Takahashi began work on the new heavy fighter in early 1940. In order to compensate for the radial engine's size—considerably larger than the Zero's Nakajima Sakae 12—Horikoshi's team extended the shaft to its three-bladed propeller and housed the engine in a long, tapered cowling with an engine-driven fan to suck in cooling air. The higher priorities afforded Zero production, and initial problems with engine cooling delayed the completion of the prototype J2M1 until February 1942, by which time the overworked and weary Dr. Horikoshi had relinquished his post as chief designer to Takahashi.

After flying it for the first time on March 20, test pilot Katsuzo Shima judged the plane's stability and controllability to be excellent, though the ailerons tended to stiffen up at speeds above 323 miles per hour. He also reported the forward view to be unacceptable and that the undercarriage would not retract at speeds exceeding 100 miles per hour. The 359-miles-per-hour speed and 7.8-minute climb to 19,685 feet fell short of the requirement; but the navy decided *that* to be rectifiable by using the more powerful 1,870-horsepower Mitsubishi MK4R-A Kasei 14 Model 23ko engine, then under development, and a Sumitomo four-bladed propeller. A modified version was therefore earmarked for production on October 13, 1942, as the J2M2 *Raiden* (Thunderbolt) Model 11, armed with two 7.7mm machine guns in the fuselage and two high-velocity 20mm Type 99 Model 2 cannon in the wings.

The production order proved to be premature. During test flights the Kasei 23*ko* engine, which had provision for water/methanol injection, emitted excessive smoke and tended to vibrate at maximum power, often to a critical degree. The latter problem was eventually alleviated with more rigid propeller blades and more resilient engine-mount shock absorbers, but further problems delayed the development of the J2M2, and not all of them had been solved when the first production machines were finally accepted by the navy in December 1943. The oscillation problems persisted, and in January 1944 the first of several incidents occurred in which J2Ms disintegrated in midair for reasons that were never conclusively explained.

In October 1943 the J2M3 appeared, replacing the fuselage-mounted 7.7mm machine guns with a wing-mounted pair of 20mm Type 99 Model 1 cannon, which had both lower muzzle velocity and rate of fire than the Model 2, but which greatly increased the interceptor's overall firepower. That type was accepted over the J2M2 in February 1944, but in June the production tempo of the later model was also reduced as a result of the navy's disenchantment with the J2M's teething troubles and its increasing interest in Kawanishi's N1K1-J *Shiden* interceptor. Although *Raiden* development continued, production of all J2M variants totaled no more than 476—too little for it to affect the outcome of the Pacific War, no matter how good it ultimately turned out to be.

The first J2M2s were flown to Toyohashi, southeast of Nagoya, and from there they were assigned to Fighter *Hikotai* (Squadron) 602 of the 381st *Kokutai*, then based at Kendari on the island of Celebes, to assist in the defense of the Balikpapan oil fields. Ten *Raidens* were on hand in February 1944, but the additional problems they were experiencing at Toyohashi dampened hopes of their completely replacing the Zeros in the 381st *Kokutai's* inventory. In mid-March, six of the *Raidens* were detached to Balikpapan via Davao, and took a direct part in air defense operations thereafter.

Hikotai 602 still had nine *Raidens* on hand—though only seven were in operational condition—on September 5, when fifty-eight Consolidated B-24 Liberators of the Fifth Air Force's 380th Heavy Bombardment Group attacked its air base at Menado, Celebes, destroying or damaging seventeen aircraft on the ground. The squadron's Zeros and *Raidens* rose to oppose them, and during the action the Japanese fighters resorted to dropping No. 3 thirty-kilogram aerial-burst phosphorus bombs into the bomber formations. Warrant Officer Keizu Kamihara of *Hikotai* 602 was credited with three of the B-24s (although two were subsequently rated only as "probables"), as well as with two of their Lockheed P-38 escorts.

At 0040 hours on September 30, twenty-four Liberators of the Thirteenth Air Force's 5th Bomb Group and twenty-four from the 307th began taking off from Kornasoren, Noemfoor, to carry out the first of five strikes on Balikpapan. Requiring a 2,600-mile round trip, this was the longest daylight mission ever undertaken by B-24s in formation, and forty-six more Liberators of the Fifth Air Force's 90th Bomb Group also participated, flying 1,243 miles to reach the target. The 5th Group was 250 miles from Balikpapan when it encountered two Japanese fighters, and soon the B-24s were running a gauntlet of antiaircraft fire and thirty more fighters. By the time the Liberators reached their target—the Pandansari refinery—they were fending off attacks by Zeros, *Raidens*, and J1N1-S *Gekko* night fighters of the 381st *Kokutai*, as well as by Zeros of the 331st *Kokutai's* *Hikotai* 309. Flak damaged fifteen of the 5th Groups' bombers, and fighters downed three of them, including that of the deputy group commander. The Japanese fighters also accounted for a B-24J of the 90th Group's 319th Squadron. In spite of intermittent cloud cover which frustrated the 307th and 90th Group bombardiers' efforts, the bomber crews reported thirty-seven direct bomb hits on the refinery, as well as nine enemy fighters shot down.

The two Thirteenth Air Force groups returned to Balikpapan on October 3, and forty Japanese fighters engaged the 307th in a seventy-minute running fight that cost the group seven of its twenty B-24s as they bombed Pandansari. The 307th's gunners made a typically exaggerated claim of having shot down nineteen enemy planes; but Lt. j.g. Keisaburo Uchiyama and Chief Petty Officer Kitami

Japanese pilots under British supervision fly two surviving Mitsubishi J2M *Raidens* of the 381st *Kokutai* on a postwar evaluation flight from Seletar, Malaya, in December 1945. (*Imperial War Museum*)

Kikuchi of *Hikotai* 602 were killed that day, though the Americans' unfamiliarity with the J2M2 makes it impossible to ascertain whether they were flying Zeros or *Raidens*, and if the latter, whether their deaths were due to combat or aircraft failure. The nineteen Liberators of the 5th Group maintained a tighter formation than the 307th as they struck at the Edeleanu refinery, claiming three enemy planes for the loss of two bombers.

For the next strike on October 10, V Fighter Command dispatched sixteen Republic P-47Ds of the 35th Fighter Group from its newly acquired base on Morotai—American Thunderbolts unwittingly stalking Japanese Thunderbolts! Preceding the bombers to the target area, the P-47s jumped twenty-five Japanese fighters and claimed a dozen of them— including three "Oscars" by Capt. William H. Strand of the 40th Squadron—for the loss of one Thunderbolt.

Fourteen Lockheed P-38Js of the 49th Fighter Group's 9th Squadron also flew ahead of the bombers and encountered six dark-green radial-engine aircraft cruising between 3,000 and 7,000 feet, which the Lightning pilots identified as Oscars of the JAAF, but which were more likely the still unknown J2M2s because one of the American element leaders, Maj. Richard I. Bong, also spotted a naval

Irving at 5,000 feet—one of the J1N1-Ss of *Hikotai* 902, the 381st *Kokutai*'s night-fighter element.

With Capt. Robert Baker following in close support, Bong did an abrupt wingover, overtook the Irving, shot it down for his twenty-ninth victory and noticed at least one of its crewmen take to his parachute. The other P-38s engaged the "Oscars," one of which was so agile that Lt. Warren Curton went into a momentary high-speed stall while trying to follow it in a turn. Swiftly recovering, Curton set another antagonist's engine ablaze, while Bong sent another down in a ball of fire. Major Wallace Jordan caught an Irving at twelve thousand feet and sent it into the sea in flames, probably killing *Hikotai* 902's division officer, Lt. Sadao Ozaki. Jordan then got one of the "Oscars" for his fifth victory before the others got into such an advantageous position that he and his wingman, Capt. Willie Williams, beat a prudent retreat. Lieutenant Edward Howes sent a Zero crashing into the ocean and then reported seeing one of two "Oscars" drop two aerial bombs at some departing B-24s, though they exploded too far aft of the bombers to cause serious damage. Chief Petty Officer Eizo Ota of *Hikotai* 602 was among the Japanese pilots killed in the day's fighting, as was Lt. Akira Tanaka, group leader of the 331st *Kokutai*'s *Hikotai* 309, and two of his pilots, Chief Petty Officer Tsumiro Tanaka and Petty Officer 1st Class Akio Fukui.

While the American fighters were trying to clear the way for them, twenty-one B-24s of the 90th Group had to fight their way through still more Japanese fighters to bomb Pandansari and Edeleanu. The group's gunners claimed sixteen of their assailants for the loss of one Liberator, which exploded after being struck by a phosphorus bomb dropped by one of the Japanese fighters. The 22nd Bomb Group's eighteen B-24s hit Pandansari next, with one of its planes claiming six enemy fighters before damage from numerous 20mm hits compelled it to crash-land on Batoedaka Island. A second Liberator exploded after being rammed by a Zero, and a third was shot down. The 43rd Bomb Group went in next, losing none of its nineteen planes and claiming thirteen of the enemy's. The 5th's twenty-four Liberators reported only halfhearted opposition as they went after Balikpapan's paraffin plant, suggesting that the Japanese fighter groups were a spent force, and the 307's twenty-five B-24s enjoyed an equally easy run against the cracking plants.

The next raid, on October 14, was also escorted, with the 35th Fighter Group's Thunderbolts preceding the bombers by fifteen minutes and again claiming to have taken a heavy toll of the defenders in advance: nineteen, including two Oscars each by Captains Strand and Alvaro J. Hunter of the 40th Squadron over Balikpapan at 1030 hours, and two more by 1st Lt. James D. Mugavero of the 41st at Manggar at 1042, for the loss of four P-47s. The "Jugs" were followed by P-38s of the 49th Fighter Group, joined by Maj. Thomas J. McGuire, the commander and leading ace of the 431st Fighter Squadron, who was obsessed with overtaking Bong's score. Again the Lightning pilots reported a mixed bag of opposition—Zeros, Oscars, and Tojos—the latter two of which again suggest confused identification of the new J2M2s. In any case, Maj. Gerald R. Johnson and Lt. Edward Cooper of the 9th Squadron and McGuire each claimed one of the "Tojos."

Also present over Balikpapan were two P-38Js of the 8th Fighter Group, flown by the group's commander, Lt. Col. Earl Dunham, and by Capt. Kenneth Ladd, commander of the 36th Squadron. As they approached the target, they encountered four "Oscars," the leader of which Ladd attacked and sent down with the Japanese pilot last seen climbing from his cockpit to bail out. Dunham saw Ladd score hits on another enemy fighter before they became separated in the melee. After the fight broke up, Dunham climbed to nineteen thousand feet and radioed Ladd but got no response. When he returned to Morotai, he heard reports of a P-38 with a smoking engine still being engaged by Japanese fighters, but nothing more was ever learned of Ladd's fate. He was the only Lightning pilot lost, while his colleagues claimed a total of sixteen enemy fighters. The Thirteenth Air Force dispatched forty-two of its own Lightnings to Balikpapan, but only six of the 68th Fighter Squadron's planes reached the target, to claim a Tojo and a Tony.

While the Thunderbolts and Lightnings were eliminating the aerial opposition, forty-nine Liberators of the 90th Group dropped two 1,000-pound and one 500-pound bomb each on Edeleanu, followed by forty Thirteenth Air Force machines. Only one of the 90th's Liberators was lost in the most effective attack on the refineries so far, which was commended as "magnificent" by Gen. Douglas MacArthur himself. The Thirteenth Air Force tried to follow up with one more strike on October 18, but foul weather prevented its planes from accomplishing much, though it also prevented the Japanese fighters from intercepting the bombers.

The Japanese claimed a total of eighty American planes during the Balikpapan raids. The Fifth and Thirteenth Air Forces had actually lost twenty-two B-24s— from which sixty crewmen were eventually recovered—as well as six P-47s and three P-38s. The total of 433 tons of bombs the Liberators dropped on Balikpapan did not cripple the complex, but did reduce its petroleum output for awhile. It cannot be positively ascertained how much of a contribution the 381st *Kokutai*'s *Raidens* made to the American casualties, but the fact that they were no longer on group strength suggests that all were either shot down or written off in the course of the five air battles.

In spite of the *Raiden*'s gallant but less than outstanding debut, Japan's position at that point was desperate enough to ensure the interceptor a role in her defense. As early as March 1944, Lt. j.g. Sadaaki Akamatsu, newly transferred to the 302nd *Kokutai* at Atsugi, was training in a new J2M3, which he would fly on and off until August 1945. In the hands of an expert like the irascible alcoholic but brilliant Akamatsu, "Jack," as the Allies called the *Raiden*, proved to be not only a formidable interceptor for bringing down a Boeing B-29 but also a fighter capable of besting a Grumman F6F Hellcat— Akamatsu accounting for four and five "probables" among the latter.

After capturing a Jack outside Manila in February 1945 and restoring it to flyable condition, the Americans tested it against their best fighters, as well as against several other Japanese and German fighters, and were impressed with its speed, outstanding climb rate, stability, and takeoff and landing qualities. Significantly, their principal criticism centered on the Achilles' heel that had prevented the J2M's becoming a significant threat from the very start—poor mechanical reliability. Nevertheless, such *Raidens* as were available were distributed among the best trained of the home-defense air groups and fought with distinction right up until Japan's surrender.

Seventeen

Dueling in the Dark

NIGHT FIGHTERS, 1940–44

Nocturnal operations, like most aspects of aerial warfare, originated during World War I. As early as 1915, reconnaissance aircraft were making stealthy flights over the lines under the cover of darkness, as were German Zeppelins when improved air defenses made it too dangerous to bomb Britain's cities by day. By 1918 modified British fighters, such as the Sopwith Comic (a Camel with the cockpit moved back and twin Lewis machine guns installed above the wing so their flash would not ruin the pilot's night vision), were playing a nocturnal game of cat and mouse with Gotha G.Vs and Zeppelin-Staaken *Riesenflugzeuge* (Giant Airplanes), while some German fighters did the same with intruding French Voisins. As with some other World War I innovations, little substantial progress was made in that form of aerial warfare at the time, but it did set a precedent for further development during the postwar years.

The real foundations for nocturnal warfare were laid on February 26, 1935, when Robert Alexander Watson-Watt and his assistant, Arnold F. Wilkins, successfully demonstrated a means of locating and pinpointing a flying object by reflecting radio waves of ultrahigh frequency off its surface. The British development of radio detection and ranging—radar—led to the establishment of stations along the English coast, which would be a decisive factor in the Battle of Britain. Inevitably, the advancing state of the art also made it possible to build radar units small enough to install in airplanes themselves, giving them the ability to navigate and detect other aircraft in the dark.

Within the first year of World War II, British and German bombers were suffering such severe losses at the hands of one another's fighters that by the autumn of 1940, both sides were making the nocturnal bombing of one another's cities a standard practice. It all began on the night of March 19, 1940, when thirty Armstrong Whitworth Whitley Mark Vs of Nos. 10, 51, 77, and 102 Squadrons RAF, followed by twenty Handley Page Hampdens of No. 5 Group, attacked the German floatplane base at Hörnum on the Frisian island of Sylt. Only one plane failed to return the next morning, but later reconnaissance flights of the area determined that this, the first deliberate British bombing attack on German soil, had not been very effective.

By what amounted to an unwritten mutual agreement, British and German cities were bombarded with nothing more than propaganda leaflets in the first months of the war, but on the night of May 11–12, 1940, Monchengladbach became the first town to suffer a bombing raid—albeit a modest one—by thirty-seven aircraft. Strategic night bombing commenced in earnest on the night of May 15–16, as a total of ninety-nine RAF bombers struck at sixteen industrial targets in the Ruhr, while twelve others attacked communications points in Belgium. Again, little damage was done, but no planes were lost.

At that point, the Germans decided to take serious steps to deal with the situation. *Oberst* Josef Kammhuber, a former Junkers Ju 88A bomber pilot, was put in charge of organizing night air defenses, including a network of radar installations and the first *Nachtjagdgeschwader*, NJG 1, under the command of *Maj.* Wolfgang Falck.

Kammhuber proved to be an outstanding organizer, and Falck's enthusiastic leadership also contributed to the rapid growth and sophistication of the Luftwaffe's nocturnal fighter component. Its primary weapon was the Me 110, which was soon proven to be ineffective in its original role as a general-purpose fighter by day, but which would prove to be an excellent night fighter, since it was faster and could climb higher than any British bomber and at the same time could be adapted to carry an increasingly advanced array of airborne radar.

Supplementing the Me 110s were the first of several bombers converted to night fighters—the Dornier Do 17Z-6 *Kauz* (Screech Owl) and Do 17Z-10 *Kauz* II, and the Ju 88C. The Ju 88, in particu-lar, would be a highly successful variant of an airframe of seemingly limitless versatility—although less maneuverable than the Me 110, it could carry just as potent an array of guns and electronics, and with a range of 1,800 miles and an endurance of five hours, it could stay in the air three times as long as an Me 110.

Ironically, none of the above-mentioned aircraft were involved in the Luftwaffe's first successful nocturnal interception. On the night of July 8–9, 1940, *Obfw.* Paul Forster of IV *Gruppe* (*Nacht*) of *Jagdgeschwader* 2 was patrolling in an Me 109E when he saw a twin-engine bomber caught in the searchlights, attacked it and sent it crashing into the sea off Heligoland. Forster, whose victim was a Whitley Mark V of No. 10 Squadron, later stated quite frankly that he had just chanced to be at the right place at the right time.

The first success for one of the *Nachtjäger* came soon after. On the night of July 20–21, *Hptmn.* Werner Streib of NJG 1 was flying an Me 110C over the Ruhr when he spotted another twin-engine plane that he at first thought to be another Me 110. As he approached it, however, he identified it as a British bomber, a conclusion which was confirmed when he closed to 250 yards abeam of the intruder, and its tail gunner opened fire at him. Turning right to position himself behind and below the enemy plane, Streib sent a volley of cannon and machine gun fire into the target and saw flames erupt from the starboard engine, followed by the sight of crewmen bailing out as their plane went down. Both Nos. 58 and 78 Squadrons lost Whitleys that night, but none of their crews survived to testify as to which had been Streib's victim.

The first German bombs to fall on London were dropped by mistake on

August 24, 1940, but they led to a reprisal raid on Berlin by the RAF the following night, which in turn led Adolf Hitler to order an all-out retaliation against Britain's cities. That shifting of aerial priorities at a time when the Luftwaffe was coming tantalizingly close to annihilating the RAF's fighter defenses was one of the decisive German blunders of the Battle of Britain. It also led to such heavy bomber losses that after September 30, the Germans resorted exclusively to night bombing. As the Germans had done, the British adapted such fighters as they could to the night interception role, the most successful, at first, being the Boulton Paul Defiant turret fighter. Better planes were needed, but they would not be long in coming.

At the forefront of British night-fighter development was Bristol. Its first such airplane was nothing more than a fighter version of the Blenheim Mark I bomber, with a ventral rack of .303-inch Browning machine guns installed in place of the bomb bay. Its first mission took the form of a daylight strafing raid, when Blenheim IFs of Nos. 25 and 601 Squadrons attacked the Luftwaffe seaplane base at Borkum on November 28, 1939. Its first use as a night fighter occurred on July 21–22, 1940, when Sgts. Arthur J. Hodgkinson and Bertram E. Dye of No. 219 Squadron took off from Redhill and made contact with an unidentified enemy plane west of Church Fenton, which they claimed to have damaged before it slipped away. The team subsequently claimed to have damaged a Junkers Ju 88 near Flamborough Head on August 15.

Given the Blenheim's obsolescence as a high-speed bomber, let alone as a fighter, the first confirmed success for the nocturnal Mark IF was impressive. On the night of September 4–5, 1940, Squadron Leader Michael J. Herrick, a New Zealander assigned to No. 25 Squadron, caught and shot down a Heinkel He 111 over Braintree at 0045 hours, followed by a Dornier Do 17 over Rendlesham twenty-five minutes later. The Blenheim IFs had further successes thereafter; but it was clear that their run of good fortune could not last.

Bristol was swift to provide a successor in the form of another even more serendipitous improvisation—a long-range fighter derived from the Beaufort torpedo bomber that was proposed in October 1938, just days after the first Beaufort flew. The Type 156, later christened Beaufighter, was to have four Hispano 20mm cannon in the fuselage, a foreshortened nose to compensate for larger propellers, and the Blenheim's four-man crew reduced to a pilot and a rear observer/navigator. The first production Beaufighter, powered by two 1,400-horsepower Bristol Hercules III engines driving de Havilland nonfeathering bracket-type airscrews, was not quite as potent as Bristol hoped, because the Hercules VI engines and Rotol constant-speed propellers the firm had wanted to use were not available; but it was deemed good enough to warrant a three-hundred-plane contract on July 3, 1939. At that point, the 1938 Munich Crisis had awakened Britain to the need to modernize its defenses in all forms, and the slow pace at which Westland was developing its Whirlwind twin-engine fighter compelled the RAF to hedge its bets with Bristol's alternative. As it turned out, the Whirlwind ended up having a disappointing career, whereas the versatile Beaufighter would serve the RAF exceptionally long and well.

Cleared for RAF service on July 26, 1940, the first four Beaufighters were allotted, one apiece, to Nos. 25, 29, 219, and 604 Squadrons on September 2. Six days later, No. 600 Squadron received its first machine. Although too late to participate in the Battle of Britain, the "Beaus" soon got the chance to show their capabilities during the "Winter Blitz" that the German bombers pursued against British cities after abandoning their all-out air offensive in October 1940.

After the first fifty Beaufighters were built, the plane's armament was increased by adding six Browning .303-inch machine guns to the wings—two port and four starboard—making it the most heavily armed fighter of the war. More significant was the experimental installation of the AI Mark IV airborne interception radar in an early production Beaufighter Mark I at the Fighter Interception Unit at Ford, followed by the modification of further aircraft from September 1940 onward.

The first operational sortie by a night-fighting "Beau" was flown by No. 29 Squadron on September 17. Although the plane's speed and climb were disappointing, crews found it more maneuverable and generally more pleasant to fly at night than their old Blenheims. The early radar proved to have some early developmental problems, however, with ground echoes limiting its range, and the first Beaufighter night-fighter pilots claimed to have had more success without the help of the radar than with it.

The first confirmed success for the Beaufighter was scored over the Kenley area on the night of October 25–26 by a machine of No. 219 Squadron piloted by Sgt. Arthur Hodgkinson. As he reported it, his first of eleven victories was, in essense, scored the old-fashioned way.

I was vectored out 170 [degrees] and back 350 onto an enemy aircraft, and I sighted the enemy at about 16,000 ft. I observed the enemy flying slightly to my north side ahead of me at a distance of 400 yards. I opened fire at 200 yards, firing approximately 200 rounds in two bursts. I gave a third burst at 70 yards but the cannon failed to fire. My AI operator observed the enemy aircraft dive steeply into cloud. The enemy aircraft returned no fire. The aircraft was definitely a Do 17 or 215, as I noticed the humped effect above the forward end of the fuselage (where the aerial is) and its high wing, as well as the twin rudders.[1]

Flying Officer John Robert Daniel Braham of No. 29 Squadron, eventually to be the highest-scoring night-fighter ace in the RAF with twenty-nine victories, got a fleeting radar contact on the night of November 18, but the first radar-assisted victory was achieved by No. 604 Squadron on November 19–20, when Flt. Lt. John Cunningham and his observer, Sgt. John R. Phillipson, destroyed a Ju 88A-5 of 3rd *Staffel*, *Kampfgeschwader* 54 north of Bridge Norton. The team struck again on the night of December 23–24, shooting down an He 111 belonging to the 3rd *Staffel* of a radar-equipped pathfinder unit, *Kampfgruppe* 100, between Alderney and Cherbourg.

Wishing to conceal the identity of the RAF's new airborne weapon as long as possible, the British press attributed Cunningham's successes to extraordinary night vision, earning him the sobriquet of "Cat's Eyes," a nickname he personally disliked. The secrecy was warranted, however, because the AI radar and the technique for using it were still evolving. "It was a long hard grind and very frustrating," Cunningham admitted. "It was a struggle to continue flying on instruments at night. The essential was teamwork."[2]

It was not until January 1941, with the introduction of GCI (ground-controlled interception), allowing controllers to plot fighters and their targets simultaneously on a PPI (plan position indicator) and vector the Beaufighters to the vicinity of the enemy, that the night fighters really came into their own. Successes grew until the night of May 19–20, when the Luftwaffe mounted its last major raid on London and lost twenty-four aircraft to RAF night fighters, compared with only two to antiaircraft fire.

The best of the British night fighters evolved from the RAF's most versatile multirole aircraft, the de Havilland D.H.98 Mosquito. Captain Geoffrey de Havilland's "Wooden Wonder" was itself a 1938 derivative of his previous Comet racer and Albatross airliner, with the original purpose of building a high-speed bomber constructed of nonstrategic material—wood—with a range of 1,500 miles and a 1,000-pound bomb load. At first the Air Ministry rejected de Havilland's proposal because he refused to arm his design, relying on speed alone for protection. On December 29, 1939, however, de Havilland was authorized to proceed, and in the course of building his prototype, he had the foresight to allow for the accommodation of cameras for photoreconnaissance duties and four 20mm cannon under the crew floor for a possible fighter variant. Less than eleven months after serious design work had begun on it, the first prototype was flown by Geoffrey de Havilland Jr. on November 25, 1940. Not only was its maneuverability impressive for a twin-engine plane, but with a speed of 382 miles per hour for the PR Mark I photoreconnaissance variant, it was the fastest combat plane at that time, and

would remain so for the next two and a half years. De Havilland had already been awarded a contract for fifty bombers on March 1, 1940, which was later altered to twenty bombers and thirty fighters. Full-scale production orders, involving manufacture in Britain and Canada, soon followed.

The Germans soon became interested in the new de Havilland, and a German spy, air-dropped near Salisbury Field on May 13, 1941, was caught the following day, just one day before the prototype night-fighter version, the NF Mark II, made its first flight. The Mosquito NF II had strengthened wing spars to stand up to high-speed maneuvering, an armament of four 20mm Hispano cannon and four .303-in. Browning guns, and an AI Mark IV radar with "arrowhead" aerial in the nose. Its maximum speed was 370 miles per hour.

The first squadron to be equipped with the night-fighting Mosquito was No. 157 Squadron, formed at Debden under the command of Wing Cmdr. Gordon Slade on December 13, 1941, then moved to Castle Camps, where it received its first T.III trainer on January 26, 1942. While the aircrews worked up, British bombers had struck the historic city of Lübeck on the night of March 28–29, 1942, and Hitler responded by ordering a series of "Baedeker" raids, specifically targeting historic or scenic cities such as Exeter and Bath. Amid this atmosphere of terror and increasing civilian casualties, three Mosquito N.F. Mark IIs of No. 157 Squadron, equipped with AI Mark V radar, flew the first operational sorties on April 27–28, joining nineteen Beaufighters and Spitfires in an attempt to intercept a Baedeker raid on Norwich and the dropping of mines by Ju 88As of KG 30 off the Norwich coast. Although some radar contacts were made, the Mosquitos

De Havilland Mosquito NF Mark II W4093 is delivered to No. 157 Squadron RAF in February 1942. (*Andrew Thomas*)

failed to accomplish anything, the eye-sight of crews being plagued by cannon flash, as well as by burning from the exhausts and cowlings. De Havilland swiftly responded to the aircrews' complaints by installing flash eliminators on the guns and shrouds over the exhausts.

Meanwhile, the Baedeker raids continued with York on the night of April 28. On the following night No. 151 Squadron flew its first Mosquito sortie, and a Mosquito of No. 157 Squadron stalked a Dornier Do 217 until the Germans spotted it in the moonlight and evaded their pursuer. On May 19, one of No. 157 Squadron's planes fell victim to engine failure and crashed at Castle Camps, killing both crewmen. There would be two other accidents in May, but the Mosquitos accomplished nothing until the night of May 29–30, when a plane of No. 151 Squadron caught and shot up a Do 217E-4 engaged in a raid on Grimsby. The Mosquito was in turn damaged by return fire, and its pilot had to fly 140 miles back to the squadron's base at Wittering on one engine. A second No. 151 Squadron Mosquito, crewed by Pilot Officer John Wain and Flt. Sgt. Thomas S. G. Grieve, damaged another Do 217E-4 from KG 2, while Squadron Leader G.

Ashfield of No. 157 Squadron claimed to have "probably" downed a Do 217E-4 south of Dover.

Official success continued to elude the Mosquitos until the night of June 24–25, when the Germans struck at Nuneaton. One of the aircraft sent up to intercept the raiders was a Mosquito piloted by Wing Cmdr. Irving Stanley Smith, the New Zealand-born leader of No. 151 Squadron, who had previously scored four victories in Hawker Hurricanes and one while flying a Defiant. At 2330 hours he detected a He 111, closed to three hundred yards, opened fire and reported seeing it dive away with fuel trailing from its port tanks. His opponent, which was counted only as "damaged," was probably an He 111H-6 of *Erprobungs und Lehr Kommando* 17, which did indeed limp back to its base in the Netherlands—minus its radio operator, *Obfw.* Paul Wilhelm Krause, who panicked and bailed out when the Mosquito's gunfire set off the bomber's flares. His body was washed ashore at Pakefield on July 5.

The night was still young for "Black" Smith. Ten minutes after attacking the Heinkel, he was vectored to a Do 217, opened fire at a range of one hundred yards and sent the bomber into the sea, where it exploded. The Mosquito had drawn first blood at last, and II./KG 40 recorded the loss of a Do 217E-4 and its crew. At 2348, Smith made radar contact with another Do 217E-4, this time from I./KG 40, flown by *Ltn.* Karl von Manowarda. Closing to three hundred yards, the New Zealander fired again, and both of the bomber's wings exploded in flames. After closing to fire one more burst at a hundred yards, Smith saw his second confirmed victory of the night— and his eighth of the war—crash into the Wash.

The team of Wain and Grieve forced an He 111H-6 down in the North Sea on the following night, and when the Germans bombed Norwich on the night of June 26–27, it cost them a Do 217E-4 piloted by *Fw.* Hans Schrödel of 3./KG 2, courtesy of another of No. 151 Squadron's Mosquitos.

While the Mosquito NF Mark II was proving its worth in the defensive role, a new Mosquito unit, No. 23 Squadron based at Ford, began flying offensive missions against German air bases in France, using NF Mark IIs with their radar deleted so they would not fall into enemy hands if one of the planes were brought down. The first such mission was flown on the night of July 5–6 by Wing Cmdr. Bertie Rex O'Bryen Hoare, who was already credited with one confirmed and one probable victory in the Blenheim IF, as well as two enemy planes shot down, four damaged, and one probably destroyed on the ground while flying Douglas Havocs. On the following night "Sammy" Hoare, who boasted a handlebar moustache that measured "six inches wingtip to wingtip," destroyed a Do 217 sixteen miles east of Chartres for the first Mosquito night intruder victory. He downed another enemy plane over Orléans on July 30–31.

From that time on, the Mosquito went on to serve as the RAF's premier night fighter throughout the war, with as much success as it enjoyed in the light bombing, pathfinder, and reconnaissance roles. One of the plane's many noteworthy proponents was Norwegian ace of aces Svein Heglund, who scored the last three of his seventeen victories in Mosquito Mark 30s of No. 85 Squadron, and who described just how different the night-fighting environment was when compared to his previous experience as a Spitfire pilot:

It was like another, fantastic world, with all sorts of lights, flares and fires punctuating the darkness. I saw V-1 jet missiles skimming along the ground, and V-2 rockets on their way to England—I sometimes tried to intercept them, only to see them go up at ever increasing speed into the stratosphere. The pathfinder aircraft dropped markers on bombing targets, in all sorts of colors. During bombing raids, huge fires would cause smoke and clouds to rise 20,000 to 30,000 feet into the air—then all would be quiet again. One saw signal rockets being fired to alert German night fighters, tracer bullets during air-to-air encounters—it was all amazing.[3]

The RAF made far more offensive use of night fighters in the intruder role than the Luftwaffe did, primarily because of Hitler's preference for the populations of German cities being able to see their attackers going down in flames. By late 1942, however, the growing use of Mosquitos to precede British bombers to the target or to prowl around the peripheries to pick off German night fighters had become a serious menace to the Luftwaffe's air defenses. The Germans responded with a new generation of specialized night fighters, including a "wooden wonder" of their own, the Kurt Tank-designed Focke-Wulf Ta 154, which was even christened with the sobriquet "*Moskito*" by its builders. The first prototype flew on July 7, 1943, and 250 were ordered in November, but a series of crashes held up the type's operational debut—it was eventually ascertained that the glue used to bond its wooden components contained too much acid, which weakened rather than strengthened the joints.

The Heinkel He 219 proved to be an excellent night fighter from its first operational test flight, in spite of official efforts to downplay its capabilities. (*Bundesarchiv*)

The closest thing to a "Mosquito-swatter" that the Germans were able to produce in quantity had originally been conceived by Ernst Heinkel A. G. as a *Zerstörer*, with provision for use as a short- to medium-range bomber or as a torpedo bomber. The design had been rejected by the *Reichsluftfahrtministerium*, but as British bombing attacks against German cities increased in late 1941, the RLM asked Heinkel to resurrect his design, this time as a night fighter. Work began at Rostock in January 1942, but just as the drawings were completed, they were destroyed in a series of RAF strikes during March and April. Heinkel transferred its development office to Vienna-Schwechat and literally went back to the drawing board. The result of that interrupted effort—the first He 219V-1 *Uhu* (Eagle Owl)—did not fly until November 15, 1942. Powered by two 1,750-horsepower Daimler-Benz DB 603A engines, the prototype showed excellent flying qualities from the outset, and in April 1943 four of Heinkel's factories were building the first of three hundred He 219A-0s which had been ordered.

By that time, concern over the Mosquito had reached the highest levels of the Nazi government, causing an urgent demand for a *Moskito-Jäger* (Mosquito-hunter). That, in turn, threatened the He 219's future, as Kurt Tank offered his economically attractive Ta 154

as the ideal solution, while Erhardt Milch, resolutely opposed to the idea of diluting German industrial efforts on the production of specialized aircraft, advocated a night-fighter version of the Junkers Ju 188. Sensing that it would take some tangible proof to save his fighter from cancellation, Ernst Heinkel rushed several prototypes, armed with a battery of four 30mm MK 108 (in the He 219A-0/R1) or MK 103 cannon (in the He 219A-0/R2) housed in a ventral fuselage tray, as well as two MG 151 machines guns in the wing roots, to 1./NJG 1's base at Venlo in the Netherlands. There he urged *Maj.* Werner Streib, by then commander of the wing's *Gruppe Stab* (Group Headquarters), to put the new fighters to the test—in combat.

The opportunity soon arose. On the night of June 11, 1943, RAF Bomber Command launched a "maximum effort" of 693 planes to Düsseldorf, with Mosquito Pathfinders marking the target. Aided by a well-established *Himmelbett* GCI system, comprised of a network of *Freya*, *Würzburg*, and *Giant Würzburg* radar stations, Streib took off in the He 219V-9 attached to *Stab*/NJG 1 and was guided to multiple targets by his *Bordfunker* (radio operator), *Uffz.* Helmut Fischer.

Handling his untried airplane with consummate skill, Streib wrought havoc among the bombers, shooting down a remarkable total of five, of which two were positively identified as a Handley Page Halifax Mark II of No. 78 Squadron and an Avro Lancaster Mark II of No. 115 Squadron. NJG 1's Me 110s took a grisly toll on the bombers as well.

As he returned to Venlo and entered his final landing approach, Streib tried to lower his flaps, but the lever failed to lock and the flaps retracted again. At that critical moment, the He 219 went out of control and the starboard engine seized. The plane came down hard on the runway, at which point the starboard engine broke away from the wing and sliced into the cockpit section, breaking the plane's back. Streib and Fischer, still strapped to their seats, skidded down the runway behind the still-intact nose wheel until they finally came to a stop about 150 feet ahead of their wrecked plane. To the amazement of both men, neither had been hurt in this spectacularly disastrous finale to a spectacularly successful combat debut.

Milch conceded that Streib had put on an impressive show, but added that he was good enough a pilot to have done the same with any other night fighter. In spite of Streib's crash landing, however, NJG 1 crews flew further combat missions in the remaining He 219s and claimed twenty more victories over the next ten days. RLM officials continued to balk at accepting the He 219, expressing misgivings about its tricycle landing gear and the close proximity of its propellers to the cockpit, although Ernst Heinkel addressed the latter complaint by installing the first ejector seat system in the plane. Production and development of the *Uhu* proceeded in the face of almost constant opposition. Virtually all of the

268 He 219s built were committed to defense of the Reich, along with all twenty preproduction aircraft, and they proved to be outstanding whenever they got the chance to fight. The vast majority of Germany's night-fighter force, however, consisted of improved versions of established designs converted to the role, mainly the Me 110G and Ju 88G. Prodigious feats were performed in them, Streib bringing his personal tally up to sixty-eight before being killed on January 21, 1944. Germany would also produce the highest-scoring night-fighter pilot in history, Heinz-Wolfgang Schnaufer, who scored his 121 victories in Me 110s. Other types were pressed into nocturnal service, such as the Do 217N and the most effective *Moskito-Jäger* of all, the Messerschmitt Me 262 jet. Belatedly committed to battle on December 17, 1944, with 10./NJG 11 under the command of *Obltn.* Kurt Welter, twenty-five Me 262A-1as and seven Me 262B-1a/U1 two-seaters equipped with *Neptun* radar managed to bring down a total of forty-eight Mosquitos by May 7, 1945. By then, however, there was no Reich left to defend.

Japan, like the other powers, adapted several of its aircraft to night-fighting tasks as necessity dictated, but the career of their most successful night fighter represented a double change in role. When the Nakajima J1N heavy fighter was originally designed for the Japanese navy in 1938, it was to have been a twin-engine long-range escort, like the Messerschmitt Me 110. Powered by twin 1,130-horsepower Nakajima NK1F Sakae 21 radial engines, which gave it a speed of 329 miles per hour at 19,685 feet, and which was fitted with long fabric-covered ailerons and slotted flaps, the J1N1 was exceptionally nimble—even able to hold its own in a dogfight with a Mitsubishi A6M2 Zero—but the navy became fixated over the practicality of the two dorsal barbettes, each housing a 7.7mm machine gun, which were to be remotely controlled by the navigator. By 1941, the single-engine Zero had proven itself more than adequate for the fighter escort role, and by the time the J1N's protracted development reached fruition in July 1942, the navy had decided to use a three-seat version, the J1N1-C, with its operational ceiling of 33,795 feet, as a high-altitude reconnaissance plane instead.

At least two J1N1-Cs were on strength with the 251st *Kokutai* at Rabaul in the spring of 1943, when the group's executive officer, Cmdr. Yasuna Kozono, came up with an idea. At that time, Boeing B-17s of the US Fifth Air Force were mounting night bombing attacks on Japanese bases throughout the Solomon Islands. The raids were more of a sleep-disrupting nuisance than a major threat, but Japanese morale was affected by the fact that nothing could be done to stop them. Kozono proposed to arm the J1N1-C with upward- and downward-firing 20mm cannon, mounted at an oblique angle behind the pilot so that the gun flashes would not affect his night vision. Navy General Headquarters dismissed his idea as absurd; but Kozono went all the way to Japan to argue his case, and eventually he persuaded them to modify two planes in accordance with his concept. After being tested at Toyohashi, the two J1N1-C-kais, as they were designated, arrived at Rabaul and were assigned to the 251st *Kokutai* on May 10. The J1N1-C-kai had no radar, so its crew had to depend on searchlights or the moon to help locate its target. The pilot and observer were given injections to aid their night vision just before a mission,

but the substance that was administered has never been identified.

The night-fighting J1N's first test came when B-17s of the 64th Squadron, 43rd Bomb Group, left their base at Port Moresby, New Guinea, just after midnight on May 21, 1943. Japanese coast watchers alerted Rabaul of the bombers' approach, and both J1N1-C-kais commenced hunting. At 0337 hours, twenty-three-year-old Chief Petty Officer Shigetoshi Kudo and his observer, Lt. j.g. Akira Sugiwara, spotted a B-17E that was caught in a searchlight. Kudo slipped underneath the unsuspecting bomber and pumped shells into its underside. The plane crashed in Saint George's Channel, killing Maj. Paul Williams and all but one of his crew. The sole survivor, Master Sgt. Gordon Manuel, later reported: "I heard one dull explosion, then a series of smaller ones. The ship wrenched to the left and shuddered."[4]

The night was still young, and at 0408 Kudo and Sugawara located another B-17 silhouetted in the thirteen-day-old moon, though they were unable to get into a good firing position before it escaped. Twenty minutes later, they encountered another B-17E of the 64th Squadron, and after Kudo fired his single gun once more, that plane, too, descended in flames. He had expended 178 rounds in the course of the night's actions.

Kozono's theories had been proven in spectacular style, although the Americans at first attributed the loss of the two bombers to antiaircraft fire or other operational causes. They should have had cause for concern after June 5, however, when two Consolidated B-24s fell victim to another of the 251st *Kokutai*'s J1Ns, crewed by Chief Petty Officer Satoshi Ono and Lt. j.g. Kisaku Hamano. On June 10 Ono and Hamano claimed another two B-17s, of which one—again from the

A Nakajima J1N1-C-kai of the 251st *Kokutai*, armed with obliquely mounted 20mm cannon, taxis for takeoff on Lakunai airfield, Rabaul. Chief Petty Officer Shigetoshi Kudo's stunning run of success with the improvisation in the summer of 1943 led to the development of the J1N1-S *Gekko* (Moonlight) night fighter (*Osamu Tagaya via James F. Lansdale*)

64th Bomb Squadron—was subsequently confirmed.

The team of Kudo and Sugawara scored again on June 11 and 13, the latter being Lt. John Woodward's B-17E *Georgia Peach* of the 65th Squadron, 43rd Bomb Group, caught in the searchlights and dispatched at 0314, to crash at Ulamona, New Britain. Of the two crewmen who got out, only one, bombardier Lt. Jack Wisener, survived his captivity; the navigator, Lt. Philip Bek, was killed by his captors.

Kudo was credited with another double victory on the night of the fifteenth and scored his seventh kill on June 26, when B-17E *Naughty But Nice* of the 65th Bomb Squadron was caught in the searchlights during an attack on Vunakanau airfield. Rabaul's antiaircraft fire, though dangerous by day, had been notoriously inaccurate at night, so the B-17's crew were little concerned until their plane was suddenly raked by 20mm fire from a J1N1-C-kai crewed by Kudo and Warrant Officer Michitaro Ichikawa. Two engines caught fire and the Fortress went into a spin. Its navigator, 1st Lt. Jose L. Holguin, was wounded in the chin and left leg and was trying to help a crewman

strap on a parachute when he was sucked out of an open escape hatch. Battered by a tree as he parachuted down, "Joe" Holguin suffered a broken back, and yet endured twenty-seven days in the jungle until natives carried him to their village. He was subsequently turned over to the 6th *Kempei Tai*—Japan's notorious military police—but survived the war in their prisoner of war camp as well. The rest of *Naughty But Nice*'s crew were killed. Kudo and Sugawara finished out the month with a B-17 on the thirtieth.

In early July the J1N1-C-kais were moved up to Ballale airfield in response to American attacks in the Buin area. The redoubtable team of Kudo and Sugawara were among them, and on July 7 he scored his final night victory when he downed a Lockheed Hudson over Buin airfield. A month later, the "King of the Night," as Kudo had come to be called, was awarded a ceremonial sword by Vice Adm. Jinichi Kusaka, commander of Rabaul's Eleventh Air Fleet, for his distinguished service. Although badly injured in a crash in May 1945, Kudo survived the war with a total of nine victories, but he died of complications resulting from his wartime injury in 1960. Satoru Ono would also survive the war with the commissioned rank of lieutenant junior grade and a total of eight enemy planes to his credit.

Kudo's success convinced the Japanese to give the J1N a new lease on life exactly as the Germans had had with their Me 110—by giving full priority to production of a night-fighter version with a smoother cabin outline and radar in the nose—the new fighter being the J1N1-S *Gekko* (Moonlight). Code-named Irving by the Allies, the *Gekko* represented the majority of the 471 J1Ns produced and was fairly successful until mid-1944, when the Boeing B-29 Superfortress outperformed it in speed and altitude.

A latecomer in dealing with the threat of night attack, the United States got its first convincing wake-up call during the Guadalcanal campaign, when Japanese aircraft attacked the US Marines at Henderson Field at night with everything from sleep-depriving nuisance raids by Mitsubishi F1M2s (nicknamed "Washing Machine Charlie" by the Marines because of the sound of their engines) to more serious attacks by Mitsubishi G4M1 Betty medium bombers. The most spectacular Japanese success of the campaign occurred near Rennell Island on the night of January 30, 1943, when flare-equipped G4M1s, soon to be dubbed "Tojo the Lamplighter" by their opponents, illuminated the heavy cruiser *Chicago*, which was then torpedoed by the bombers and finished off by more G4M1s the following day.

As with the other warring powers, the United States Army Air Forces responded by converting a variety of day fighter and bomber types to the night-fighter role, most notably the Douglas A-20 Havoc (which in its radar-equipped night-fighter

guise was known as the P-70) and the Lockheed P-38M Lightning. In November 1943, the US Navy enjoyed some success by using a radar-equipped Grumman TBF-1C Avenger to vector two F6F-3 Hellcats to their targets, but it would do better in the following year by installing a radar in one of the Hellcat's own wings, to produce the Grumman F6F-5N. The Vought F4U-2 Corsair also sported radar in one wing, as would its postwar successors, the F4U-4N and F4U-5N. One F4U-5N pilot, Cmdr. Guy P. Bordelon Jr. of composite squadron VC-3, shot down five enemy planes to become the Korean War's only American night-fighter ace, as well as its only US Navy ace.

Even while such improvisations tided over the American air bases and carriers, a new, more specialized airplane was in the works—in fact, it was the only World War II plane designed from the onset as a night fighter to achieve production status. In 1944, the USAAF was ready to unleash the Northrop P-61A Black Widow.

During the autumn of 1940, US Army observers in Britain had taken note of the nocturnal missions already being carried out by the RAF and the Luftwaffe. As a result, on October 21, 1940, the Army Air Corps Materiel Command presented John K. Northrop with a requirement for a specialized night fighter. Two weeks later, on November 5, he and his assistant, Walter J. Cerny, were presenting their proposal at Wright Field, Ohio. After undergoing some revisions, the design was approved for development on January 30, 1941. On May 26, 1942, Vance Breese took the XP-61 prototype up for its first test flight.

Jack Northrop's creation was an all-metal, twin-boom, twin-tail monoplane, 66 feet 3/4 inch in wingspan and 49 feet 7 inches in length. With an empty weight of 24,000 pounds and a maximum loaded weight of 32,400 pounds, it was three times heavier than a P-51 and almost twice as heavy as a P-47. As with other large American fighters, all that weight was compensated for by the engines— two eighteen-cylinder Pratt & Whitney R-2800-65 twin-row air-cooled radials generating more than 2,000 horsepower each. Double-slotted flaps reached along almost the full span of the wings, which also featured small ailerons and lateral-control spoilers, an overall arrangement that was years ahead of its time and endowed the big plane with astonishing agility. With a range of 1,000 miles, it could cover a wide area, and with a maximum speed of 366 miles per hour at 20,000 feet, it could fly as fast as many single-engine fighters.

The first of thirteen preproduction YP-61s flew on August 6, 1943, and provided much critical data that would go into the P-61A, which entered service in March 1944. Early Widows had four 20mm cannon housed in a ventral fairing and a dorsal barbette with four .50-caliber machine guns, though buffeting problems encountered when elevating or rotating the latter resulted in the barbettes being removed after the first thirty-seven P-61A-1s left the Northrop assembly line in October 1943. Equally important to the plane's success was the new SCR-720 AI radar housed in the nose of the fuselage nacelle.

In March 1944, personnel of the 422nd Night Fighter Squadron (NFS) arrived at Scorton, England, and began training in their new P-61A-5s and A-10s, to be joined in June by the 425th NFS. At about the same time, Widows arrived in the Pacific to replace P-70s in the 6th NFS on May 1, and those of the 419th NFS, two days later.

As much as the European P-61 crews were dying to match wits with the Luftwaffe's best, it would be in the Central Pacific that the Black Widow would bite first. Like most P-61 pilots, 2nd Lt. Dale F. Haberman had undergone his initial training in the new type with the 481st Night Fighter Operational Training Group at Orlando Field, Florida, before being assigned in March 1944 to the 6th NFS, Seventh Air Force, at John Rogers Airport, Oahu, Hawaiian Islands. From there, Haberman and eight other pilots and their ROs (radar operators) were detached from the squadron and sent to Isley Field on the newly taken island of Saipan, arriving on June 21.

Flying with Flight Officer Raymond P. Mooney as his RO in a plane christened *Moonhappy*, Haberman claimed a "probable" about a week before scoring his— and the Black Widow's—first confirmed victory. On June 30, Haberman and Mooney took off to investigate an unidentified plane, which they located about fifty miles east of Saipan. He subsequently described what followed:

> We were vectored on the bogey at 7,000 feet, with Mooney insisting that we had two targets in formation on our scope. We chased them from 7,000 to 22,000 over a 25-minute time period, never quite reaching their altitude. I discovered later that I had forgotten the radio, leading in on "Transmit," so the whole intercept was broadcast over the entire Saipan area.
>
> I finally got a visual on the two bogeys. A Betty with a Zero on her right wing were turning toward the island and at a much higher altitude than we were. I pulled the "Widow" up on her tail and nailed the Betty with a five-second burst. The Betty burst into flame behind the left engine. The flame trailed for some time, then the aircraft exploded and was seen crashing into the ground.

A Northrop P-61 Black Widow of the 6th Night Fighter Squadron patrols the Pacific waters off Saipan in July 1944. On the night of June 30, 2nd Lt. Dale F. Haberman of the 6th NFS and his radar operator, Flight Officer Raymond P. Mooney, scored the first P-61 victory when they destroyed a Mitsubishi G4M2. (*National Archives*)

Haberman then performed a high-speed stalling turn and went into a power dive with the Zero in pursuit. Although the P-61 reached 475 miles per hour, the Japanese fighter closed in on its tail. Glancing back, Mooney yelled to Haberman: "Twist port or starboard! Let's get the hell out of here! The wingman is right there!" Haberman saw tracers flash by his plane and wondered if his wings would come off, but the Black Widow held together and he managed to lose the Zero in a cloud.[5]

Pulling out at 1,500 feet, Haberman returned to Saipan, only to be fired on by his own troops, who were not familiar with the new airplane. After having to dodge his own flak at treetop level, Haberman finally brought *Moonhappy* down safely on Isley Field. Haberman would later add two more to his score, and his colleagues did yeoman service in protecting the vital Boeing B-29 air bases

of Saipan, Guam, and Tinian from further nocturnal attacks by the Japanese.

Back in England, the P-61 aircrews became livid with rage to learn that they were being barred from operations because their aircraft had been declared inferior to the Mosquito. The 422nd NFS demanded a fly-off, and when the RAF granted its request, 1st Lt. Donald J. Doyle went up to defend the Widow's honor against an equally determined RAF champion on July 7. The result was that the P-61 outran, outclimbed, and outturned the Mosquito at 5,000, 10,000, and 20,000 feet. Besides that convincing show, the appearance of growing numbers of V-1 pulse-jet-powered guided bombs over Britain since mid-June led to the lifting of the ban on P-61 operations. The Widow crews' first job would be to join the RAF fighters in "anti-diver" missions against Hitler's latest "vengeance weapon."

On the night of July 14–15, 2nd Lt. Herman E. Ernst, with Flight Officer Edward H. Kopsel as his RO, were cruising over the Channel at an altitude of 7,500 feet when they spotted a V-1 at 2,000 feet. Ernst dived after the "diver," which he noted was racing along toward England at 340 miles per hour. "All of a sudden there was a loud boom and a tremendous amount of noise in the cockpit," Ernst recalled. "Kopsel was screaming into the intercom, but I could not understand a word. My first thought was that a German night fighter had gotten in behind us and had shot us down . . . and on our very first mission!"

Ernst's plane was still responding to the controls, so he brought it back to his base at Ford to discover that the Plexiglas tail cone had disintegrated due to the air pressure. A flat piece of Plexiglas was installed in its place, and on the following night Ernst and Kopsel went up again. Again they encountered a V-1, and once more Ernst dived after it. "This time we closed the gap and fired several 20mm rounds," he reported. "They found their mark all over the propulsion unit and the bomb lost power, nosed over and went into the sea."[6]

The 422nd NFS had scored the first victory for the P-61 in Europe—over a pilotless flying robot. The Black Widows would have ample opportunities to strike against Luftwaffe aircraft in the months to come, however. In spite of—or perhaps because of—its late arrival in the night-fighting arena, the Black Widow's fighting career was as distinguished as it was brief. Sixteen squadrons operated P-61s on all fronts in which the USAAF operated, and at least three pilots—one in the Pacific, two in Europe—would "make ace" in Black Widows. Herman Ernst would be one of them, adding

three Ju 87s, a Ju 188, and an Me 110 to his score by May 1945.

Eighteen

Odd and Ends

IMPROVISATIONS AND DEVELOPMENTAL
DEAD ENDS, 1940–45

War induces combatants to seek advantages by any means possible. In regard to World War II aircraft, the quest for an edge encompassed numerous aspects, such as size, bomb capacity, speed, rate of climb, altitude, maneuverability, and potency of armament. Even though the fundamental configurations of aircraft had been established by the end of World War I, the quest for an advantage in the air brought a spate of new variations into the sky. For every successful innovation, such as the radar-equipped all-weather fighter, there were interesting failures, such as the turret fighter and the lightweight interceptor. And for every evolved or carefully conceived design, there was a wartime improvisation that occasionally worked—though not always as originally intended.

The British should have known better than to develop the turret fighter. The two-seat fighter from which it evolved, the Bristol F.2B of 1917, had achieved its phenomenal success by being flown as a single-seater with a sting in the tail, rather than relying primarily on the rear gunner's weapon. The F.2B's successor in 1931, the Hawker Demon, differed little from it in armament, but the problems encountered by the gunner in handling a .303-inch gun in the open cockpit of an airplane flying at nearly twice the Bristol's speed led the Air Ministry to seek a more advanced weapons system.

One solution to the problem was offered by Boulton Paul Aircraft Ltd., which had been subcontracted to build Demons for the RAF, and which had also obtained the rights to produce an electro-hydraulically operated turret—invented by French engineer Joseph Bernard Antoine de Boysson—capable of traversing 360 degrees and incorporating either a 20mm cannon or four .303-inch machine guns. After seeing the turret demonstrated in the nose of a Boulton Paul Overstrand bomber, the Air Ministry issued a specification for a fighter armed with four machine guns in the de Boysson turret and capable of flying as fast as the Hawker Hurricane fighter. Since Hurricanes were expected to protect the turret fighter from enemy fighters while it attacked enemy bombers from the side or below, the specification limited armament to the turret. Such a measure saved weight, but in essence it made the pilot nothing more than a chauffeur for his gunner—hardly a role that went over well with aggressive fighter jockeys.

Designed by John Dudley North, the Boulton Paul Defiant was a commendably clean and compact airplane, powered by the same 1,030-horsepower Rolls-Royce Merlin III engine used in the Hurricane and the Supermarine Spitfire. A retractable fairing helped to smooth the airflow behind the rear turret when it was not in use, and in spite of the drag that the turret still imposed on it—as well as a gross weight of 8,600 pounds compared to the Hurricane's 6,218—the Defiant managed a maximum speed of 302 miles per hour at 16,500 feet compared to the Hurricane's 316. It took the Defiant 11.4 minutes to climb to that altitude, however, whereas the Hurricane could reach it in only 6 1/2 minutes. First flown on August 11, 1937, the Defiant was approved for production, but because Boulton Paul was then relocating from Norwich to a new plant at Wolverhampton, the first operational Defiants were not deployed with No. 264 Squadron until December 1939. When the Germans invaded the Low Countries on May 10, 1940, the unit moved from its training base at Martlesham Heath to Duxford; from there, A Flight flew to Horsham Saint Faith and B Flight returned to Martlesham, where it would operate alongside the Spitfires of No. 66 squadron.

The Defiants did their intended job fairly well in their first combat. On May 11, No. 264's commander, Squadron Leader Philip A. Hunter, and Pilot Officer Michael H. Young flew an evening convoy patrol as far as the Happisburgh lighthouse. The next day Flt. Lt. Nicholas G. Cooke led A Flight on a patrol off the Dutch coast, accompanied by six Spitfires of No. 66 Squadron. They soon encountered enemy aircraft, and the Defiants

Boulton Paul Defiants of No. 264 Squadron fly a tight formation in July 1940. The foreground plane, L7026 PS-V, was flown by Pilot Officer Eric G. Barwell, while N1636 PS-A behind it was piloted by Squadron Leader Philip A. Hunter. (*Imperial War Museum*)

drew first blood five miles south of The Hague as a Junkers Ju 88A fell victim to Hunter and his gunner, Sgt. Frederick H. King, while a second was claimed by Young and Leading Aircraftman Stanley B. Johnson. Cooke added a third victory to the squadron's opening tally when he caught an He 111 six miles south of The Hague and his gunner, Cpl. Albert Lippett, shot it down.

On the following morning, six Defiants of B Flight, accompanied by six Spitfires of 66 Squadron's A Flight, were flying another sweep over the Dutch coast when they spotted Junkers Ju 87Bs dive-bombing a railway line and attacked. Between them, the British claimed ten of the Stukas—four of which were credited to the Defiants—before themselves coming under attack by Messerschmitt Me 109Es of the 5th *Staffel* of *Jagdgeschwader* 26. Flight Lieutenant Kenneth McLeod Gillies, a Spitfire pilot who had shot down a Stuka east of Rotterdam, damaged an Me 109 before 66 Squadron disengaged. One Spitfire fell victim to *Ltn.* Hans Krug, but its pilot managed to force land his damaged plane in Belgium.

The fight had a much grimmer outcome for 264 Squadron. In their first encounter with enemy fighters, the six

Defiant crews found themselves unable to evade the Me 109s, and five were shot down in short order, although only one German, *Fw.* Erwin Stolz, identified his adversary as a Defiant at the outset; the other victors—*Ltn.* Eckardt Roch (who claimed three), *Leutnant* Krug, *Uffz.* Hans Wemhöhner, and *Fw.* Wilhelm Meyer— all claimed Spitfires, before subsequently learning the true identity of their adversaries. The sole Defiant pilot to return, Pilot Officer Desmond Kay, claimed that five German fighters went down in the course of the massacre, and they were duly credited to the squadron.

In actuality, the Defiants had managed to shoot down only one of their assailants, who—contrary to popular misconception—already knew what he was up against and fell victim to overconfidence, rather than from mistaking the turret fighter for a single-seater. As *Ltn.* Karl Borris himself recorded it in his diary:

> Enemy contact with a mixed British formation . . . I bank toward a Defiant, I can clearly see the four machine guns in its turret firing; however, I do not think they can track me in a dogfight. I approach closer, and open fire at about seventy meters range. At this moment, something hits my aircraft, hard. I immediately pull up into the clouds and examine the damage. The left side of my instrument panel is shot through; a round had penetrated the Revi [reflex gunsight]; and a fuel line has obviously been hit—the cockpit is swimming in gasoline. The engine coughs and quits, starved of fuel. I push a wing over and drop from the clouds. Unbuckle, canopy off, out![1]

Borris parachuted onto a dike wall near the mouth of the Rhine River and made his way back to 5./JG 26 four days later. Having lived to profit from this reminder

of the price one pays for cockily dismissing any armed opponent, he would survive the war with forty-three victories.

For the next ten days, No. 264 Squadron refrained from operations, but on May 23, its Defiants joined in the RAF's desperate effort to cover the evacuation of Allied troops from Dunkirk. By the end of the month, the squadron had claimed forty-eight victories—thirty-seven on May 29 alone—but lost nine planes, including that of Cooke and Lippett, killed on the thirty-first. Occasionally, the Defiant's superficial resemblance to the Hurricane did mislead German fighters into attacking it from the rear, with sometimes fatal results for the attackers. Soon the Germans learned to distinguish between the Defiant and its single-seat stablemates, however, with results that spoke for themselves. A second Defiant unit, No. 141 Squadron, had a disastrous combat debut on June 28, when nine of its planes tangled with Me 109Es and lost seven while claiming only four victories. On July 19, nine more Defiants of 141 Squadron encountered Me 109Es of JG 51 and again lost seven planes, while one of the two surviving crews, Flt. Lt. Hugh N. Tamblyn and Sgt. S. W. N. "Sandy" Powell, claimed one of the enemy in return. In August the Defiant units' air bases were moved farther north, but the RAF's need for anything which could fly and fight at that time kept them engaged—and suffering mounting losses. By late 1940, the Defiant Mark Is were being relegated to the night-fighting role, and a radar-equipped version, the Defiant Mark II, was introduced. As such, they did well, being in fact the most successful night fighters of 1941 until sufficient numbers of Bristol Beaufighters and de Havilland Mosquitos became available to phase them out of first-line service.

Another concept which persisted throughout World War II was that of the lightweight interceptor, inspired by the matter of mass production amid the exigencies of war. The lightweight fighters' exact roles varied as much as did their builders' approaches to achieving them; but they all sought to wring the highest possible performance from the smallest, lightest possible airframe, built using the greatest amount of easily available materials (usually wood) in lieu of strategic materials (such as aluminum). Another thing that most of them held in common was failure. Of the many lightweight interceptors created just before or during the war, only three attained production status, and only one could truly be called successful.

During the mid-1930s, the *Service Technique de l'Aéronautique* of the French *Armée de l'Air* laid down a specification for a lightweight interceptor that was influenced by the monoplanes, which were then attracting much publicity in speed competitions. The chosen design, the Caudron-Renault CR.710, was designed by Marcel Riffard and was based on his sleek C.460, which between 1933 and 1936 had been outperforming larger, more powerful aircraft in international competitions. (During which time Renault had bought up the Caudron firm in 1933.) Like the racing plane, Riffard's CR.710 fighter was of wooden stressed-skin construction and characterized by a long, slim fuselage. Its power plant was a 500-horsepower Renault 12R 01 twelve-cylinder inverted-V air-cooled engine. The first prototype had fixed, spatted landing gear and oval-shaped vertical tail surfaces when it first flew on July 18, 1937, and later carried two wing-mounted drum-fed 20mm Hispano-Suiza HS-9 cannon. The second prototype, CR.710-

02, featured more angular vertical tail surfaces, and the CR.713 introduced retractable landing gear. A third version, the CR.714, first flew on July 6, 1938, and differed from the CR.713 primarily in armament, the cannon being replaced by four 7.5mm MAC M39 machine guns, housed in two underwing trays.

After some final modifications, the CR.714, also known as the Cyclone, was ordered into production on November 5. The production CR.714 featured an improved 12R 03 engine, which had a carburetor that allowed negative-G maneuvers. It had maximum speed of 286 miles per hour at 16,450 feet, and climbed to 13,125 feet in 9 minutes and 40 seconds.

The original production order was for twenty CR.714s, with an option for a further hundred eighty, but once the fighter entered service, the *Armée de l'Air* judged it to be unsuitable for combat. Six were sent to Finland but arrived too late to take part in the Winter War, and desperate though the Finns were for any combat aircraft, they never used the Caudrons in battle. The other Cyclones were assigned to two training squadrons at Lyon-Bron, made up of expatriate Polish pilots. By June 2, a total of thirty-nine CR.714s had been delivered to the Poles, who flew them operationally as *Groupe de Chasse I/145*, also known as the 1*ère Groupe Polonaise de Varsovie*, under the joint command of *Commandant* Józef Kepinski and his French advisor, *Commandant* Lionel A. de Marmier, a six-victory World War I ace.

For all the Cyclone's racy looks, the Poles soon became disenchanted with their new mount. It required a long take-off and landing run; the landing gear release often jammed; the variable-pitch propeller mechanism was prone to failure; the rate of climb was slow and so was the aileron response. Worst of all was the 12R 03 engine, which had trouble starting, was plagued by a weak crankshaft, had a tendency to overheat, and suffered from fuel and oil leaks. *Sous-Lieutenant* Witold Dobrzynski was killed in a crash on May 19, and three other Caudrons were written off in landing accidents on May 25. After inspecting GC.I/145 on May 25, Air Minister Guy La Chambre considered grounding the interceptors. Kepinski chose to keep them in spite of their faults, however. His men wanted to fight, and with the German offensive in the West under way, they had little choice but to make do with the fighters they had until their intended replacements, Bloch MB.152s, became available.

A Caudron-Renault CR.714 of *Groupe de Chasse* 1/145, whose Polish pilots tried to make the most of their less-than-ideal fighters in June 1940. (*Service Historique de Défense—Air*)

On June 2, GC.I/145's Cyclones flew from Villacoublay to the former RAF airfield at Dreux. Combat was joined on June 3, when *Commandant* de Marmier, *Lt.* Tadeusz Czerwinski, and *Sous-Lt.* Aleksy Zukowski dived on three He 111s and shot down two over Villacoublay. The unit carried out further patrols, but its next fight did not occur until June 8, when a flight led by *Capitaine* Antoni Wczelik engaged at least fifteen Messerschmitt Me 110Cs over Rouen. One Caudron was damaged, but the French confirmed the destruction of two Me 110s by Czerwinski, one each by Wczelik and Zukowski, and one shared between *Sous-Lt.* Jerzy Godlewski and *Caporal* Piotr Zaniewski. Kepinski and *Sous-Lt.* Czeslaw Glówczynski, who was already credited with three and a half enemy planes during the German invasion of Poland, scored probable victories over another two Me 110s.

Perhaps inevitably, GC.I/145's luck took a turn for the worse the next day, when seventeen Cyclones encountered twenty-five Dornier Do 17s escorted by twenty Me 109Es. Malfunctioning radios

prevented the CR.714 pilots from making a coordinated attack, and while Wczelik's flight hurled itself at the bomber formation, other Poles found themselves engaged in individual duels with the German fighters. Glówczynski was credited with one of the Messerschmitts, along with probable credits for a second Me 109 and a Do 17 (he would add one more German to his score on December 30, 1941, as a Spitfire pilot in the RAF). *Sous-Lieutenant* Jerzy Czerniak and *Sgt.* Mieczyslaw Parafinski were credited with one Me 109 each, while Wczelik, *Lt.* Julian Kowalski, and *Sgt.* Antoni Markiewicz shared in the destruction of another of the bombers (the Germans reported no Do 17 losses, but *Fw.* Fritz Specht of II *Gruppe*, *Kampfgeschwader* 54 returned to his base at Köln-Butzweilerhof on one engine with the tail and rudder of his Heinkel He 111P badly damaged after being attacked by enemy fighters over Evreux). *Lieutenant* Jan Obuchowski, *Sous-Lt.* Lech Lachowicki-Czechowicz, and *Caporal* Edward Uchto were killed, however, and Kowalski was wounded in the right arm, although he managed to land his damaged plane at Bernay. In addition, the riddled Caudrons of *Commandant* Kepinski and *Sous-Lts.* Jerzy Godlewski and Bronislaw Skibinski crash-landed in the Norman countryside, Czerniak crash-landed his shot-up plane at Dreux, and most of the other CR.714s returned in various damaged states.

Twelve of the group's thirteen remaining CR.714s were operational as they attacked fifteen Do 17s and twelve Me 109s over Étampes on June 10. The Poles' radios failed again, as de Marmier led them in a head-on attack against the bombers. One Dornier fell to de Marmier, a second to Czerniak, and Zukowski downed a third, while *Capitaine* Piotr Laguna accounted for an Me 109 over Henonville following a long pursuit. Kepinski was wounded in a lung by Me 109s, but in spite of a considerable loss of blood, he managed to make a wheels-up landing in a field. *Capitaine* Juliusz Frey, *Lt.* Waclaw Wilczewski, and *Lt.* Zdislaw Zadronski were also compelled to force land their shot-up planes.

Kepinski's executive officer, *Capitaine* Laguna, took command of what remained of GC.I/145, but there was little left to take charge of. On June 11, French technicians removed the instruments from eleven of the group's defective Caudrons and then burned them. The remaining twelve Cyclones were withdrawn to Sermaize, from whence eight of GC.I/145's pilots were assigned to GC.I/1 and eight to GC.I/8, both of which were equipped with MB.152s. The Poles continued to fly missions until June 18, when they learned of France's capitulation. Released from French service, they departed by ship from La Rochelle on the twentieth, to carry on their fight in Britain. Using hit-and-run tactics to make the most of their faulty fighters, the aggressive Poles of GC.I/145 had managed to shoot down twelve German aircraft and probably downed two others in the course of the Caudron-Renault CR.714's brief fighting career.

The United States produced two lightweight fighters during World War II. One, the Bell XP-77, was not accepted for production, the other, the Curtiss-Wright CW-21, was, but not by the Americans.

In the 1920s, the Curtiss-Wright Company established an independent division in Saint Louis, Missouri, under Vice President of Engineering George Page. One of its early products was an all-metal, two-seat monoplane, the CW-19L, small quantities of which were sold to China and Cuba in 1937 as the A-19R military trainer. Although criticized for vicious stalling characteristics and a tendency to ground-loop, the CW-19 had an outstanding rate of climb—1,890 feet per minute and 23,000 feet in fifteen minutes with an 820-horsepower Wright engine. That capability inspired Page to design a lightweight interceptor. His specialized concept, which completely ignored dogfighting or tactical capabilities, ran counter to what the US Army Air Corps was looking for in a fighter, but he proceeded with the CW-21's development with an eye on foreign customers.

Begun in 1938 under chief engineer Willis Wells, the CW-21 used a lot of CW-19 components, but was trimmed down and strengthened. The landing gear retracted into two clamshell-shaped fairings under the wing. Its engine was a Wright Cyclone R-1820-G5 radial with a takeoff rating of 1,000 horsepower. The original intended armament of four .30-caliber machine guns firing from inside the engine cowling was changed to two .50-caliber guns, effective beyond the range of a bomber's .30-caliber defenses.

Using an Alclad semimonocoque fuselage devoid of armor or fuel tank protection, the CW-21 weighed only 3,050 pounds empty and climbed at the phenomenal rate of 4,800 feet per minute, leading Curtiss-Wright to tout it as the "mile-a-minute interceptor." Its first flight was on September 22, 1938, with Ned Warren at the controls.

The first logical customer was China, where Curtiss-Wright was a shareholder in the Central Aircraft Manufacturing Company (CAMCO) at Nanking. On January 24, 1939, the first CW-21 prototype arrived by ship in Rangoon, Burma, then made its way up the Irrawaddy River by barge and finally by truck to Loiwing, where CAMCO had moved its facilities following the fall of Nanking to the Japanese in December 1938. Painted with blue and white Chinese tail stripes and demonstrated by Curtiss-Wright test pilot Robert Fausel in March, the plane proved to be surprisingly nimble as well as a fast climber—in a comparative dogfight with a Polikarpov I-152, Fausel managed to get on the biplane fighter's tail by means of a steep climb followed by a wingover. The Chinese were impressed, but had reservations about the plane's $70,000 price tag, and were concerned that it might be too hot for the average Chinese pilot to handle. They also requested that two .30-caliber machine guns be added to the existing armament, along with provision for belly tanks and the replacement of the wraparound Pyrolin plastic windshield, which yellowed with exposure to weather, with a three-piece armor-glass assembly, although that would reduce speed by an estimated 4 percent.

While contract negotiations dragged on, Fausel flew a patrol with a Chinese I-152 squadron on March 29. He later claimed that when Japanese bombers attacked Chunking on April 4, he took off just as the first bombs exploded, and two minutes later he attacked a formation of Italian-built Fiat BR.20s at 10,000 feet. On his third burst, Fausel's machine guns

jammed in the "on" position until they ran out of ammunition, but he saw one of the bombers smoking, and later learned that it had belly-landed in Chinese territory, where its crew were taken prisoner. Fausel's account notwithstanding, the Japanese army did not record any bombing missions to Chungking that day.

Turned over to the Chinese at Kwang Yong Pa, the demonstrator aircraft is believed to have crashed in June 1939; but three more CW-21s, with increased armament and auxiliary fuel tanks as requested by the Chinese, but retaining the original windshield, were soon produced at Saint Louis to serve as models for the assembly of twenty-seven more by CAMCO. Shipped to Burma in May 1940, the CW-21s were to be flown from there to Kunming by pilots of the newly formed American Volunteer Group— Ken Merritt, Lacey Mangleburg, and Eric Shilling—on December 23, 1941. Soon after taking off from Lashio, however, all three planes developed engine trouble, probably because of dirty fuel, and crash-landed on a mountain slope. Shilling and Merritt survived, but Mangleburg was killed. In China, two assembled CW-21s were almost ready for flight testing when CAMCO, learning that advancing Japanese forces were only sixty miles away, burned them on May 1, 1942, and evacuated its equipment to India. By then, the CW-21 was demonstrably obsolete, and the type never saw use over Asia.

Meanwhile, Curtiss-Wright had produced another trainer, the CW-23, which used landing gear that retracted inward and flush under the wing, rather than into underwing fairings. First flown in April 1939, the CW-23 failed to find any customers, but its undercarriage was easily adapted to the CW-21, along with hydraulic wing flaps in place of the original chain-driven ones. Weight was

Ground personnel work on a Curtiss-Wright CW-21B of *2e Jachtafdeling* of *Vliegtuiggroep* IV at Andir, Java, in February 1942. (*Royal Netherlands Air Force*)

increased and the climb rate reduced by three hundred feet per minute, but the reduced drag produced by the new undercarriage increased the CW-21B's level speed by eighteen miles per hour.

During a sales visit to the Netherlands in January 1940, E. C. "Red" Walton managed to interest the Dutch in ordering thirty-six CW-21Bs, although that order was later reduced to twenty-four. The German conquest of the Netherlands on May 14, 1940, followed by the reestablishment of the exiled Dutch government in England, led to the CW-21Bs being shipped to the Dutch East Indies. By February 1941, twenty CW-21Bs had been assembled and were assigned to the *2e Jachtafdeling* (Fighter Squadron) of *Vliegtuiggroep* (Aircraft Group) IV, or 2-VlG-IV, at Andir air base near Bandoeng, Java. Operational losses had lowered their number to seventeen by the time the Japanese invaded the East Indies in January 1942.

The CW-21B's first chance to fight occurred on February 3, when twelve of the fighters scrambled up from Perak airfield and spread out in three flights to intercept Japanese bombers heading for Soerabaja naval base. They did not find any bombers, but they did encounter some of their escort—twenty-seven Mitsubishi A6M2 Zeros of the 3rd *Kokutai*. In the dogfight which ensued, the outnumbered Dutch learned that their CW-21Bs were no better protected than the Zeros, but that the Japanese fighters were faster, more maneuverable, and, with two 20mm cannons in the wings, better armed. In one flight, *Vaandrig* (Ensign) J. Hogenes and *Sergeant-Vlieger* (Sergeant) Robert C. Halberstadt died in flames, but *2e Luitenant-Vlieger* (Second Lieutenant) J. Kingma thought he shot two Zeros down before his own plane was set on fire. Kingma bailed out and survived, although he was badly burned. *Sergeant* Henk M. Haye also shot down a Zero before force landing his bullet-riddled CW-21B at Ngoro airfield.

Elsewhere, *1e Luitenant-Vlieger* W. A. Bedet was wounded but managed to force land at Perak. *Vaandrig* D. Dekker and *Sergeants* O. B. Roernimper and J. Brouwer force landed their damaged planes. The commander of 2-VlG-IV, *1e*

Luitenant Ricardo A. D. Anemaet, found no enemy planes, but as he returned to Perak he discovered that the field had already been attacked, as his CW-21B crashed into a bomb crater. *Vaandrig* A. W. Hamming and *Sgt.* Nicolass Dejalle landed safely, but *Sgt.* F. van Balen was killed by Zeros as he made his landing approach.

The three Zero claims by Kingma and Haye may have been accurate—the 3rd *Kokutai* did lose three pilots, Petty Officer 2nd Class Hatsuma Yayama and Petty Officers 3rd Class Shoichi Shoji and Masaru Morita, in the February 3 raid. The Dutch had been decimated, however, losing eight CW-21Bs in thirty minutes.

The CW-21Bs fared no better in subsequent run-ins with such formidable adversaries as the Tainan *Kokutai* and, later, the Nakajima Ki.43-equipped 59th and 64th *Sentais*. After the Dutch surrendered on March 9, the Japanese recovered and flight-tested a number of intact Allied aircraft, including one CW-21B that was found in Singapore after the war.

In addition to France and the United States, Britain, Italy, Germany, and the Soviet Union also tested a variety of lightweight fighters. Ironically, the only country to succeed, the Soviet Union, had not given the lightweight concept serious consideration until 1941, when the *Voyenno-Vozdushny Sili* began pursuing two different directions of development with the Yakovlev Yak-1, a new fighter that held great promise but was handicapped by an unsatisfactory power-to-weight ratio. The *tyazhely* (heavy) variants took advantage of the new boosted Klimov M-105PF engine, leading to the Yak-7A and the famous Yak-9. The other, *legky* (light), option involved taking radical steps to compensate for the existing engine with the smallest, simplest, cleanest airframe

possible. Unlike Western designs, the *legky* fighter was intended to achieve local air superiority over the battlefield, rather than be confined to point defense. Unlike the similarly lightened Mitsubishi Zero, however, the Soviet fighter did not have to satisfy a long-range requirement, since it would be operating just over the lines during the sweeping land battles that could—and would—be fought on Russia's western steppes.

When the Germans invaded the Soviet Union on June 22, 1941, the *legky* project had to be shelved, while production priority concentrated on the existing Yak-1. Work on the "lightened Yak" resumed in the late summer of 1942, resulting in the Yak-1M (*Modifikatsirovanny*, or "Modified"), which incorporated aerodynamic refinements such as a shallower oil cooler intake and, most significantly, a cut-down rear fuselage with a three-piece canopy, affording the pilot a much-improved rearward view.

Even while that derivative was under way, one of Yakovlev's design team leaders, Konstantin V. Sinelshchikov, was working on a more extensive redesign. His airframe looked like a Yak-1M, but the wingspan was decreased from 32 feet 9 1/2 inches to 30 feet 2 1/5 inches, and the aspect ratio was also reduced. The canopy was shallower and had a one-piece frameless windshield. The ventral radiator was moved farther aft, a retractable tail wheel was installed, and the radio mast discarded. The oil cooler was initially extended under the nose but was later relocated in the left wing root.

Structurally, the Yak-3 was identical to the Yak-1M, with a two-spar wooden wing skinned with highly polished plywood, and a 27-foot-10 1/2-inch-long fuselage consisting of a chrome-molybdenum steel-tube frame, covered by duralumin engine panels forward and plywood

Russian-born French pilot Léon Ougloff of the 1ère *Escadrille*, *Régiment de Normandie*, flies a Yak-3 over Königsberg in March 1945. (*Service Historique de Défense-Air, Château de Vincennes*)

or fabric aft. All control surfaces were metal framed and fabric covered. Empty weight was 4,641 pounds and the normal maximum takeoff weight 5,862 pounds, making the Yak-3 the lightest non-Japanese fighter to see combat during World War II.

The Yak-1M was test-flown in late 1942, and its performance exceeded expectations, including a speed of 422 miles per hour at 12,140 feet. At the same time, the first prototype of Sinelshchikov's Yak-3 was well under way. Meanwhile, Vladimir Klimov had developed a new, more efficient and powerful four-valve version of his triple-valve M105, the M-107A; but since that engine was then being tested for use in the new Yak-9U, an M-105PF-2 engine, with an output of 1,244 horsepower on takeoff, was installed in the Yak-3. Armament consisted of a 20mm ShVAK cannon firing through the propeller shaft along with two nose-mounted 12.7mm Berezin UB machine guns.

During one of the Yak-3's first test flights early in 1943, the left wing broke away during a snap roll. The pilot, Sergei N. Anokhin, bailed out at low altitude and parachuted into a marsh, which cushioned his landing enough to save his life. Despite the destruction of the first proto-type, a strengthened Yak-3 was factory tested successfully in April and passed its official V-VS trials in October 1943. Maximum speed was 407 miles per hour at 10,170 feet, and with an initial climb rate of 3,800 feet per minute, it could reach 16,405 feet in 4.1 minutes.

The first production Yak-3 left the factory on March 1, 1944, and the type entered frontline service with the 91st *Istrebitelny Aviatsionny Polk* in June 1944. Still-crude manufacturing standards reduced the plane's speed by as much as 12.4 miles per hour and added half a minute to its climb to 16,400 feet in comparison with the prototype, but pilots still found the Yak-3's performance impressive. Commanded by *Podpolkovnik* (Lieutenant Colonel) A. R. Kovalyov, almost half of the 91st IAP's personnel were replacements who were seeing combat for the first time. In a way, that was fortunate for them, since they would be approaching the Yak-3 with open minds and would be less prone to judge its flying characteristics against those of its predecessors. Its high wing loading took some getting used to at low speeds—its stalling speed was high; it had a tendency to drop a wing during a slow landing approach; and a tendency to ground-loop in the hands of an inexperienced pilot. During a high-speed low-level dogfight, however, the Yak-3 was in its element and was demonstrably superior to both the Me 109G and the Fw 190A. It could do a 360-degree turn in only 18.5 seconds, and could perform tight rolls and snap rolls with remarkable accuracy and smoothness.

Commencing operations over the Lwow area of Poland, the Yak-3's debut stood in marked contrast to those of its forebears, even the best of which could claim only qualified successes in their first encounters with the Luftwaffe. On June 16, 1944, eighteen Yak-3s of the 91st IAP charged headlong into twenty-four equally game German fighters, and in the twisting aerial battle royal that ensued, the Soviet pilots claimed no less than fifteen victories for the loss of one Yak destroyed and one damaged. Although it is likely that the regiment's enthusiastic neophytes made exaggerated or duplicated claims, the virtual cessation of Luftwaffe activity over the front on the following day suggests that the Germans found the new Soviet fighter an unpleasant surprise.

From then on, the Yak-3s were usually sent over the front to strike at enemy airfields or to engage enemy fighters about ten minutes ahead of the Ilyushin Il-2 or Petlyakov Pe-2 ground-attack planes, at altitudes that seldom exceeded 11,500 feet. In the course of their next 431 missions, the pilots of the 91st IAP accounted for twenty more German fighters and three Ju 87s, while two Yak-3s were lost in aerial combat, and three others were damaged by antiaircraft fire but returned to base.

Such early successes gave the Yak-3s instant popularity as more arrived at the front. When offered their choice of the latest fighters, the French volunteers of the Normandie-Niemen Regiment unani-mously requested Yak-3s, which started to reach them in mid-July 1944. During a ten-day rampage in October, the regiment's four *escadrilles* claimed 119 victories out of their wartime total of 273. By late summer 1944, Luftwaffe units were receiving directives to "avoid combat below 5,000 meters with Yakovlev fighters lacking an oil cooler under the nose."[2]

Early in 1944, the Yakovlev team unveiled an all-metal version of their lightweight wonder, the Yak-3U (*Usilenny*, or "strengthened"). Powered by the 1,650-horsepower M-107A engine, the Yak-3U had a speed of 447 miles per hour at 18,045 feet and the ability to climb to 16,405 feet in 3.9 minutes—the highest performance of any piston-engine Yak. Armament consisted of two fuselage-mounted 20mm B-20 cannon. Nicknamed *Ubiytsa* (Killer) by its pilots, the Yak-3U was rushed into production in the autumn of 1944, but did not reach operational units before Germany surrendered. By the time the Yak-3U was phased out of production in early 1946, a total of 4,848 Yak-3s of all models had been built. If the Yak-3 was exceptionally successful for a lightweight fighter, it was also an exception that proved the rule, for unlike its more original—and less successful—contemporaries, it was really a conventional fighter reduced in size and stripped to bare essentials.

The Curtiss-Wright CW-21 was not the only fighter to be developed from a trainer. The crippling of the US Pacific Fleet at Pearl Harbor on December 7, 1941, and the equally disastrous losses of the British battleship *Prince of Wales* and battle cruiser *Repulse* in the Gulf of Thailand three days later, left Australia largely isolated and feeling literally left to its own devices. The only major aircraft factory in

Australia at that time was the Commonwealth Aircraft Corporation (CAC) at Fisherman's Bend, outside of Melbourne, Victoria, then engaged in license production of a version of the North American NA-33 single-engine two-seat trainer called the CA-3 Wirraway (Aboriginal for "Challenge"). With neither the time nor the resources to build a new fighter from scratch, CAC's design team, headed by Fred David, proposed a single-seat fighter based on the Wirraway's airframe, and received the go-ahead for the project on December 21, 1941.

Delivered in March 1943, CAC Boomerang CA46-47 served in No. 83 Squadron RAAF at Strathpine, Queensland, with the code letters MH-Q. It is shown in its later livery serving with 2 Operational Training Unit. (*Australian War Memorial*)

Ironically, the chief designer behind the Boomerang was not Australian but Austrian—Fred David had worked for Heinkel and Aichi before his Jewish religion forced him to seek sanctuary Down Under. Even while supervising the project, he was still technically classed as an enemy alien and had to report weekly to the immigration authorities. Although the CA-12 was conceived to use as many Wirraway components as possible, by the time it was completed only 2 percent of its 60,000 components—mainly at the tail—were Wirraway compatible. The forward fuselage was strengthened to accept the largest engine then available— the license-built fourteen-cylinder, 1,200-horsepower Pratt & Whitney R-1830-S3C4-G Twin Wasp that powered Government Aircraft Factory-produced Bristol Beaufort torpedo bombers. The outer wing panels were reduced to a span of 36 feet 3 inches, and the fuselage shortened to 25 feet 6 inches. A 1 1/4-inch armor-glass windshield was installed along with armor plate behind the pilot's seat. Four .303-inch Browning machine guns and two 20mm cannon were installed in the wings. The Hispano cannon was not being produced in Australia at that time, but a single example of the

weapon was located, and using it as a pattern, Harland Engineering Ltd. established a production line to manufacture the weapon. In addition to the guns, a five-hundred-pound bomb could be carried under the CA-12's fuselage. There was also provision for light bomb racks under the wings.

Such importance was placed on the CA-12 Boomerang, as the new fighter was known, that the designers worked seventy-hour weeks to complete it. The fighter was already in production form when Ken Frewin flew it for the first time on May 29, 1942, just five months after work on it had begun—a remarkable achievement for a company that had never built an original airplane before.

Considering its rushed genesis, the Boomerang performed remarkably well, with a maximum speed of 296 miles per hour and an initial climb rate of 2,940 feet per minute. The Australian fighter's normal 930-mile range could be extended to 1,600 miles with the addition of a 70-gallon drop-tank. Tested against the Curtiss P-40E Kittyhawk and the Bell P-39 Airacobra, the CA-12 outclimbed the P-40E and could outmaneuver both of its American counterparts, although they both possessed an edge in diving and level speed. One later example, designated the CA-14, was tested with an 1,850-horsepower Pratt & Whitney R-2800 engine, and superchargers were considered as a means of improving the Boomerang's poor high-altitude performance. By that time, however, the crisis that had brought the Boomerang into being had passed.

Some minor teething troubles were experienced with the first CA-12s, but on April 4, 1943, No. 84 Squadron, Royal Australian Air Force, got its first ten fighters and began operations from Horn Island and across the Torres Strait at Merauke, New Guinea. By April 30, Nos. 83 and 85 Squadrons had each received their first six CA-12s for home defense. Meanwhile an improved Boomerang, the CA-13, was in production at Fisherman's Bend.

On May 16, 1943, two CA-12s from No. 84 Squadron were performing a typical one-hour standing patrol over Torres Strait when they encountered three Mitsubishi G4M1s of the 753rd *Kokutai* on a mission to bomb a 2,000-ton transport at Merauke. Pilot Officer Robert W. Johnstone and Sgt. M.F.J. Stammer closed to 250 yards when electrical failure caused one fighter's guns to jam, while the other managed to loose only one short burst at the bombers before they disappeared into a cloud. The Japanese turret and tail gunners returned fire but scored no hits, either. To add injury to insult, one of the Boomerangs, serial number A46-24, cracked up upon landing at Merauke, damaging its propeller, left wing, and left landing gear. The first interception by Boomerangs was not entirely in vain, however—the Japanese, unable to line up for a proper attack on the ship, went for a secondary target at Kimaam and returned to report an inconclusive combat with two enemy aircraft.[3]

On the night of May 20, two Boomerangs of No. 85 Squadron took off to intercept a Kawanishi H8K2 flying boat of the 851st *Kokutai* that was reconnoitering the Gulf of Carpentaria, while a second H8K tried to bomb the US Navy submarine support facilities in Exmouth Gulf. Pilot Officer Roy F. Goon in CA-12 A46-61 reported some sightings, but failed to make positive contact and came home empty-handed. The H8K on the reconnaissance mission reported two indecisive encounters with enemy aircraft, at 2103 and 2113 hours. On the following night Boomerangs sought out another H8K that was trying to reconnoiter an airfield reported to be at Onslow. They failed to make contact, while the Japanese, failing to find any airfield, bombed Exmouth on the way home.[4]

Swift though its development had been, by the time the Boomerang commenced operations in mid-1943, the RAAF had acquired enough British and American fighters to handle the threat of Japanese air attack. The Boomerang did, however, find another niche as a close-support aircraft. Jinking about over the mountainous rain forests of New Guinea, the Solomon Islands, Bougainville, and Borneo, Boomerang pilots—known as "Boomer Boys" to the troops they supported ("Boomer" being a reference to a male of the largest species of kangaroo)—became adept at ferreting out and eliminating enemy strongpoints.

It was just as well that the Boomerangs never came up against Zeros, for the few times that they encountered aerial opposition did not end well for the Aussies. On September 6, 1943, CA-13 A46-112 of No. 4 Squadron was on a tactical reconnaissance mission during the Australian campaign to take Lae and Salamaua when it ran afoul of an area sweep by the 14th Flying Brigade of the Japanese Fourth Air Army, involving Ki.43 *Hayabusas* of the 24th and 59th *Sentais*, and Kawasaki Ki.61 Hiens of the 68th and 78th *Sentais*.[5] Last seen engaging Japanese fighters, A46-112 went missing until November 12, 1948, when it and the remains of its pilot, Flying Officer Thomas J. Laidlaw, were found near the Busa River.

On November 15, CA-13 A46-136 of No. 4 Squadron was returning from a sortie when it was attacked by a P-38 whose pilot, 1st Lt. Gerald R. Johnson of the 9th Squadron, 49th Fighter Group, mistook it for an Oscar. Flying Officer Robert McColl Stewart managed to belly-land his burning Boomer in some scrub near Salamaua, emerging without serious injury. After profuse apologies, Johnson added an Australian flag to the sizeable collection of rising suns gracing his Lightning's nacelle.

The last air-to-air loss occurred on November 26, when two Boomers of No. 4 Squadron, on a tactical reconnaissance mission over the Finschhafen area with an escort of four P-39s of the 41st Squadron, 35th Fighter Group, were jumped by seven "Zeros," and both were shot down over the Sanga River. The Japanese army dispatched ten Kawasaki Ki.48s to bomb Allied artillery positions near Finschhafen that day, and their escorts, thirty-seven Ki.43s of the 59th and 248th *Sentais* and Ki.61s of the 68th and 78th, reported twenty engagements with Allied fighters, claiming two "P-40s," one of which was credited to Captain Ryoichi Tateyama, commander of the 78th *Sentai*'s 3rd *Chutai*, for the loss of five Japanese pilots.[6] As of this writing, CA-13s A46-109 and A46-132 and their respective pilots, Flt. Sgt. Alan J. Salter and Flying Officer Hector C. Munro, remain missing in action.

The only indigenously designed Australian aircraft to see service during World War II, a total of 250 Boomerangs were built, serving in five RAAF squadrons, by the time production was

terminated in January 1945. Of that number, 105 were CA-12s, 95 CA-13s, 49 CA-19 photoreconnaissance planes, and the single prototype CA-14.

Australia's Japanese adversaries came up with a rather unique improvisation of their own when they went over to the defensive. The floatplane fighter, originally developed and employed with some success in World War I, declined in importance during World War II. Most such aircraft were land fighters modified to operate from island bases where airfields were unavailable, at the sacrifice of performance due to the weight and drag of the floats that replaced their retractable undercarriage. Such aircraft seemed to make sense to the Japanese, who after World War I had acquired a far-flung collection of island mandates across the Pacific. Two of their floatplane fighters—the two-seat Mitsubishi F1M2 biplane and the single-seat A6M2-N, a Nakajima-built floatplane version of the Zero—even enjoyed a modest degree of success in the Pacific War's first year. However, a more ambitious follow-up project by Kawanishi Kokoki K.K. ended up taking an unexpected turn: what began as a purpose-built floatplane fighter to supersede the A6M2-N evolved into a land fighter.

Begun by an engineering team led by Elizaburo Adachi, Toshiharu Baba, Hiroyuki Inoue, and Shizuo Kikuhara, the new Kawanishi floatplane was to feature a new laminar-flow wing installed at mid-fuselage. Undercarriage comprised a central float and stabilizing floats under each wingtip. Using a 1,463-horsepower Mitsubishi Kasei-14 fourteen-cylinder radial engine—partly because it was easier to maintain under primitive conditions than a water-cooled inline engine, but mainly because it was readily available—

the Kawasaki team tried to compensate for its drag by using two contra-rotating propellers and a large spinner. In that form, the N1K1 prototype first flew on May 6, 1942.

As early as December 1941, Kawanishi's experimental shop had been investigating the possibilities of building a land-based version of the new floatplane. The Japanese navy rejected its proposal at that time, but Kawanishi went ahead with the project as a private parallel venture. In addition to using an 1,820-horsepower Nakajima Homare 11 eighteen-cylinder radial driving a four-bladed propeller in place of the Kasei, Kawanishi's Model X-1 experimental land-based fighter required extra-long landing gear to compensate for its midwing configuration, a problem that the designers tried to remedy with a mechanism which would extend the undercarriage leg upon landing and contract it during retraction.

Troubles with the Kasei 14's contra-rotating installation resulted in the substitution of the more conventional Kasei 13 with a single three-blade airscrew in later prototype N1K1s, as well as in the final production version of the N1K1 *Kyofu* (Mighty Wind), the first of which was delivered in July 1943. Service trials held up the new fighter's introduction until December, by which time the navy's priorities had changed from offensive floatplane operations to land-based defense.

With a top speed of 304 miles per hour at 18,700 feet, the *Kyofu* was judged capable of intercepting a bomber if it could avoid the escort, so the first N1K1s were sent to defend the oil refineries at Balikpapan, Borneo, while the rest were kept at home, serving in the Otsu *Kokutai* from Lake Biwa on the isle of Honshu. In January 1944, the Singapore-based 934th *Kokutai* received nine N1K1s and then set out via Java to Ambon Island, from which

base it flew its first interception mission on January 16. On that occasion, the *Kyofus* engaged six Consolidated B-24s, one of which was claimed by Warrant Officer Kiyomi Katsuki, an exceptionally aggressive floatplane pilot who had scored his first two victories in F1M2s (the second, a Boeing B-17E of the 72nd Bomb Squadron being destroyed on October 4, 1942, by ramming, after which Katsuki and his observer bailed out of their plane), and at least three in A6M2-Ns before striking the first blow for the *Kyofu*. The 924th *Kokutai* engaged another twenty B-24s two days later, claiming one bomber destroyed and another probably shot down for the loss of one plane and pilot.

The Mighty Wind did not persist for long, however. On March 1, 1944, the 924th *Kokutai* was disbanded, and its personnel returned to Soerabaya. By then, its pilots had claimed twenty-nine enemy aircraft and seven probables—including five victories and shares in twelve others by Hidenori Matsunaga—for the loss of five pilots killed. Only ninety-seven N1K1s were completed before production of the Mighty Wind petered out. The *Kyofu* had performed about as well as could have been expected of a floatplane fighter, but the concept on which it had been based was no longer relevant.

Fortuitous it was, then, that Kawanishi had persevered in ironing out the problems with its land fighter: installing an improved 1,990-horsepower Homare 21 engine; increasing the armament to two cowl-mounted 7.7mm machine guns, two 20mm Type 99 Model 2 cannon in the wings, and two more cannon in underwing gondolas; and making provision under the fuselage for a 400-liter auxiliary fuel tank. The firm's efforts were finally rewarded when the navy accepted the heavy land fighter for production as the N1K1-J *Shiden* (Violet Lightning) in December 1943.

Gearing up for production was slow, but by October 1944 Kawanishi's Naruo and Himeji plants had built 662 *Shidens* and completed another 106 in that month. The first operational N1K1-Js went to the 341st, or "*Shishi*" (Lion) *Kokutai,* commanded by Capt. Motoharu Okamura. The group was divided into three *hikotais*, or squadrons—the 401st commanded by Lt. Ayao Shirane, the 402nd led by Lt. Iyozo Fujita, and the 701st under Lt. Kunio Iwashita. All three leaders were experienced pilots, Shirane having scored his first of nine victories in the first Zero action on September 13, 1940, and Fujita having flown Zeros at Pearl Harbor (where he destroyed a Curtiss P-36), Midway (where he was credited with three personal and seven joint victories on June 4, 1942), Guadalcanal, and Rabaul. Between August 31 and mid-September, *Hikotai* 402 was dispatched to Takao (now Kaohsiung) on Formosa, where its forty-two planes were to intercept American aircraft en route to and from China. Opportunities for combat were few, which was just as well because the new *Shidens* were still proving to be mechanically unreliable.

The N1K1-J got its first real taste of combat on October 12, when the US Third Fleet struck at Formosa. *Hikotais* 401 and 701 were moved to southern Kyushu and attached to the 2nd Air Fleet to assist in the island's defense, while the 402nd, at that time led by Lt. Masaaki Asakawa, joined in from its base at Kaohsiung. Fielding thirty-one *Shidens* and one Zero, *Hikotai* 402 alone claimed ten victories that day, but lost fourteen planes and two pilots, Lts. j.g. Shigemi Wakabayashi and Katsumi Yamaguchi. Another of the 341st *Kokutai*'s pilots,

A Kawanishi N1K1 Ja *Shiden* attached to the 201st *Kokutai* to escort suicide planes to their targets was disabled at Clark Field, Luzon, and subsequently captured when the Americans retook the airfield on January 31, 1945. (*National Archives*)

Chief Petty Officer Tadashi Sakai, was killed in action the following day.

It remains difficult to determine which Americans first met the N1K1-Js in battle that day, though cases where "Zekes," "Oscars," and "Tojos" are encountered at the same time (a suspicious mixing of navy and army types) may suggest that *Shidens* were engaged by Grumman F6F-5s of VF-15 from carrier *Essex*, VF-18 from *Intrepid*, and VF-20 from *Enterprise*. The Americans claimed a total of 500 Japanese planes destroyed in the air or on the ground between October 12 and 16 for the loss of 71 carrier planes. Japanese records claimed 112 American aircraft in that time, while admitting to the destruction of 312 of their own.

On October 17, US Army troops landed at Leyte in the Philippines, and twenty-three *Shidens* of the 341st *Kokutai* were dispatched south to reinforce the 201st *Kokutai*, based at Clark Field, Luzon. Battle was joined on the twenty-fourth, as eighteen *Shidens* of *Hikotai* 402 met a similar number of Hellcats, the Japanese claiming seven F6Fs but losing ten N1K1-Js and two of their pilots, Warrant Officer Minoru Shibamura and Chief Petty Officer Rokusaburo Shinohara. Undercarriage failures, oil leaks, and strafing attacks by American fighters took their toll, and by the end of the day, only four N1K1-Js remained airworthy at Clark Field. Another was lost in action the following day, along with its pilot, Lt. j.g. Masaru Sometani of *Hikotai* 401, and on October 29 Chief Petty Officer Kyoji Handa was killed over Manila.

At that point *Hikotai* 701, with Shirane commanding on behalf of the ailing Iwashita, was dispatched to Mabalacat Airfield on Cebu Island. Attrition on the ground continued, punctuated by occasional losses in the air, such as Ensign Munesaburo Takahashi, a veteran of service with the 13th and 12th *Kokutais*, as well as aboard the carriers *Soryu* and *Hiyo*, killed over Tacloban on November 18. A crippling blow to flagging morale in the 341st *Kokutai* occurred on November 24, when Shirane was killed along with Chief Petty Officer Sadao Koike during a fight between *Hikotai* 701 and Lockheed P-38s of the 433rd Squadron, 475th Fighter Group, near Ponson Island off the western coast of Leyte. Iwashita, recovered from his illness, reassumed command of *Hikotai* 701 and led twelve *Shidens* during attacks on an American ship convoy off Mindoro, during which Lt. Sumio Arikawa and Chief Petty Officer Toshiharu Kagami of *Hikotai* 701 were killed on December 14, and Lt. Seiya Nakajima, division leader of *Hikotai* 402, died the next day. By the end of December, eight N1K1-Js remained operational, and some of the 341st's pilots were being selected to carry out kamikaze attacks on American shipping.

Thirteen new *Shidens* arrived from Japan on January 3, 1945, but on the following day they were caught on the airfield by two Republic P-47Ds, which in moments set eight of them on fire and killed four pilots and five maintenance crewmen. When an American invasion force entered Lingayen Gulf on January

7, Iwashita led the 341st's four remaining *Shidens* on missions from Tuguegarao in northern Luzon, until the last of them was destroyed a few days later. Most of the group's personnel were subsequently killed in ground fighting for Luzon, although a few, led by Fujita, managed to fight their way through a gauntlet of Filipino guerrilla units back to Tuguegarao, from whence they were evacuated by air to Japan.

Generally, "George," as the Allies came to call the N1K1-J, had not made an impressive name for itself in its first actions. Its engine and undercarriage extending mechanism were both chronically prone to failure, and Fujita summed the new fighter's overall performance up in two words: "No good!"[7] Nevertheless, Kawanishi was committed to production and eventually built 1,007 machines, including the N1K1-Ja, on which two of the wing cannon were transferred from underwing gondolas to the wings themselves.

Even while gearing up to produce the *Shiden*, Kawanishi's engineers were taking steps to address its shortcomings, by repositioning the wings at the lower fuselage and enclosing the engine in a cleaner cowling. The resulting plane, which first flew on December 21, 1943, also had a longer fuselage, redesigned tail surfaces, and a shorter, simpler, sturdier undercarriage. The N1K2-J *Shiden-Kai* (Violet Lightning, Modified) had a maximum speed of 369 miles per hour at 13,375 feet—23 miles per hour slower than the army's Nakajima Ki.84 *Hayate*, but 6 miles per hour faster than the *Shiden* and 4 miles per hour faster than the Mitsubishi J2M3 *Raiden*. It was also superior in maneuverability, visibility, range, and overall performance to both navy types. In spite of the Homare 21 engine's persistent problems, the N1K2-J was lauded by its pilots and accepted by the Japanese navy as its principal interceptor, though only 428 would be built between December 1943 and August 1945.

As had been the case with the original *Shiden*, the Japanese could not afford to spend much time equipping and training its units for action in the *Shiden-Kai*. The 343rd *Kokutai*, a group that had been annihilated in Yap, Guam, and Saipan in July 1944, was reconstituted on December 25 under the command of Capt. Minoru Genda. At the end of January 1945 its three forty-eight-plane fighter *hikotais*—301, 701, and 407— commenced training in N1K1-Js at Matsuyama, Oita, and Izumi air bases, respectively. In addition, the 4th *Hikotai*, with twenty-four Nakajima C6N1 *Saiun* (Iridescent Cloud) reconnaissance planes, were attached to the group to provide advance intelligence on enemy activity. N1K2-Js finally began arriving in mid-February, by which time the type was having its baptism of fire with another unit.

On February 16 carriers of US Task Force 58, conducting a preemptive strike in support of the invasion of Iwo Jima, attacked air bases in the Tokyo area—the first appearance of American aircraft over the Japanese capital since the Doolittle Raid of April 18, 1942. Among the Zeros and other fighters that rose to challenge them were a handful of *Shiden-Kai and* J2M *Raiden* interceptors attached to the evaluation section of the Yokosuka *Kokutai*, Japan's oldest naval air group, which since 1916 had been responsible for testing new aircraft types and training new pilots. At the controls of an N1K2-J was Ensign Kaneyoshi Muto, a twenty-eight-year-old veteran of exceptionally short stature who nevertheless was described by his comrade and fellow ace, Saburo Sakai, as the "toughest fighter pilot

in the Imperial Navy."[8] Joining other aircraft from the Yokosuka *Kokutai*'s fighter element, Muto took part in a fierce engagement with F6F-5s of VF-82 from the carrier *Bennington* over Atsugi, during which four Hellcats were shot down.

After two days of mammoth air battles, the Americans departed, claiming the destruction of 332 Japanese planes in the air and another 177 on the ground for the loss of 52 F6Fs and 16 F4Us to all causes, some forty-nine being destroyed in combat. The Japanese navy, in turn, claimed that its airmen downed at least ninety-eight American planes, while admitting to the loss of seventy-eight fighters, with fourteen of their pilots killed. The defenders, who included some of the best pilots Japan had left, had done a creditable job; but amid the deteriorating wartime situation, it was not good enough for the Japanese press, which seized on Muto's successful first fight in the *Shiden-Kai* and beefed it up for public consumption. In consequence, Japanese propaganda described Muto as taking on twelve Hellcats single-handed, dispatching each of his four victims with a single burst in a display of martial skills that the press compared with those of seventeenth-century samurai and master swordsman Miyamoto Musashi. In so doing, they had inflated Muto's actual exploit to the realm of myth—a myth that still persists more than seventy years later.

By mid-March 1945, there were enough *Shiden-Kais* for the three *hikotais* of the 343rd *Kokutai* to operate formations of sixteen each. Then, on March 19, Task Force 58 returned, this time striking at the Kure naval base. Ready or not, the 343rd *Kokutai* would have to fight.

Their morale buoyed by a long series of successes in the past year, the Americans were confident that they were facing a spent force when F6F-5s of VBF-17 from the carrier *Hornet* ran into elements of *Hikotais* 407 and 701. A vicious melee ensued, during which Lt. j.g. Byron A. Eberts claimed two Georges, Ensign Robert A. Clark downed another George, and Lt. Edwin S. Conant claimed two "Franks." Conant's victims, along with a "Zeke" claimed by Eberts, were more likely also *Shiden-Kais*, matching the six actually lost by the 343rd *Kokutai* in that fight. VBF-17, in turn, lost six planes—the Hellcat had literally met its match.

Hikotai 301, commanded by Lt. Naoshi Kanno, was even more active that morning, starting with a wild engagement with Marine F4U-1Ds of VMF-123, also operating from *Bennington*, north of Kure. After disengaging off the western coast of Shikoku Island and reassembling his squadron, Kanno sought out more trouble. He found it—or rather, it found him—in the form of two F4U-1Ds of *Intrepid*'s VBF-10, flown by Lt. Robert Hill and Ensign Roy Erickson, who jumped the Japanese fighters from behind and sent their leader down in flames. Kanno bailed out, and a Corsair passed by close enough for Erickson to notice the astonished look that was still on Kanno's face. Suffering burns to his hands and face, Kanno parachuted into a field near Matsuyama Castle and was menacingly approached by an elderly farmer armed with a pitchfork who thought Kanno to be American until he heard the curses leveled at him in Japanese. Kanno then appropriated a bicycle and pedaled back to his base.

With Kanno out of the fighting, Chief Petty Officer Shoichi Sugita, a veteran of fighting over the Solomons, took charge of a flight and quickly accounted for three Hellcats. Another member of *Hikotai* 301, twenty-one-year-old Chief Petty Officer Katsue Kato, was involved in several fights that resulted in the destruction of nine Hellcats, all of which were subsequently credited to him alone by the Japanese press. Even in such desperate times, Kato's superiors would not officially confirm such a fantastic score, although they did cite him by name in the Naval All Units Proclamation afterward.

By morning's end the 343rd *Kokutai* had claimed fifty-three Hellcats and Corsairs, as well as four Curtiss SB2C Helldivers, for the loss of thirteen pilots. Any impact the *Shiden-Kai* may have had on its American opponents was somewhat overshadowed by the dive-bombing and near loss of the carrier *Franklin* that day, but as far as the Japanese were concerned, the N1K2-J was an unqualified success. Superb fighter though it was, however, it had appeared too late to stave off their ultimate defeat. One by one, even its greatest exponents were overwhelmed and killed—Sugita and his wingman, Petty Officer 2nd Class Toyomi Miyazawa, by Lt. Cmdr. Robert Weatherup, an F6F-5 pilot of VF-46 from *Independence* on April 16; Kato on April 17 by Hellcats of *Bataan*'s VF-47; Muto in a fight with Corsairs and Hellcats on July 24; and Kanno following an engagement with B-24s on August 1. Nevertheless, the final variant of what began as a floatplane fighter was one of the war's outstanding land-based fighters—and, given its genesis, one of the most unique.

When one thinks of unique, however, there can be little argument that the Messerschmitt Me 163 is the all-time champion, being the only tailless fighter to serve during World War II and the only rocket-powered manned aircraft ever to see combat under the stratosphere. Its roots lay both in Dr. Alexander Lippisch's tailless glider experiments during the

1920s and in Hellmuth Walter's development of a rocket engine suitable for aircraft in 1936. In March 1938, the two elements were combined in a conception by the *Deutsches Forschunginstitut für Segelflug* (German Sailplane Research Institute), the DFS-194. Development of the design was transferred to Messerschmitt in January 1939, and after the poor showing made by another rocket-powered design, the Heinkel He 176, its Walter R I-203 engine was installed in the tailless, sweptwing airframe of the DFS-194 and flown with encouraging results—a speed of 342 miles per hour, a fast rate of climb and, most notably, pleasant handling characteristics. In the spring of 1941 gliding trials began with the Me 163V1, the only snag encountered being its tendency to keep gliding, even when the pilot was trying to bring it down for a landing. Using the rocket propellant between July and September 1941, test pilot Heinrich Dittmar pushed the speed envelope ever higher until he reached a maximum of 623.85 miles per hour on October 2—a world record. In December, Lippisch and his design team began work on an operational fighter version.

Although none of the other Western powers had come anywhere near the Germans' level of rocket aircraft development, they were not quite alone in such endeavors. Scientists in the Soviet Union had also been experimenting with liquid rocket propellants in the 1930s, and in the spring of 1941 the Viktor Bolkhovitinov design bureau had begun work on an airplane to use the D-1-A rocket motor, fueled by kerosene with concentrated nitric acid as the oxidizer. An airframe was designed by Aleksandr Y. Bereznyak and Aleksei M. Isayev, and given the high fuel consumption of the power plant, it was intended as a short-

The first rocket interceptor to become operational in a home defense squadron in mid-1942, the Bereznyak-Isayev BI-1 was quickly retired after killing one of the Soviet Union's top test pilots. (*National Archives*)

range point interceptor—essentially the same role that would later be envisioned for the Me 163. Combining a bullet-shaped fuselage with conventional straight wings and tail surfaces, all constructed of plywood and fabric, the Bereznyak-Isayev BI fighter made its first powered flight on May 15, 1942. Captain Grigori Y. Bakhchivandzhi reported that everything went smoothly and that the plane handled the same as any other.

Armed with two nose-mounted 20mm cannon, a handful of BI-1 rocket interceptors were built and assigned to a squadron, making it the first rocket-powered fighter to see military service. On March 27, 1943, however, Bakhchivandzhi put the third prototype through a test flight; he accelerated to a speed of 497 miles per hour at 6,500 feet and suddenly went into a dive from which he was unable to recover. Testing of the BI continued in the hands of Konstantin Gruzdev and Boris Kudrin, but the death of Bakhchivandzhi—who was posthumously awarded the Gold Star of a Hero of the Soviet Union—left V-VS personnel uneasy about the "devil's broomstick," as they came to call the BI, and about the rocket fighter concept in general. The BI was never used operationally and the

Soviet Union, the first nation to develop a rocket fighter, also became the first to abandon it.

Even while the Bereznyak-Isayev fighter was being readied for its first test flight, the Me 163V3 was being completed in May 1942, using Walter's HWK R.II engine, which used *T-Stoff* (hydrogen peroxide and water) as a fuel and calcium permanganate as a catalyst. The latter tended to clog the jets, however, so Walter devised another motor, the HWK 509, which employed *C-Stoff* (hydrazine hydrate and methyl alcohol) as the catalyst. Using the latter power plant, the Me 163V3 made its first powered flight at Peenemünde in August.

As the Allied bombing campaign against Germany increased, the new plane was ordered into production at Messerschmitt's Regensburg plant, and at the end of June 1944 the first Me 163B-1 *Komet* fighters were ready to commence operations with 1st *Staffel* of JG 400. The first interceptor mission by the new plane, however, had already been flown a month earlier.

In May 1942 a test unit, *Erprobungskommando* 16, had been established at Bad Zwischenahn under the command of *Obltn*. Wolfgang Späte, a prewar glider pilot and crack fighter ace who already had seventy-two victories to his credit when he was recalled from the Eastern Front. After two years of test flying and training new pilots on the new plane, the newly promoted *Major* Späte was notified by his superiors that since the actual combat unit was still undergoing training, he would demonstrate the new interceptor's capabilities on May 13, 1944. When he entered the hangar that morning, he was astonished to find the only combat-ready machine, an Me 163V41, call letters PK+QL, overpainted entirely in tomato red. His own mechanics had finished it in

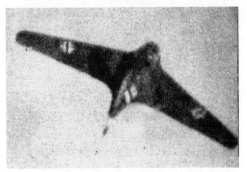

During an Eighth Air Force raid on Merseburg-Leuna on October 7, 1944, a Messerschmitt Me 163B of I *Gruppe, Jagdgeschwader* 400 was attacked by Capt. Elmer A. Taylor and 1st Lt. Willard G. Erfkamp of the 385th Squadron, 364th Fighter Group. After force landing, *Fw.* Rudolf Zimmermann just got clear before the P-51s strafed his plane to destruction. (*US Air Force*)

Manfred von Richthofen's colors in hopes that it would bring him luck in the revolutionary machine. Späte did not appreciate the gesture, declaring that at least the Baron had scored some victories before having his plane painted red—and, on a more practical level, the extra paint added forty pounds to an airplane whose rocket motor gave limited endurance as it was. At least the unit's morale was in the right place, though, and when ground radar detected enemy fighters in the sky, Späte took off, hoping that its garish finish or its long exhaust trail would not attract every other enemy plane in the area.

Once he reached his operating altitude, Späte leveled off and spotted two Republic P-47Ds ahead of him and two other Thunderbolts above. Switching on his Revi 16 gunsight and cocking his 20mm MG 151 cannon, Späte began a tight weaving maneuver to let the higher P-47s get ahead of him, during which he inadvertently released the control column, causing his engine to flame out. He then had to go into a shallow glide for two minutes, helpless until he could restart the motor. Once he did, however, he found that none of the American pilots were aware of his presence. Two of the P-47s were three miles distant, so Späte slid the throttle forward to 100 percent power, bringing his airspeed past 550 miles per hour, and soon had one of the silver fighters in his sight. He was about to open fire when the Me 163's left wing abruptly snapped down and the airplane went into a dive, vibrating dangerously and creating negative Gs that caused the motor to flame out again. Späte eased back the throttle and regained control, and the shaking gradually abated.

Chastened and frustrated, Späte brought his *Komet* down safely at Bad Zwischenahn, having still gone completely unnoticed by his quarry. He had been undone by an encounter with the compressibility threshold known as the sound barrier, a sometimes fatal phenomenon that occurred more often with diving aircraft, rather than one engaged in roughly level flight as his had been. Späte, transferred out of EK 16 after May, would never get another chance to claim an enemy plane while flying the Me 163, but he would later have the privilege of claiming five victories in the jet-powered Me 262, as *Gruppenkommandeur* of III./JG 7.

On May 31, a photoreconnaissance Spitfire of the RAF, flying at altitudes of 37,000 to 41,000 feet, provided the Allies with the first report of an airplane that rose quickly, leaving three white trails behind it, which seemed to be "nearly all wing," possibly with a marked sweepback. At about that same time, the Germans had discovered why the Me 163 tended to cut out as it leveled off at the summit of its rapid climb—the fuels would slop around in their tanks, which in turn activated a safety device that would automatically cut the motor off, since any change in the ratio of its two unstable fuels might cause an explosion.

The installation of additional baffle plates to the fuel tanks partially, but never completely, solved that problem.

Another flaw lay in the motor's running time, which the Walter firm had claimed would be twelve minutes, but which in practice barely lasted four minutes before all the fuel was consumed. For example, I./JG 400's first combat mission, in which *Uffz.* Konrad Schiebeler tried to intercept a Lockheed F-5 reconnaissance plane on July 7, ended in frustration when his fuel ran out. While Walter labored on to increase the fighter's endurance, the Me 163B-1 was physically confined to the point-interception role, using its rocket motor to shoot it above an incoming Allied bomber stream, then glide through the formation firing its two wing-mounted 30mm MK 108 cannon, or using its supplemental underwing armament of R4M rockets, its pilot hoping to have conserved enough of a fuel reserve for an accelerated getaway if enemy escort fighters should intervene. By late July, the newly expanded I *Gruppe* of JG 400 was assigned to defend the oil refinery at Merseburg-Leuna, in spite of the protests of its commander, *Hptmn.* Rudolf Opitz, that its base at Brandis, east of Leipzig and forty kilometers from the target area, made it more difficult to carry out the mission.

On July 29, the Eighth Air Force dispatched 569 Boeing B-17s to bomb the Merseburg-Leuna petroleum complex. As the bombers passed over Brandis, seven Me 163s, including those flown by *Leutnante* Hans Bott and Hartmut Ryll, and by *Fw.* Siegfried Schubert and Hans Glogner, shot up to engage them. At 0945 hours crewmen of the 96th, 388th, and 452nd Bomb Groups began calling out positions of strange aerial projectiles approaching them. Schubert, whose early mastery of the Me 163 had earned him delegated leadership of the formation, brought his flight above the bombers, then dived through them. The Germans misjudged their closing speed, however, and found themselves with no time to select a target.

Colonel Avelin P. Tacon Jr. was leading eight North American P-51s of the 359th fighter Group at 25,000 feet when he reported: "One of my pilots called in two contrails at six o'clock high some five miles back at 32,000 feet. I identified them immediately as jet-propelled aircraft. Their contrails could not be mistaken and looked very dense and white, somewhat like an elongated cumulus cloud some three-quarters of a mile in length." His flight turned 180 degrees and went to engage five of the Germans, three of which were gliding down with their motors cut off. The two whose rockets were on turned toward the bombers, but at 3,000 yards distance turned toward the oncoming Mustangs instead. "In this turn they banked about 80 degrees but their course changed only about 20 degrees," Tacon reported. "Their turn radius was very large but their rate of roll appeared excellent. Their speed I estimated was 500 to 600 miles per hour." The *Komets* passed 1,000 feet under the Mustangs, but when the Americans went after them one shot up toward the sun, its rockets bursting intermittently "as though it was blowing smoke rings," as one of Tacon's pilots described it. The other glided down, and by the time Tacon turned his attention to that German, "he was five miles away at 10,000 feet."[9]

Six Me 163s rose to intercept a force of 647 bombers attacking Merseburg on July 29, but again failed to draw blood. One B-17 of the 100th Bomb Group was returning from the mission at 11,000 feet when an Me 163 came up behind it near Wesermünde at 1145 hours. Captain

Arthur F. Jeffrey and three other Lockheed P-38J pilots of the 434th Squadron, 479th Fighter Group, had been watching over the straggler, and now he and his wingman, 2nd Lt. Richard G. Simpson, dived to attack the rocket, which began to climb and weave before breaking off to the left. Cutting inside its turn, Jeffrey got in a deflection shot from 300 yards, observed hits, and then tried to follow the rocket in a vertical dive until his P-38 reached a speed of 500 miles per hour, and he was compelled to pull out at 1,500 feet, blacking out in the process. An examination of Jeffrey's camera film led to his being awarded credit for destroying the enemy plane, but German records indicated no Me 163 losses that day.

The remaining two P-38s of Jeffrey's flight had remained 16,000 feet above and behind the bomber when a second Me 163 appeared at 1148 and dived out of the sun at the B-17. It made a seventy-degree firing pass but failed to score hits, then dived into a cloud deck before the startled Lightning pilots could even begin pursuing it.

In spite of days of interception attempts, the Me 163s did nothing to prevent the substantial crippling of Merseburg's fuel-producing facilities. On August 2 Lt. Gerald M. Adams of the 14th Squadron, 7th Photo Reconnaissance Group, was flying a Lockheed F-5 to assess the damage when he found himself being pursued at 33,000 feet by an extraordinarily fast airplane, which swiftly closed the range to 1,000 yards and started firing at him. "This was a very startling revelation to find something that much faster than the P-38," Adams recalled, "so I pointed the nose straight down toward a low cloud deck at about 8,000 feet and fortunately made it, but the old P-38 was bucking like a bronc all the way. The Me 163 followed me all the way down making firing passes, but was never able to maneuver into a good tail chase position, fortunately. The cloud deck extended far enough to lose the Me 163, or his fuel was short."[10]

The Eighth Air Force struck at Merseburg again on August 5, and aircrews of both the 94th and 352nd Bomb Groups reported seeing ten of the rockets. The 486th Group spotted eight, and the 490th reported "bat-type-wing" fighters leaving smoke trails at 28,000 feet. Sergeant Charles M. Nevaskil, radio operator in the *Royal Flush*, a B-17G of the 100th Bomb Group, reported seeing three Me 163s "about 35,000 feet high and about 9 o'clock to our course. They flew in the direction of our formation, then turned left to attack three P-51s, which were flying approximately 3,000 feet above and to the left of our formation. The 'Jets' dived on to the P-51s, which were about 8 o'clock to our formation at the time. The Me 163s were in trail as they swooped down on the P-51s. The attacks were pressed to almost point-blank range, then the enemy aircraft zoom-climbed into the clear blue sky above. I saw each of the three P-51s catch fire and dive earthward."[11] In spite of what Nevaskil reported seeing, although VIII Fighter Command did lose three Mustangs in combat, each of them came from a different squadron, and their losses were attributed to Me 109Gs; moreover, JG 400 made no P-51 claims that day.

Thus far, the Me 163 had been making its enemies uneasy, but for all the Sturm und Drang, it had done little material damage. Then, on August 16, 1,096 B-17s and B-24s set out to bomb Zeitz, Rositz, Leuna, Boehlen, Halle, Dresden, and Koethen. Only five Me 163s were operational, but all five were launched, and at 1045 hours they began making rear-end

attacks on B-17s of the 91st and 305th Bomb Groups. *Feldwebel* Herbert Straznicky made a determined attack on 2nd Lt. Donald M. Waltz's B-17G *Towering Titan* of the 305th Group, closing to 50 yards before peeling off. Equally determined, however, was the bomber's tail gunner, Sgt. Howard J. Kaysen, who began shooting at 1,000 yards and kept firing bursts until he saw the rocket go down, emitting smoke from 23,000 feet. Wounded in the left arm and thigh, Straznicky bailed out and parachuted safely near Brandis.

Near Waltz's plane, 1st Lt. W. E. Jenks's B-17G became the target of another gliding Me 163. Both the upper turret gunner, Technical Sgt. H. K. Tubbs, and the navigator, 2nd Lt. W. G. McGregor, fired back at the *Komet* between 800 and 200 yards distance, then saw it pass under their bomber and dive away. A third *Komet*, flown by *Leutnant* Ryll, attacked 2nd Lt. Charles J. Lavadiere's B-17, pressing his attack so aggressively that the tail gunner bailed out. Closing to point-blank range, Ryll struck both inboard engines, the flaps and tail, and killed a waist gunner. He then turned and resumed his attack from two o'clock low, blasting the B-17's ball turret and its gunner out of the plane. The Germans subsequently credited the Fortress to Ryll, but in fact Lavadiere managed to bring his severely mauled plane back to base.

Shooting away from the 305th's formation, Ryll went after fresh prey in the form of the *Outhouse Mouse*, a B-17G of the 91st Bomb Group piloted by 1st Lt. W. Reese Mullins, limping along with two superchargers out after a previous attack by Fw 190A-8s of IV.(*Sturm*)/JG 3. During that earlier attack, 20mm shells had struck the tail gunner, Staff Sgt. M. D. Barker, in the right leg and severed the ammunition train to his right gun, while in the top turret, Technical Sgt. Carl A. Dickson was hit in the face with shell splinters. Firing back with his one working weapon, Barker saw an Fw 190 go down on fire, but "Moon" Mullins had been compelled to drop out of formation, jettison his bombs to lighten the plane and head for home. The *Outhouse Mouse* was therefore not in the best of shape when Ryll came at it, but the B-17's plight had not gone unnoticed by friendlier eyes, either.

Lieutenant Colonel John B. Murphy of the 359th Fighter Group and his wingman, 1st Lt. Cyril W. Jones Jr., were surveying three boxes of B-17s when he looked over his shoulder and saw Ryll's contrail. Murphy knew about the new German fighter and doubted that he could intercept it until he noticed the *Outhouse Mouse* two miles to his right at 25,000 feet. Expecting the ailing Fort to attract enemy attention, Murphy led Jones in a dive toward it. Even at a speed well past 400 miles per hour, however, he and Jones were unable to reach the B-17 before Ryll did.

The *Outhouse Mouse*'s radioman, Technical Sgt. James R. Knaub, had dragged Barker to the waist gunner's position to treat his leg wound, and the bomber's crew were preparing for the homeward flight when Staff Sgt. Robert D. Loomis, who had taken Barker's place at the tail gun, remarked in his Montana drawl: "One of them things they been tellin' us about is messin' around back here." "If he starts to come at us, let me know," replied Mullins over the radio. Seconds later Loomis calmly announced, "Here he comes," and Mullins and his copilot, 2nd Lt. Forrest B. Drewry, manhandled their B-17 into a three-second dive with alternating rudder action. Loomis assured the crew that "he's shootin' at us and I'm shootin' at him,"

but the B-17's desperate evasive maneuvers prevented Ryll from doing any further serious damage before he broke sharply to the right.[12]

For more than half a minute the Me 163 flew alongside the *Outhouse Mouse*, then to its crew's surprise and relief it dived away. The reason became clearer to them when two P-51s flashed by. Closing to 1,000 feet, Murphy opened fire just as the *Komet* was pulling out of its dive, observing hits on the tail and left side of its fuselage before overshooting it and being forced to pull up violently to the left. Jones, trailing Murphy by 1,000 feet, came at the Me 163 from below and slightly to the left, but as he opened fire he saw the German go into a split-S, rolling inverted and then diving. Going down after the rocket, Jones scored some hits on its canopy before his Mustang hit the enemy plane's backwash, and he blacked out. He recovered at 14,000 feet, by which time Murphy had rejoined the chase, staying with the turning, descending *Komet* to an altitude of 8,000 feet. Firing a continuous burst as he closed to 100 feet, he saw multiple hits along the fuselage and pieces fly off until the entire left side of the fighter, from the cockpit back, blew away. As he passed through the smoke, Murphy noticed "a strange chemical fume" seeping into his cockpit. At that point, he saw another Me 163 about two miles off, and because he was running low on fuel, he and Jones decided to set course for England and home. Aboard the *Outhouse Mouse*, its ventral ball turret gunner, Staff Sgt. Kenneth L. Blackburn, saw Murphy's victim spiral into the ground, followed by an explosion and a tall column of smoke. The Germans subsequently determined that Ryll, the first combat casualty in a rocket fighter, had suffered fatal head and chest wounds before crashing just west of Brandis at 1052 hours.[13]

At 0100 hours two other Mustang pilots of the 359th Group, Capt. C. W. Hipscher and Lt. James Shoffit, saw a rocket fighter ten to twelve miles away, swiftly ascending to 32,000 feet to attack a formation of B-17s. Shoffit went after it, and as he approached he got into a head-on gun duel with the Me 163. Neither plane scored hits, but the *Komet* turned away from the bombers, and Shoffit continued to engage the rocket until the German applied full power and left the Mustang behind as if it were standing still.

Another of the 305th Group's Forts came under attack at 1102. The ball turret gunner, Sgt. J. D. Adriano, saw an Me 163 approach from nine o'clock low, fired at 800 yards as it passed under his formation, and saw it dive into a cloud, smoking.

At Brandis, the pilots of I./JG 400 assessed the results of fifteen intense minutes with mixed feelings. The frustration of the past month had finally been broken by Hartmut Ryll's B-17 kill (understandably, nobody was aware that his adversary would actually survive the punishment he had inflicted on it), but the cost had been Ryll himself, cut down while trying to score a second success. The fighting on August 16, 1944, would not be typical. The Germans built 364 Me 163s before production ceased in February 1945, but their pilots were credited with no more than sixteen aerial victories, a number far exceeded by their losses. The vast majority of Me 163 fatalities were not in combat, but in the act of taking off or landing, when a sharp bump could be enough to make the plane's unstable fuel explode. All of the Allied powers acquired Me 163s to evaluate after the war, but while it was unquestionably a spectacular weapon, the *Komet* would remain the only rocket-powered fighter ever to be used in combat.

One other German fighter design that saw brief service involved a mismatch of concepts. As the war turned irretrievably against the Third Reich, *Reichsmarschall* Hermann Göring proposed a *Volksjäger*, or "People's Fighter"—a cheap, simple jet made of nonstrategic materials and capable of being flown even by hastily trained Hitler Youth. "*Projekt Salamander*," issued on September 8, 1944, was not beyond the technical capabilities of Germany's advanced aircraft industry at that time, but the requirement that the design be drafted in twelve days and the prototype flying within ninety days placed it in the realm of fantasy. *Generalmajor* Adolf Galland, chief of the *Jagdwaffe*, opposed the whole idea, and only the Ernst Heinkel Flugzeugwerke made a serious effort to fulfill it. Working round the clock, the Heinkel designers managed to come up with a basic layout in four days. Of mixed construction, the Heinkel He 162V1 had a fuselage of light steel with a wooden nose cone, a single-piece wooden wing with metal down-angled tips, and twin vertical stabilizers at the ends of a V-shaped tail plane. A BMW 003E axial-flow turbojet engine, producing a maximum of 2,028 pounds of thrust, was mounted above the fuselage aft of the cockpit.

First test-flown by Gotthard Peter on December 6, 1944—incredibly, within the required time frame—the He 162V1 displayed a good turn of speed but marked instability along its longitudinal axis. During a subsequent flight on December 10, it reached a speed of 560 miles per hour at 20,000 feet—faster than an Me 262—but during a final high-speed run at low altitude, the He 162 suddenly rolled hard to the right and crashed in a fireball, killing Peter. An investigation concluded that the glue used in the plane's wing had failed under the aerodynamic stresses of high-speed flight.

By that time, Göring had already committed Heinkel to the production of his pet project, so its engineers did their best to make it more stable, increasing the wing area and the span of its tail plane. Because the twin 30mm MK 108 cannon intended for the He 162A-1 placed excessive stress on its light airframe, two fuselage-mounted 20mm MG 151s had to be substituted in the He 162A-2. Given the location of the engine intake, waiting to suck in the pilot if he had to bail out, an ejection seat, activated by an explosive cartridge, was added. The definitive He 162A-2 model was rushed into production, but a two-seat version was never built, and training consisted of veteran pilots getting a cockpit briefing at *Erprobungskommando* 162 at Rechlin-Roggenthin, and then going up to take its measure solo. In February 1945, the He 162 *Spatz* (Sparrow) entered service in JG 1, commanded by *Oberstleutnant* (Lieutenant Colonel) Herbert Ihlefeld.

Ihlefeld, a 123-victory ace who had scored his first nine flying Bf-109Bs over Spain, was not impressed with this latest Heinkel product. "The jet was rubbish," he said in a telephone interview in 1989. "Wood and glue. In very cold temperatures it became brittle. The shaking from the engine made it a death trap, but when the weather warmed up it became flexible. In my opinion it was a flying torch. Just a sad waste of resources, really."[14]

By mid-March, JG 1 had twenty-five combat-ready He 162s, but the loss of nine experienced pilots killed and five injured in accidents, against only one to enemy action, gave a clue as to how the *Volksjäger* would have fared in the hands of a neophyte *Hitlerjugend*. The earliest combat loss, Harald Bauer, was ferrying an He

A Heinkel He 162A-2 No. 120222, allegedly flown by *Oberst* Herbert Ihlefeld, commander of JG 1 at Leck airfield, awaits evaluation by its new American owners after Germany's surrender in May 1945. (*National Archives*)

162 to Varel on March 24 when he was spotted by the lead navigator of a B-17G of the 390th Bomb Group, who directed escorting Mustangs to shoot him down. Wounded in the leg and discounting bailing out as too chancy an option, Bauer belly-landed and found himself surrounded by American soldiers of the 2nd Armored Division, who promptly took him prisoner. Oddly, no corresponding claim has been found among VIII Fighter Command's Mustang pilots for that day.

On April 15, *Ltn.* Rudolf Schmitt of 1./JG 1 claimed to have had a brief, bloodless run-in with a Spitfire. On the nineteenth, a captured Royal Air Force pilot claimed to have been shot down by a jet, and *Oberst* Ihlefeld attributed the victory to *Fähnenjunker-Feldwebel* (Sergeant /Officer Cadet) Günther Kirschner of the 3rd *Staffel* (3./JG 1). Kirschner himself was unavailable to testify, however, because shortly afterward he crashed to his death. He was apparently the victim of Flying Officer Geoff Walkington, a Tempest V pilot of No. 222 Squadron RAF, who claimed to have attacked a small twin-tailed jet fighter he'd never seen before, and although unable to catch up at 360 miles per hour, he saw it make the fatal effort of turning sharply to the right, allowing him to close the range and shoot it down.

With casualties mounting, Ihlefeld, joined by Galland and other veteran officers, confronted Göring at his estate at Karinhall. As Ihlefeld recalled it:

> Galland was very much against the *Volksjäger*, claiming that it was a waste of time, money and lives. He argued that it was stupid to place these young boys in this thing. I agreed, and this was in front of Göring, in April. Göring waved away our concerns, and he told Galland, "You will do as you are told, and so will you, Ihlefeld." Well, I told him that unless I saw that a pilot was well qualified, I would not be responsible for killing young men. Galland then said that he would never follow that order. Göring turned red and became angry, but said, "Then do what you can." We left Karinhall with the others, and everyone just looked at me. [Günther] Lützow said, "What will you do, Herbert?" I told him that as far as I was concerned, my new pilots would all be unable to pass the flight physical.[15]

JG 1 formally began operations against low-flying Allied fighter bombers on April 21. Three days later, *Hptmn.* Paul Heinrich Dähne, *Gruppenkommandeur* of II./JG 1 and a *Ritterkreuz* recipient with ninety-nine victories previously scored over the Eastern Front, lost control of his He 162 and crashed near Warnemünde, dying due to a malfunction of the ejection system. On the twenty-sixth, *Uffz.* Helmut Rechenbach of 3./JG 1 claimed a de Havilland Mosquito, with his *Staffelführer*, *Obltn.* Emil Demuth, and *Stabsintendant* (Staff Director) Siegfried present to bear witness, but again the victor fatally crashed soon afterward, before he could submit a combat report.

On May 1, JG 1 was based at Leck in Schleswig-Holstein, from which *Ltn.* Rudi Schmitt was flying his fifth mission on May 4 when he caught what he reported as a low-flying Hawker Typhoon and sent it crashing to earth. In fact, Flying Officer M. Austin of No. 486 Squadron, Royal New Zealand Air Force, was shot down, bailed out of his Tempest, and was taken prisoner. Unfortunately for Schmitt, however, his claim was disputed by a local flak unit, which got the credit instead. Any further hopes for the *Volksjäger* to balance some success against all the sweat and blood which had gone into its hurried creation ended that evening, as British troops occupied Leck.

Because of Göring's crash program, numerous He 162s were available for the Allies to evaluate—and later to preserve in their museums. While the Me 262 pointed to the future, however, the He 162 endures merely as an aeronautical curiosity, a monument to desperation, and a far-fetched deviation along the road to jet-plane development.

Nineteen

Dawn of a New Era

MESSERSCHMITT ME262, GLOSTER METEOR,
AND THE P-80 SHOOTING STAR, 1944–45

In the early morning hours of August 27, 1939, Ernst Udet, chief of the *Technisches Amt* of the *Reichsluftfahrtministerium* (RLM), was awakened by a telephone call from the Heinkel works' airfield at Rostock-Marienehe.

"Good morning," said Ernst Heinkel. "I just wanted to inform you that *Flugkapitän* (Flight Captain) Warsitz has just successfully flown the world's first jet plane, the Heinkel He 178, and landed safely."

Udet paused a moment to gather his drowsy thoughts, then congratulated Heinkel and asked that he be allowed to get back to sleep.[1]

At the moment—and for some time thereafter—neither Udet nor the Luftwaffe was quite aware of the significance of what had occurred. In fact, Udet's rude awakening had been amid the dawn of a new era in aviation, for the Heinkel He 178 that Erich Warsitz had test-flown to a speed of 373 miles per hour was indeed the first turbojet-powered aircraft to take to the air. And just six days later, Adolf Hitler's invasion of Poland would plunge Germany into a war that would compel the Luftwaffe to accel-

erate the development of Heinkel's experimental He 178 into a practical weapon.

Heinkel, aided by physicist Hans-Joachim Pabst von Ohain, had also built an airplane powered by a liquid-fueled rocket, the He 176, which Warsitz had successfully test-flown at the Peene-münde experimental rocket station in June 1939. It was the jet, however, that proved to have the greater potential, and Heinkel proceeded with the development of a twin-jet fighter, the Me 280, which made its first powered flight on March 20, 1941. By then, however, Heinkel was not alone—his rivals at the Messerschmitt plant at Augsburg were working on a jet fighter of their own. And in Britain, Frank Whittle had long been working on a jet engine and an airframe on which to demonstrate it—the Gloster E.28/39, which finally got into the air on May 15, 1941.

Jet development proceeded apace over the next two years, with Heinkel and Messerschmitt vying for a production order until March 27, 1943, when the German chief of aircraft procurement, *General-Luftzeugmeister* (General of Air

Production) Erhard Milch, informed Heinkel that his He 280 had been passed over in favor of Messerschmitt's Me 262. Although the He 280 was faster, had a higher rate of climb and a higher service ceiling, its twin vertical tail surfaces were suspect, and its range was only two-thirds that of the Me 262.

First proposed to the RLM in the summer of 1938 as a research aircraft to use the new BMW P 3302 gas turbine engine, Messerschmitt's *Projekt* P.1065 was developed into an airframe by a design team led by Dr. Woldemar Voigt. Intended as an interceptor from the outset—even though the Luftwaffe requirement had not specified that role for it—the original Me 262 had a tail wheel and a slight sweepback that was incorporated with the center of gravity, rather than aerodynamic qualities, as the causal consideration. The airframe, powered by a single 690-horsepower Junkers Jumo 210 piston engine, was taken up for its first test flight from Messerschmitt's airfield at Augsburg by *Flugkapitän* Paul Wendel on April 18, 1941.

Although Hitler is popularly blamed for holding up the Me 262's development with his much-publicized fixation on using it as a high-speed fighter-bomber, the real delaying factor was the engine. The Bayerische Motoren Werke had claimed that its P 3302 engine could be made to produce 600 kilograms (about 1,300 pounds) of thrust by the end of 1939, but when one of the engines was bench-tested at the end of 1940, it only produced 260 kilograms (570 pounds). Meanwhile, Heinkel had developed his own engine, capable of producing 500 kilograms of thrust, to power his He 178 on its historic first flight. The first of BMW's P 3302 engines—redesignated BMW 003s—did not arrive at Augsburg until November 1941, and shortly after

Wendel took the Me 262 up for its first jet-powered flight on March 25, 1942, both of them failed, requiring him to return to using the piston engine.

By then, the Junkers Jumo 004 turbojet had been developed and tested, producing 1,000 kilograms (2,200 pounds) of thrust. Two of the engines were installed in the wings of the third prototype, and on July 18, 1942, Wendel took off from Leipheim in the Me 262V3 for a successful twenty-minute flight. Even after plane and engine went into full production, however, the Jumo 004 would be an Achilles' heel for the Me 262. Chromium and nickel, essential ingredients for the steel alloys necessary to operate at a jet engine's high temperatures, were in short supply and substitute metals, such as ordinary steel with a spray coating of aluminum, were prone to burn and adversely affected engine life. At the end of the war, the average engine life of an Me 262 was only twenty-five hours, with a time between overhauls of ten hours.

So secret was the new plane's development that *General der Jagdflieger* (Commanding General of Fighters) Adolf Galland knew nothing of it until he finally got his chance to fly the fourth prototype on May 22, 1943. He was instantly impressed, declaring that flying the jet felt "as if an angel were pushing," and recommended that Me 109 production be halted so that Messerschmitt could concentrate on the all-out manufacture of the new jet fighter. His influence did speed things up—seventy-two hours later, Milch ordered the Me 262 into series production. The first hundred Me 262s to be built would be issued to special test units, or *Erprobungskommandos*, that would give the fighter its first operational exposure to the enemy and simultaneously identify and iron out any shortcomings as they arose.[2]

In mid-1943, German air defenses had been holding their own against British and American bombers. At the same time, however, the failed British raid on Dieppe in August 1942, followed by the successful Allied invasions of North Africa in November 1942, of Sicily in July 1943, and of Italy in September, had alerted Hitler to the danger of an Allied invasion of France. Concluding that the best way to repel such an invasion would be a series of lightning air strikes on the beachhead, he became adamant about developing a high-speed "Blitz Bomber." Such was the situation on November 2, 1943, when Hermann Göring, while visiting the Augsburg plant, first asked Willy Messerschmitt if the Me 262 could be adapted to the bombing role. "*Herr Reichsmarschall*," Messerschmitt replied, "from the very outset we have provided for the fitting of two bomb pylons so it can carry bombs—either one 500 kg. or two 250 kgs. [2,200 pounds]." He went on to state his confidence that the payload could be doubled and that such a modification could be completed in a couple of weeks.[3]

On November 26, the Me 262 was demonstrated to Hitler at Insterburg, and he, too, broached the question. Again, Messerschmitt answered affirmatively, and the *Führer* blissfully assumed that his wish would be carried out. Unknown to him, however, Messerschmitt proceeded with the Me 262 as an anti-bomber interceptor, with a quartet of four 30mm Mark 108 low-velocity cannon in the nose.

The Luftwaffe accepted its first sixteen preproduction Me 262A-0 fighters, which had been lying at Lechfeld waiting for engines, between April 18 and 29, 1944, and at the end of that month *Erprobungskommando* 262 was formed at Lechfeld, Bavaria, under *Hptmn.* Werner

Thierfelder. Even while they gained experience in the new type, the special test unit's pilots were writing an operating manual for the Me 262A-1a fighter.

On May 23 Hitler summoned Göring, Milch, Galland, and other *senior* Luftwaffe officers, as well as Albert Speer and officials from his armament ministry, to Berchtesgaden to discuss fighter production. The *Führer* listened somewhat lackadaisically to Milch's statistics, but when he mentioned progress on the Me 262 fighter program, Hitler interrupted him: "I thought the 262 was coming as a high-speed bomber? How many of the 262s already manufactured can carry bombs?"

"None, *mein Führer*," Milch replied. "The Me 262 is being manufactured exclusively as a fighter aircraft." There was an awkward silence, then Milch added that extensive design changes would be necessary to convert the jet into a bomber, and even then it would not be able to carry more than five hundred kilograms.

"Never mind!" Hitler exclaimed. "I wanted only one 250-kilo bomb." Losing his composure, he demanded precise weight statistics on the fighter's armor, guns, and ammunition. "Who pays the slightest attention to the orders I give?" he railed. "I gave an unqualified order, and left nobody in any doubt that the aircraft was to be equipped as a fighter-*bomber*."[4]

Hitler's confidence in Milch was irreparably shattered by the meeting in Berchtesgaden, and Milch would be progressively stripped of his authority in the weeks that followed. Göring was made personally responsible for the implementation of the Blitz-Bomber program.

On May 27, a still furious Hitler ordered that the Me 262 not be regarded as a fighter, but to enter service exclusively as a fighter-bomber. He relented somewhat a few days later, allowing testing of the fighter to continue, but with the proviso that the first operational units must be equipped with the bomber. Messerschmitt responded by mounting two pylons, each capable of carrying a 550-pound SC 250 bomb under the nose of the tenth prototype, and fitting an extra fuel tank with a capacity of 600 liters (132 Imperial gallons) in the rear fuselage. To compensate for the weight, two of the MK 108 nose cannon and most of the armor plating from the cockpit were removed.

While the Me 262A-2a *Sturmvogel* (Storm Bird) was hastened into production, a unit of *Kampfgeschwader* 51, commanded by *Maj*. Wolfgang Schenck, was detached and sent to Lechfeld to train on the new Blitz Bomber. On June 8, two days after the Allied landings in Normandy, members of the 3rd *Staffel* (3./KG 51) assembled at Lechfeld, and by the end of the month seven jets were at their disposal, although only one may have been an Me 262A-2a. After about a month of conversion training, the first nine Me 262A-2as of *Erprobungskommando* /KG 51 were ordered to Châteaudun, France, on July 17. The next day, the town of Saint Lô fell to US Army troops of the 29th Infantry Division, ending weeks of stalemate in the hedgerow country of the Cotentin peninsula. Hitler, however, still did not realize that his original hope of using Blitz Bombers to drive the Allies from their beachhead was long gone. He was still convinced that Normandy was a feint and that the main Allied landing was yet to come at Calais, for which event the *Sturmvogel* would surely be ready.

On July 20, *Erprobungskommando*/KG 51, also known as *Kommando* Schenck, arrived at Châteaudun—while at Hitler's Wolf's Lair, a bomb placed by *Oberstleutnant* Claus Schenk *Graf* von Stauffenberg injured but failed to kill the *Führer*. The transition to fighter-bombers

Messerschmitt Me 262A-1a No. 130017 "White 4" of *Erprobungskommando* 262, in which *Ltn*. Alfred Schreiber was credited with scoring the first jet victory in history on July, 24, 1944, although he actually did not shoot his quarry down. (*Bundesarchiv*)

was still proving to be anything but smooth, as reports arose of a need for a strengthened undercarriage to accommodate the bombs and auxiliary fuel tanks, and of a lack of suitable bomb sights. Meanwhile, several pilots were killed in the course of training for *Jabo* operations in the jets.

While the Blitz-Bomber program slowly proceeded, the fighter pilots found it impossible to wait. *Erprobungskommando* 262 opened its account on July 26, when *Ltn*. Alfred Schreiber, flying Me 262A-1a No. 130017, call number "White 4," attacked de Havilland Mosquito PR Mark XVI MM273 of No. 544 Squadron RAF just as it was completing a photoreconnaissance mission over Munich at 28,000 feet. The British pilot, Flight Lieutenant Albert E. Wall, was about to head back to his air base at Benson, England, when his navigator, Pilot Officer Albert S. Lobban, reported a "bandit" at six o'clock and closing fast. Wall raised speed to 430 miles per hour, only to hear from Lobban that the enemy was still closing. He then put MM273 into a shallow dive, but at its not-to-be-exceeded speed of 470 miles per hour, the Mosquito was vibrating dangerously when the German caught up and opened fire. The British managed to evade five passes before Schreiber scored a hit and saw a piece fall off the plane. At that point Wall dived into a cloud and lost his pursuer. Although Schreiber was credited with the victory, what he'd seen fall off was the outer door of the pressurized Mosquito, but the damage was enough for Wall to abandon hopes of returning to Benson. Lobban instead directed him to the nearest Allied airfield available, Fermo on the Adriatic coast of northern Italy. With engine cylinder temperatures running dangerously high, Wall gently nursed his plane over the Alps to make an emergency landing at Fermo and reported his harrowing encounter with an unnervingly swift new enemy fighter that had no propellers.

A former Me 110 pilot of *Zerstörergeschwader* 26, "Bubi" Schreiber claimed Supermarine Spitfires engaged in PR missions on August 2 and 26. In between, *Ltn.* Joachim Weber destroyed another 540 Squadron Mosquito on August 8. Schreiber downed a Spitfire PR Mark XI of the U.S. Army Air Forces on September 5, and on October 28 he shot down a Lockheed F-5E-3-LO, photo Lightning, whose pilot, 2nd Lt. Eugene Selzer Williams of the 22nd Squadron, 7th Photoreconnaissance Group, bailed out and was taken prisoner.

Schreiber's impressive run of luck in the Me 262 ended on November 26, as the new ace was taking off with a comrade on an intercept mission, and his engine suffered a flameout. Schreiber tried to land but ran into a trench and was crushed to death when his plane turned over on top of him.

After moving bases several times in the face of Allied advances in France, *Kommando* Schenck finally flew its first Blitz-Bomber mission on the morning of August 25, when four Me 262s took off from Juvincourt and attacked targets along the Seine River, northwest of Paris. The same quartet flew a second mission that day, but one was lost in an emergency landing near the front. The Führer's wishes were finally being carried out, for all the good it would do. On that same day, *General der Infanterie* (Lieutenant General) Dietrich von Choltitz, disobeying Hitler's order to raze Paris to the ground rather than give it up, surrendered the French capital to the Allies.

By the autumn of 1944, Me 262A-1a fighters were making their presence felt among the American bomber streams and—to a considerably lesser degree— among the Allied ground forces. In either role, they were too little and too late to affect the course of the war in, or over, Europe. The presence of the long-ranging North American P-51D Mustang and the steady Allied advance across the Continent brought Me 262 air bases within range of an increasing number of Allied fighters. If the jets were too fast to catch in the air (though some were shot down by a handful of lucky pilots), their poor acceleration rate made them vulnerable to ambush as they took off or landed, a weakness of which the Allied fighter pilots took frequent advantage.

While the Me 262 pilots were putting up a gallant but futile fight against overwhelming numerical odds, another jet fighter was entering service on the Allied side—the Gloster Meteor.

Since the early 1930s, while Ernst Heinkel worked on his jet engine, Frank Whittle had been engaged in a double struggle—with the problems of perfecting a gas turbine driving a series of enclosed impellers, and with trying to interest the RAF in his project. In 1939, Whittle joined creative forces with

George Carter, chief designer for the Gloster Aircraft Company, and at about that same time the Air Ministry finally took an interest in the jet concept, issuing a contract to Gloster for an experimental airframe that could be adapted for operational use with minimum modification.

The result of Whittle's and Carter's efforts, the single-engine Gloster E28/39 finally flew at RAF Cranwell on May 15, 1941. Although the maiden flight lasted little more than a quarter of an hour, test pilot Phillip E. G. Sawyer emerged from the cockpit praising the plane, declaring that the jet was indeed the way of the future. The RAF was already convinced of that, for by then it had issued Specification F9/40, calling for five hundred twin-engine fighters using Whittle's engine.

Eight developmental aircraft were built, the fifth of which, DG206, made the F9/40's first flight from RAF Cranwell on March 5, 1942, with Michael Daunt at the controls. Whittle's WSB engines were still being built at the Rover plant in Coventry, so de Havilland H1 Halford engines were substituted for the first flight. Although the plane exhibited a few problems, including a tendency to yaw violently as its speed approached 230 miles per hour, the Air Ministry found it promising enough to continue development, starting with redesigned larger tail surfaces. Trials were switched to Newmarket Heath, then to Barford Saint John, and finally to Moreton Valence, once a hardened runway was completed there. Policemen closed the roads whenever the Meteor, as the new jet was then being called, flew. Test flights were usually conducted when there was low cloud cover, to reduce the odds of unauthorized eyes seeing the top-secret fighter.

Meteor DG205/G, fitted at last with the Whittle-developed W2B engine, made

its first flight on June 17, 1943. Development continued at a rather slow pace until the first production Meteor F Mark 1, EE210/G, flew from Moreton Valence on January 12, 1944. The production aircraft used Rolls-Royce-built W1B Welland I engines, but apart from a modified canopy and the installation of four 20mm Hispano Mark V cannon in the nose, it differed little from the prototype. Rated at 1,600 horsepower, the Welland I was a reliable and tractable power plant, but because of the Meteor's size, rather than its modest weight, the plane was only able to reach a speed of about 390 miles per hour at sea level and a maximum of 415 miles per hour at higher altitudes, with a service ceiling of 40,000 feet. Given that less than exhilarating performance, some RAF people suggested that it was fit only to serve as a trainer. By then, however, the Air Ministry was aware of the Me 262 and of its imminent introduction to service, and consequently judged it psychologically important that the RAF have a jet of its own in frontline squadrons.

Another psychological factor arose that would serve as the ultimate call to arms for the Meteor. On June 12, 1944, the first V-1 (*Vergeltungswaffe*, or Vengeance Weapon) was launched against London. A pilotless flying bomb packing a ton of high explosive in the nose, powered by a simple, externally mounted pulse-jet engine, guided by gyros, and made to fall on its target when its predetermined allotment of fuel ran out, the V-1—or "Buzz Bomb," as it came to be called by its intended victims—was meant to terrorize the British home front for the first time since the Battle of Britain, in reprisal for the Allied landings in Normandy six days earlier. To the RAF, it seemed almost a matter of destiny that its first jet should be among the aircraft mobilized to defend

England against these swift, small, jet-propelled flying robots.

The unsuspecting first recipient of the Meteor was No. 616 Squadron, Auxiliary Air Force, which in June 1944 was flying Spitfire Mark VIIs from Culmhead, Somerset, on escort missions for bombers striking at German tactical targets in France. Rumors had been rife since the spring that the squadron was to be reequipped, but most pilots believed that the replacements would be Rolls-Royce Griffon-engine Spitfire Mark XIVs, two of which arrived at Culmhead in June. Shortly afterward, Squadron Leader Andrew McDowall and five other pilots were summoned to Farnborough to acquaint themselves with the new aircraft, but when they returned, they announced that 616 Squadron's replacements would be jets—and unanimously added that once they had gotten accustomed to their tricycle landing gear, they were delightful machines to fly.

The first two Meteor Is, EE213 and EE214, began flying with 616 Squadron on July 12, 1944. Five more Meteors arrived on the fourteenth, and by the twenty-fifth squadron strength had reached a dozen, completely replacing its Spitfire VIIs. On July 21 two of the Meteors, escorted by the squadron's Spitfires, flew to Manston airfield in Kent, followed two days later by five more jets. From there, newly promoted Wing Commander McDowall, Wing Cmdr. Hugh J. Wilson, Squadron Leader Leslie W. Watts, Flying Officers William H. McKenzie and T. J. Dean, and Warrant Officer Wilkes commenced operations against the V-1s.

At 1430 hours on July 27, 1944, Bill McKenzie flew the first Allied combat mission in a jet, patrolling around Ashford without incident. More "diver patrols," as the RAF code-named its efforts to stop the flying bombs, were flown that afternoon by Dean and Watts, the latter catching up with an incoming V-1 over Ashford. With the "doodlebug" in his gunsight, Watts pressed the trigger button on the control column, but nothing happened. His guns had jammed, and the V-1 escaped and went on to hit its target. After that setback, it was decided that patrols against the vengeance weapons should be carried out by pairs of aircraft, since the odds of both Meteors' guns jamming were unlikely. In order to increase their time in the air, the squadron moved its Meteors to a dispersal aerodrome near Ashford, reducing the distance they had to fly to reach the V-1s' expected routes.

Finally, on the evening of August 4, "Dixie" Dean was only minutes from takeoff at Ashford when he spotted a V-1 ahead of and below him, heading toward Tunbridge Wells. Going into a shallow dive, Dean increased his speed to 385 miles per hour, then got in a brief burst of his guns before they jammed. Dean then brought the Meteor alongside the V-1 as close as he felt safe, slid his wing under the V-1's and slowly pushed his control column to the left. As Dean's plane banked, the force of air lifted the robot's wing, unbalancing its autopilot until it abruptly flicked over on its back and dived into the ground, exploding harmlessly in the open countryside. Contrary to popular belief, the Meteor's wingtip did not actually touch the V-1 during these "tip-and-run" tactics, since there was too much chance of damaging the fighter or even the loss of valuable pilots and aircraft. Air pressure sufficed to do the job.

Within minutes of Dean's success, Flying Officer J. K. Rodger closed on another "diver" over Tunbridge and opened fire with his four 20mm cannon. This time, the weapons did not jam, and the V-1 went down in open countryside near Tenterden.

After that, 616 Squadron relayed two-plane patrols throughout the day, each flight lasting about thirty minutes. By August 10, Dean had added two more divers to his score. On August 16 and 17, the Meteors accounted for five more of the robot bombs. A total of thirteen V-1s were destroyed in one way or another by 616's "Meatboxes"—modest in number, but a great boost to public morale.

Aside from its guns, the Meteor gave no trouble, and its two engines needed less servicing than the single engine of a Spitfire. Indeed, the only thing bad that happened to No. 616 Squadron was when one of its Meteors was almost shot down in error by a Spitfire and had to land under control of the elevator trimmers.

For the remainder of 1944, 616 Squadron's Meteors operated from Debden, where they were used to acquaint RAF and USAAF units with the characteristics of jets, and to help them develop tactics to counter the Me 262As that were starting to take their toll of bomber formations. After being subjected to hit-and-run strikes by the Meteors, the American P-51 and P-47 pilots concluded that the only way to protect the bombers was to increase their numbers 5,000 feet above them, allowing the fighters time to build up speed to intercept the German jets. Such tactics required split-second timing at high speeds, but they paid off, as a number of Americans added Me 262s to their scores while on escort duty.

On December 18, No. 616 Squadron received its first two Meteor F Mark 3s, EE231 and EE232, which were powered by Rolls-Royce Derwent engines in revised nacelles. In addition to improved performance, the F3 had a larger fuel capacity, which gave it an hour's longer endurance; an enlarged, more stream-lined windscreen; and a rear-sliding bulged bubble in place of the F1's side-hinged canopy. Three more Meteor 3s were on strength by January 1945, when the unit moved to Colerne, Wiltshire. There, the squadron exchanged the last of its Mark 1s for the newer type, and on January 20 one of the flights was dispatched across the Channel to Melsbroek, near Brussels, Belgium, to join No. 84 Group of the Second Tactical Air Force. For some weeks, the Meteors flew patrols over local Allied airfields, primarily to acquaint them with the new jet's silhouette. While there, the squadron had its closest brush with its technological counterparts on March 19, when the air base was raided by four Arado Ar-234B-2 jet bombers of II *Gruppe*, *Kampfgeschwader* 76. The rest of the squadron arrived on March 31, and in early April it resumed offensive operations from Gilze-Rijen in the Netherlands as part of No. 122 Wing.

As was the case with the German jet fighter units, 616 Squadron's personnel now included some skilled veterans, including Wing Cmdr. Warren Edward Schrader from Wellington, New Zealand, who had previously flown Hawker Tempests in No. 486 Squadron and had eleven victories, plus two shared, to his credit. An identical victory tally had been scored by Wing Commander McDowell from Kirkenner, Scotland, but he had done so in Spitfires with No. 602 Squadron. Much to the disappointment of the pilots, however, no contact was made with the Luftwaffe in the course of their short patrols, and in consequence they were employed in the armed reconnaissance and ground-attack roles.

On April 13, No. 616 Squadron moved to Nijmegen, and on the fourteenth, Flt. Lt. Mike Cooper became the first "Meatbox" pilot to fire his guns in anger over the Continent when he spotted a

Gloster Meteor F.1 and F.3 jet fighters of No. 616 Squadron refuel at Manton, Kent, in January 1945, not long before departing to the Continent. (*Imperial War Museum*)

large German truck near Ijmuiden and in a single firing pass sent it careening off the road, to burst into flames seconds later. On the twenty-fourth, McDowell, flying Meteor F1 YQ-A, led four others on a strike against an enemy airfield at Nordholtz, Germany. Diving out of the sun from 8,000 feet, he destroyed a Ju 88 on the ground and shot up a vehicle. Flying Officer Ian T. Wilson set two petrol bowsers on fire and used up the rest of his shells on other airfield installations. Flying Officer H. Moon roamed the perimeter of the field, strafing a dozen railway trucks and destroying a flak post. Flying Officer T. Gordon Clegg attacked a large vehicle full of German troops who, thinking the twin-engine jet to be one of their own, waved and cheered until Clegg opened fire.

Up to that time, 616 Squadron had taken no casualties, but that unblemished record came to a tragic end on April 29, when Squadron Leader Watts, who had been with the unit since August 1943, collided with Flt. Sgt. Brian Cartmell in a cloud bank. The two planes exploded, and both pilots were killed.

On May 2, Wing Cmdr. "Smokey" Schrader replaced McDowell as commander of No. 616 Squadron. On the same day, one of the "Meatbox" pilots encountered a Fieseler Fi.156 *Storch*, but the nimble liaison plane was able to outmaneuver the fighter and landed—after which the Meteor strafed it to destruction. On May 3 Schrader, flying Meteor YQ-F, led the squadron in an attack on Schönberg air base near Kiel, during which six aircraft were destroyed on the ground—Schrader personally accounting for an Me 109, an He 111, and a Ju 87. On another occasion, four Meteors encountered some Fw 190s, but their hopes of adding some air-to-air victories to the squadron tally were again frustrated when some Spitfires and Hawker Tempests mistook the British jets for Me 262s and prepared to attack, compelling the Meteors to abandon their attempt against the Germans. On the following day, 616 Squadron's pilots destroyed one locomotive, damaged another, knocked out ten vehicles and two half-tracks, and strafed a number of installations. At 1700 hours, the unit was ordered to suspend offensive operations. Four days later, Germany surrendered.

It is probably fortunate for the Meteor pilots that they never had to do battle with the Me 262s—all other things being even, neither their aircraft nor their own level of expertise would have matched the performance of the German fighter, nor of the crack *Experten* who were flying it in the final weeks of the war. Nevertheless, the "Meatbox" proved to have considerable development potential, and on November 7, 1945, a Meteor F3 piloted by Wing Cmdr. "Willie" Wilson reached a record speed of 606 miles per hour. Progressively improved marks were to follow, serving in the fighter, reconnaissance, and night-fighter roles until September 1961, by which time a total of 3,875 Meteors had been built.

One more jet fighter waited in the wings for its fighting debut in the war's last months. Witnessing the earliest British jet flights in 1941, US Army Air Corps general Henry H. "Hap" Arnold returned home impressed and determined to get the United States into the race. At his urging, General Electric studied Frank Whittle's engine and set about developing its own turbojet, while Laurence Bell got a contract on October 3 to design a jet fighter, which was ready for testing a year later as the XP-59A Airacomet. With a maximum speed of 413 miles per hour at 30,000 feet, however, it lacked the performance to be competitive, and the fifty Airacomets built served only as trainers for pilots making the transition into jet flight.

In May 1943 USAAF representatives approached Lockheed's chief engineer, Clarence L. "Kelly" Johnson, who had been playing around with turbojet designs of his own since 1940. Assembling a

design team in a secret facility in Burbank, California, which they called the "Skunk Works," Johnson had a mockup ready by July and a prototype ready for testing within the 180 days he promised the Army that he would. On January 8, 1944, Lockheed test pilot Milo Burcham took the XP-80 up for its first flight from Muroc Dry Lake, California. Powered by a GE I-40 engine generating 4,000 pounds of thrust, the sleek plane performed perfectly, diving at 490 miles per hour and roaring over the observers at 475.

While the USAAF was looking forward to Lockheed producing 450 P-80s per month by 1946, development of the fighter continued, necessitated by inevitable flaws such as fuel pump failure and flameouts, which killed Burcham on October 20, 1944, and which would similarly claim the life of American ace of aces Maj. Richard I. Bong on August 6, 1945. After three prototypes, Lockheed produced thirteen preproduction YP-80As, and selected four of them for Operation Extraversion. This, in response to the appearance of German jets over the Reich, involved their being assigned to the Eighth Air Force in England and the Fifteenth Air Force in Italy. Giving demonstration flights at various air bases, they would reassure uneasy bomber crews and apprehensive fighter pilots that an American response to the German jets was on the way.

In January 1945 two P-80A Shooting Stars, as they were being called, arrived in England, and another two at Lesina, Italy, the latter being assigned to the 94th Squadron of the 1st Fighter Group "for testing under combat conditions in a remote location." The program suffered a setback during a demonstration flight over RAF Burtonwood on January 28, when an engine fire caused P-80A 44-

83026 to crash, killing the pilot, Maj. Fredrick Austin Borsodi.[5]

Given the P-80's limited range, that should have been the end of it, but for the appearance of enemy jets over Italy. In late February, a detachment of three Ar 234B-1s led by *Obltn*. Erich Sommer arrived at Udine-Campoformido airfield, and on March 15 he flew its first of some twenty photoreconnaissance sorties over Ancona and San Benedetto del Tronto. This brought the prospect of a jet-versus-jet encounter tantalizingly within the 1st Fighter Group's bailiwick, and Maj. Edward F. LaClare is known to have flown two operational sorties in the P-80A, though there is no known instance of he or anyone else making contact with the enemy.

The two Lockheed P-80As do a pass by Mount Vesuvius in 1945. Assigned to the 1st Fighter Group, they missed a chance to test their mettle against Arado Ar 234B-1s operating in northern Italy that April. (*National Archives*)

The American jets proved to be the least of *Kommando* Sommer's problems. On March 29, Spitfires strafed Lonate-Pozzolo (now Milano-Malpensa) airfield just as *Fw.* Walter Arnold was taking off in an Ar 234, forcing him to abort his mission. On April 11, *Ltn.* Günther Gniesmer's Arado was jumped by two P-51Ds of the 52nd Fighter Group, and 1st Lt. Benjamin I. Hall III sent it crashing in flames in the Comacchio valley. Gniesmer bailed out, but struck his horizontal stabilizer and died of his injuries two days later. A landing crash on April 20 destroyed Sommer's plane, and on the thirtieth, with Italy lost, Sommer and Arnold burned the last Ar 234 at Holzkirchen and started down the road for Austria.

Having missed their fleeting opportunity for combat, the 94th Fighter Squadron's P-80As spent their time being photographed flying past Mount Vesuvius and engaging in comparative mock dogfights with P-38s and P-51s. The Shooting Star's problems would be solved by 1950, when the redesignated F-80C went to war anew in Korea, where it would finally get its postponed chance to face another jet in action—in the form of sweptwing MiG-15s flown by Soviet pilots. But that, as they say, is another story for another time.

It had been a remarkable thirty years. At the start of 1915, the fighter plane did not exist, and airmen were trying to figure out how to fire a machine gun around the propeller. By the end of 1945, there would no longer be a propeller in the way, as fighters took to the sky using means of propulsion that few would have envisioned three decades earlier. A new, frightening, and exhilarating era of aerial combat had begun.

NOTES

CHAPTER ONE

1. Carson, *Flight Fantastic*, 35.
2. Durkota, Darcey, and Kulikov, *Imperial Russian Air Service*, 202.
3. Ibid., 204.
4. Cuneo, *Winged Mars*, 424.
5. Lawson and Lawson, *First Air Campaign*, 102.
6. Werner, *Knight of Germany*, 136.
7. Guttman, *Groupe de Combat 12*, 11.
8. Jammer28, March 26, 2009 (8:36 a.m.), "Charles Nungesser," Combat Ace Forum, http://combatace.com/topic/38892-charles-nungesser.
9. Mason, *High Flew the Falcons*, 126.
10. Jankowski, *Verdun*, 113.
11. VanWyngarden, *Jagdstaffel 2*, 12.
12. Kulikov, *Russian Aces*, 70–74.
13. Revell, *British Fighter Units*, 8.
14. Guttman, *Pusher Aces*, 22.

CHAPTER TWO

1. Grosz and Ferko, "Biplanes for the *Fliegertruppe*," 57.
2. Ibid.
3. Grosz, *Halberstadt Fighters*, 5.
4. VanWyngarden, *Jagdstaffel 2*, 13.
5. Kilduff, *Red Baron: The Life and Death of an Ace*, 74.
6. Guttman, *Spad VII Aces*, 41.

CHAPTER THREE

1. Klaeylé and Osché, *Guynemer*, 5.
2. Austin and Dicks, *Great Fighting Planes*, 28.
3. Ibid.
4. Rimell, *Zeppelin!*, 111.
5. Ibid., 110.
6. Ibid., 111.

7. Franks, Giblin, and McCrery, *Under the Guns*, 96.
8. Bowyer, *Sopwith Camel*, 13.
9. Bruce, *De Havilland D.H.5*, 9.
10. Bruce, *Bristol M.1*, 7.
11. Lighthall, "Royal Air Force," 171.
12. Bruce, *Bristol M.1*, 6.
13. Lighthall, "Royal Air Force," 171–72.
14. Alegi, *Hanriot HD.1/HD.2*, 2.
15. Ibid., 6.
16. Bruce, "Spad Story," pt. 1, 240.
17. Bruce, "Spad Story," pt. 2, 290.
18. Hylands, *Georges Guynemer*, 17.
19. Bruce, "Spad Story," pt. 2, 290.
20. Grosz, Haddow, and Schiemer, *Austro-Hungarian Army Aircraft*, 442.
21. Ibid., 142.
22. Ibid., 119.

CHAPTER FOUR

1. Bruce, "Sopwith's Pedigree Pup," 188.
2. Ibid., 190.
3. E. Rochford Grange, letter to author, May 25, 1982.
4. Kilduff, *Richthofen*, 149–50.
5. Ibid., 150.
6. Lee, *No Parachute*, 108.
7. Kilduff, *Richthofen*, 151.
8. Ibid.
9. Ibid., 153.
10. Hylands, *Werner Voss*, 8.
11. McCudden, *Flying Fury*, 221.
12. Hylands, *Werner Voss*, 16.
13. Ibid.
14. Ibid.
15. Revell, *High in the Empty Blue*, 161.
16. Hylands, *Werner Voss*, 20.

CHAPTER FIVE

1. *Adj.* Walter John Shaffer, letter to mother, Plessis-Belleville, February 10, 1918, unpublished memoir, copy in my possession courtesy of Jack Eder, 234–35.
2. Shaffer, unpublished memoir, letter to mother, somewhere in France, March 14, 1918, 253.
3. Shaffer, unpublished memoir, letter to mother, Secteur Postal 127, May 21, 1918, 324.
4. Van Wyngarden, Jagdstaffel 2, 87–88.
5. Gray, Siemens Schuckert D III & IV, 7.
6. Gray, Fokker D.VII, 3.
7. Kilduff, *Richthofen*, 213.
8. Stark, *Wings of War*, 149–50. Originally published as *Die Jagdstaffel unsere Heimat: Ein Fliegertagebuch aus dem letzten Kriegsjahr* (London: John Hamilton, 1933).
9. Kilduff, *Red Baron Combat Wing*, 232.

CHAPTER SIX

1. Forsyth, *Aces of the Legion Condor*, 19.
2. Ibid., 20–21.
3. Ibid., 21.
4. Ibid., 30.
5. Guttman, "One-Oh-Nine," 40.
6. Forsyth, *Aces of the Legion Condor*, 58.
7. López, *Spanish Republican Aces*, 33.
8. Forsyth, *Aces of the Legion Condor*, 62.

CHAPTER SEVEN

1. Campbell, *Messerschmitt Bf 110 Zerstörer*, 11.
2. Weal, *Junkers Ju 87 Stukageschwader*, 22
3. Guttman, "Poland's Circus Master," 34–36.
4. Kuipers, "Un spectacle fascinant," 106.
5. Morton, "5 jours de folles battaille," 80.
6. Ibid.

CHAPTER EIGHT

1. Bishop, *Hurricane*, 24.
2. Robertson and Wilson, *Scotland's War*, 11–12.
3. Ibid., 12–13.

CHAPTER NINE

1. Shores and Cull, *Bloody Shambles*, vol. 1, *December 1941–April 1942*, 145.
2. Ibid., 301.
3. Shores and Cull, *Bloody Shambles*, vol. 2, *December 1941–May 1942*, 36.
4. Ibid., 276.
5. Millman, *Ki-44 'Tojo' Aces*, 13.

CHAPTER TEN

1. Guttman "Finland's Ace of Aces," 43.
2. Ibid.

3. Scutts, *Mustang Aces*, 10.
4. Ibid., 11.
5. Ibid.

CHAPTER ELEVEN

1. Francillon, *American Fighters*, 11.
2. Chloe, *Aleutian Warriors*, 199.
3. Ibid., 201.
4. Haugland, *Eagles War*, 113.
5. Ibid., 117.
6. Ibid., 115.
7. Hall, *1000 Destroyed*, 25.
8. Ibid., 26.
9. Haugland, *Eagles War*, 116.
10. Ibid.
11. Robert S. Johnson, in discussion with the author, June 1996.

CHAPTER TWELVE

1. Shores and Cull, *Malta*, 60.
2. Ibid., 60–61.

CHAPTER THIRTEEN

1. Caldwell, *JG 26*, 98–99.
2. Shores and Cull, *Malta*, 2–4.
3. Ibid., 11.
4. "Oriental Swallow," 77.
5. Sakaida, *Japanese Army Air Force Aces*, 55.
6. Cornelius and Short, *Ding Hao*, 455.
7. "High Wind from Ota," 46.

CHAPTER FOURTEEN

1. Green, "End of an Era," 10.
2. Khazanov and Medvedev, *MiG-3 Aces*, 14.
3. Morgan, *Soviet Aces*, 55.
4. Gordon and Khazanov, *Yakovlev's Piston-Engined Fighters*, 16.
5. Gordon and Khazanov, *Soviet Combat Aircraft*, 199.
6. Khazanov and Medvedev, *La-5/7 vs. Fw 190*, 10–12.
7. Bergström, *Stalingrad*, 76–77.
8. Khazanov and Medvedev, *La-5/7 vs. Fw 190*, 58.

CHAPTER FIFTEEN

1. Styling, *Corsair Aces*, 13.
2. Tillman, *Corsair*, 30.
3. Styling, *Corsair Aces*, 9.
4. Ibid., 10.
5. Tillman, *Hellcat Aces*, 8–9.
6. Jensen, *Carrier War*, 55.
7. Robert Duncan, letter to author, February 25, 2000.

8. Jensen, *Carrier War*, 55.
9. Ewing and Lundstrom, *Fateful Rendezvous*, 210.
10. Ibid., 212.
11. Ibid., 214.
12. Polmar, *Aircraft Carriers*, 359.

CHAPTER SIXTEEN

1. Courteville, "Bloch's Fighters," 205.
2. Thomas, *Typhoon and Tempest Aces*, 10–11.
3. Franks, *Typhoon Attack*, 23.
4. Thomas, *Typhoon and Tempest Aces*, 11.
5. "Slayer of Dragons," 227.
6. Ford, *Flying Tigers*, 315.
7. Ibid., 316.
8. Ibid., 317–18.
9. Ibid., 317.

CHAPTER SEVENTEEN

1. Thomas, *Beaufighter Aces*, 6.
2. Ibid., 9.
3. Guttman, "Norway's Ace of Aces," 47.
4. Sakaida, *Imperial Japanese Navy Aces*, 50.
5. Cooper, "A Black Widow strikes at night," 62.
6. Thompson, *P-61 Black Widow Units*, 17–18.

CHAPTER EIGHTEEN

1. Caldwell, *JG 26*, 20.
2. Pilawskii, *Soviet Air Force Fighter Colours*, 194.
3. 753rd *Kokutai* combat log, courtesy of Osamu Tagaya.
4. 851st *Kokutai* combat log, courtesy of Osamu Tagaya.
5. *Senshi Sosho*, vol. 7, "Eastern New Guinea Army Air Operations," courtesy of Osamu Tagaya.
6. Ibid.
7. Sakaida, *Imperial Japanese Navy Aces*, 110.
8. Ibid., 93.
9. Ethell and Price, *German Jets*, 119.
10. Ethell, *Komet*, 108.
11. Ibid., 110.
12. Ibid., 11–13.
13. Ibid., 13.
14. Herbert Ihlefeld, interview with Colin D. Heaton, October 7, 1989.
15. Herbert Ihlefeld, interview with Colin D. Heaton, Wennigsen, Germany, May 29, 1984.

CHAPTER NINETEEN

1. Morgan and Weal, *German Jet Aces*, 6.
2. Ibid., 9.
3. Ethell and Price, *German Jets in Combat*, 14.
4. Ibid., 17.
5. Mullins, *Escort of P-38s*, 157.

BIBLIOGRAPHY

ADF Serials. Australian & New Zealand Military Aircraft Serials & History: RAAF A46 CAC Boomerang, CA-12/CA-13/CA-14/CA-14A/CA-19, last modified July 5, 2013, http://www.adf-serials.com.au/2a46.htm.

Alegi, Gregory. *Hanriot HD.1/HD.2.* Windsock Datafile 92. Berkhamsted, UK: Albatros, 2002.

Alexander, Captain Richard L., USAAF (Ret.). *They Called Me Dixie*, 1st ed. Hemet, CA: Robinson Typographics, 1988.

Ashley, Glenn. *Meteor in Action*. Carrollton, TX: Squadron/Signal, 1995.

Austin, Alan, and Anthony Dicks. *Great Fighting Planes: World War I to the Present Day*. London: Military Press, 1986.

Bailey, Frank W., and Paul Chamberlain. "*L'Escadrille de Chasse* Spa.57," *Cross & Cockade Journal* 26, no. 1 (Spring 1985): 1–23.

Barker, Ralph. *The RAF at War*. Time-Life Series "The Epic of Flight." Alexandria, VA: Time-Life Books, 1981.

Barnett, Corelli. *Engage the Enemy More Closely: The Royal Navy in the Second World War*. New York: Norton, 1991.

Beedle, J. *43 Squadron, Royal Flying Corps, Royal Air Force: The History of the Fighting Cocks, 1916–66*. London: Beaumont Aviation Literature, 1966.

Bergström, Christer. *Stalingrad: The Air Battle, 1942 through January 1943*. Hinckley, UK: Midland, 2007.

Bernád, Dénes, Dmitriy Karlenko, and Jean-Louis Roba. *From Barbarossa to Odessa: The Luftwaffe and Axis Allies Strike South-East, June–October 1941*. Vol. 1, *The Air Battle for Bessarabia: 22 June–31 July 1941*. Hinckley, UK: Ian Allan, 2007.

Birdsall, Steve. *Log of the Liberators*. Garden City, NY: Doubleday, 1973.

Bishop, Edward. *Hurricane*. Washington, DC: Smithsonian Institution, 1986.

Bond, Charles R. Jr., and Terry H. Anderson. *A Flying Tiger's Diary*. College Station: Texas A&M University Press, 1984.

Botquin, Gaston. "Un bon chasseur pour la 'Drôle de Guerre.'" *Le Fanatique de l'Aviation*, no. 103 (June 1978).

Bowman, Martin. *Mosquito Fighter/Fighter-Bomber Units of World War II*. Botley, Oxford, UK: Osprey, 1998.

Bowyer, Chaz. *Sopwith Camel: King of Combat*. Falmouth, UK: Glasney Press, 1978.

Brown, Captain Eric. "The Fortuitous Fulmar." *Air Enthusiast* 13, no. 2 (August 1977): 74–79.

Bruce, J. M. *The Bristol M.1*. Aircraft in Profile No. 193. Leatherhead, Surrey, UK: Profile, 1967.

——. "Bristol's Fighter Manqué." *Air Enthusiast* 32 (December 1986–April 1987): 1–21, 73–75.

——. *The De Havilland D.H.5*. Aircraft in Profile No. 181. Leatherhead, Surrey, UK: Profile, 1967.

——. *Hanriot HD.1*. Windsock Datafile 8. Berkhamsted, UK: Albatros, 1988.

_____. *Morane-Saulnier Types N, I, V*. Windsock Datafile 92. Berkhamsted, UK: Albatros, 1996.

_____. *RAF SE5*. Windsock Datafile 30. Berkhamsted, UK: Albatros, 1991.

_____. *Sopwith Pup*. Windsock Datafile 2. Berkhamsted, UK: Albatross, 1987.

_____. "Sopwith's Pedigree Pup." *Air Enthusiast Quarterly* 4 (1976): 187–207.

_____. *Spad 7.C1*. Windsock Datafile 12. Berkhamsted, UK: Albatros, 1988.

_____. *Spad 13.C1*. Windsock Datafile 32. Berkhamsted, UK: Albatros, 1992.

_____. "Spad Story," Pts. 1 and 2. *Air International* 10, no. 5 (May 1976): 237–42; no. 6 (June 1976): 289–96, 310–11.

_____. *War Planes of the First World War*. Vol. 1, *Fighters*. Garden City, NY: Doubleday, 1965.

_____. *War Planes of the First World War*. Vols. 2 and 5, *Fighters*. Garden City, NY: Doubleday, 1968.

Caldwell, Donald L. *JG 26: Top Guns of the Luftwaffe*. New York: Orion Books, 1991.

_____. *JG 26: Luftwaffe Fighter Wing War Diary*. Vol. 1, *1939-1942*. London: Grub Street, 1996.

Campbell, Jerry L. *Messerschmitt Bf 110 Zerstörer in Action*. Warren, MI: Squadron/Signal, 1977.

Carson, Annette. *Flight Fantastic: The Illustrated History of Aerobatics*. Newbury Park, CA: Haynes, 1985.

Cattaneo, Gianni. *The Fiat CR.42*. Leatherhead, Surrey, UK: Profile, 1965.

Christienne, Charles, and Pierre Lissarague. *A History of French Military Aviation*. Translated by Francis Kianka. Washington, DC: Smithsonian Institution Press, 1986.

Cieslak, Krzysztof, Wojciech Gawrych, and Andrzej Glass. *Samoloty mysliwskie wrze?nia 1939*. Warsaw: Aerohobby, 1987.

Cloe, John Haile. *The Aleutian Warriors: A History of the 11th Air Force and Fleet Air Wing*. Pt. 1. Missoula, MT: Pictorial Histories, 1991.

Cooksley, Peter. *Bristol Fighter in Action*. Carrollton, TX: Squadron/Signal, 1993.

Cooper, Ann. "A Black Widow strikes at night to down a Japanese Betty bomber over the Pacific during World War II." *Aviation History* 7, no. 4 (March 1997): 62–64.

Cornelius, Wanda, and Thayne Short. *Ding Hao: America's Air War in China, 1937-1945*. Gretna, LA: Pelican, 1980.

Courteville, Lt. Col. Pierre. "Bloch's Fighters: Viewed from the Cockpit." *Air International* 14, no. 4 (April 1978): 179–89, 204–5.

Cuneo, John R. *Winged Mars: The Air Weapon, 1914-1916*. Vol. 2. Harrisburg, PA: Military Service, 1942.

Cynk, Jerzy B. *The Polish Air Force at War: The Official History*. Vol.1, *1939-1943*. Atglen, PA: Schiffer, 1998.

Davis, Donald A. *Lightning Strike: The Secret Mission to Kill Admiral Yamamoto and Avenge Pearl Harbor*. New York: St. Martin's Press, 2005.

Duiven, Richard. "Das Königliches Jagdgeschwader Nr.II." *Over the Front* 9, no .3 (Fall 1994): 196–229.

Durkota, Alan, Thomas Darcey, and Victor Kulikov. *The Imperial Russian Air Service: Famous Pilots and Aircraft of World War I*. Mountain View, CA: Flying Machines Press, 1995.

Ethell, Jeffrey, and Alfred Price. *The German Jets in Combat*. New York: Jane's, 1980.

Ethell, Jeffrey L. *Komet: The Messerschmitt 163*. New York: Sky Books, 1978.

Ewing, Steve, and John B. Lundstrom. *Fateful Rendezvous: The Life of Butch O'Hare*. Annapolis, MD: Naval Institute Press, 1997.

Ferguson, S. W., and William K. Paskalis. *Protect and Avenge: The 49th Fighter Group in World War II*. Atglen, PA: Schiffer, 1996.

"Firefly: A Masterpiece for the Matelots." *Air Enthusiast* 2, no. 3 (March 1972): 139–47.

Ford, Daniel. *Flying Tigers: Claire Chennault and His American Volunteers, 1941–1942*. Washington: Smithsonian Institution Press, 2007.

Forsyth, Robert. *Aces of the Legion Condor.* Botley, Oxford, UK: Osprey, 2008.

_____. *Me 262 Bomber and Reconnaissance Units.* With the assistance of Eddie J. Creek. Botley, Oxford, UK: Osprey, 2012.

Francillon, Dr. René J. *American Fighters of World War Two.* Vol. 1. Garden City, NY: Doubleday, 1969.

_____. *Kawanishi Kyofu, Shiden and Shiden Kai Variants.* Aircraft in Profile No. 213. Windsor, UK: Profile, 1971.

Franks, Norman. *Typhoon Attack.* London: Grub Street, 1993.

Franks, Norman, Frank W. Bailey, and Russell Guest. *Above the Lines.* London: Grub Street, 1993.

Franks, Norman, Hal Giblin, and Nigel McCrery. *Under the Guns of the Red Baron.* London: Grub Street, 2007.

Franks, Norman, Russell Guest, and Gregory Alegi. *Above the War Fronts.* London: Grub Street, 1997.

Freeman, Roger A. *The Mighty Eighth.* New York: Jane's, 1970.

Goetz, Thomas J. "Birth of the American Jet Age." *Aviation History* 17, no. 1 (September 2006): 22–29.

Gordon, Yefim, and Dmitriy Khazanov. *Soviet Combat Aircraft of the Second World War.* Vol. 1. Hinckley, UK: Midland, 2002.

Gordon, Yefim, and Dmitriy Khazanov. *Yakovlev's Piston-Engined Fighters.* Hinckley, UK: Midland, 2002.

Goworek, Tomasz. "The U.S. Army Air Service's first air-to-air victim may have arranged his own capture." *Military History* 11, no. 4 (October 1994): 10–16.

Gray, Peter L. *The Fokker D. VII.* Aircraft in Profile No. 25. Leatherhead, Surrey, UK: Profile, 1967.

_____. *The Siemens Schuckert D III & IV.* Aircraft in Profile No. 86. Leatherhead, Surrey, UK: Profile, 1966.

Green, William. "End of an Era . . . Polikarpov's *Chaika*." *Air Enthusiast* 1, no. 1 (June 1971): 9–15.

_____. *Famous Fighters of the Second World War.* Garden City, NY: Doubleday, 1969.

_____. *War Planes of the Second World War.* Vol. 1, *Fighters.* Garden City, NY: Doubleday, 1965.

Green, William, and Gordon Swanborough. "The Agile Asian . . . Japan's Type 97 Fighter." *Air Enthusiast* 6 (March–June 1978).

Green, William, and Gordon Swanborough. "The Era of the Gull: The Chronicles of the Pulawski Fighter Line." *Air Enthusiast,* no. 28 (July–October 1985): 35–53, 80.

Green, William, and Gordon Swanborough. "Nakajima Demonology: The Story of the Shoki." *Air International* 3, no.1 (July 1972): 17–25.

Green, William, and Gordon Swanborough. "Of Chaika and Chato." *Air Enthusiast* 11 (November 1979–February 1980): 9–29.

Green, William, and Gordon Swanborough. "Soviet Flies in Spanish Skies: The Operational Use of I-16s in the Spanish Civil War and Thereafter." *Air Enthusiast Quarterly,* no. 1 (December 1975): 1–16.

Green, William, and Gordon Swanborough. "The Zero Precursor . . . Mitsubishi's A5M." *Air Enthusiast,* no. 19 (August–November 1982): 26–43.

Grosz, Peter M. *Fokker D. VIII.* Windsock Datafile 25. Berkhamsted, UK: Albatros, 1991.

_____. *Halberstadt Fighters.* Classics of WWI Aviation. Vol. 1. Berkhamsted, UK: Albatros, 1996.

_____. *SSW D.III-D.IV.* Windsock Datafile 29. Berkhamsted, UK: Albatros, 1991.

Grosz, Peter M., and A. E. Ferko. "Biplanes for the *Fliegertruppe.*" *Air Enthusiast* 14 (December 1980–March 1981): 57–67.

Grosz, Peter M., George Haddow, and Peter Schiemer. *Austro-Hungarian Army Aircraft of World War One.* Mountain View, CA: Flying Machines Press, 1993.

Gunston, Bill. "Birth of the Jet Fighter." *Royal Air Force Yearbook 1984,* 51–60.

Guttman, Jon. *Bristol F2 Fighter Aces of World War I.* Botley, Oxford, UK: Osprey, 2007.

_____. "Finland's Ace of Aces: Interview with Eino Ilmari Juutilainen." Military History 15, no. 6 (February 1999): 42–48, 80.

_____. "The first cannon-armed fighter owed its modest success to outstanding pilots." *Aviation Heritage* (September 1991): 12–16.

_____. *Groupe de Combat 12 'Les Cigognes': France's Ace Fighter Group in World War 1.* Botley, Oxford, UK: Osprey, 2004.

_____. *Nieuport 28.* Windsock Datafile 36. Berkhamsted, UK: Albatros, 1992.

_____. "Norway's Ace of Aces: Interview with Maj. Gen. Svein Heglund." *Aviation History* 9, no. 6 (July 1999): 42–48.

_____. "One-Oh-Nine: Messerschmitt's Killing Machine." *Aviation History* 9, no. 5 (May 1999): 38–46.

_____. "Plumage: Spa.156: L'Escadrille des Deux Martinets." *Over the Front* 9, no. 1 (Spring 1994): 71–74.

_____. "Poland's Circus Master: Interview with Stanislaw Skalski." *Aviation History* 10, no. 2 (September 1999): 34–40.

_____. *Pusher Aces of World War 1.* Botley, Oxford, UK: Osprey, 2009.

_____. *Sopwith Camel vs Fokker DrI.* New York: Osprey, 2008.

_____. *Spad VII Aces of World War I.* Botley, Oxford, UK: Osprey, 2001.

Guttman, Robert. "Arming a fighter plane with nothing more than a four-gun turret proved unique—but not brilliantly successful." *Aviation History* 7, no. 1 (September 1996): 12.

_____. "A speedy fighter of 1916 was rejected out of hand because it had only one wing in a biplane era." *Aviation History* (November 1996): 8–12.

Hall, Grover C. Jr. *1000 Destroyed: The Life and Times of the 4th Fighter Group.* Fallbrook, CA: Aero, 1978.

Hata, Ikuhiko, and Yasuho Izawa. *Japanese Naval Aces and Fighter Units in World War II.* Translated by Don Cyril Gorham. Annapolis: Naval Institute Press, 1989.

Haugland, Vern. *The Eagle's War: The Saga of the Eagle Squadrons Pilots, 1940–45.* Blue Ridge Summit, PA: TAB Books, 1992.

"The High Wind from Ota." *Air International* 10, no. 1 (January 1976): 22–29, 43–46.

Holmes, Tony. *Hurricane Aces, 1939-40.* Botley, Oxford, UK: Osprey, 1998.

Hooton, Edward R. "Air War Over China." *Air Enthusiast Thirty-four* (September–December 1987.

Howson, Gerald. *Aircraft of the Spanish Civil War, 1936–1939.* Washington, DC: Smithsonian Institution Press, 1990.

Huggins, Mark. "Falcons on Every Front: Nakajima Ki.43-I *Hayabusa* in Combat." *Air Enthusiast*, no. 131 (September/October 2007): 32–38.

Hylands, Dennis. *Georges Guynemer.* Aces & Aeroplanes 2. Berkhamsted, UK: Albatros, 1986.

_____. *Werner Voss.* Aces & Aeroplanes 1. Berkhamsted, UK: Albatros, 1986.

Ichimura, Hiroshi. *Ki-43 'Oscar' Aces of World War 2.* Botley, Oxford, UK: Osprey, 2009.

Jackson, Robert. *Spitfire: The Combat History.* Osceola, WI: Motorbooks International, 1995.

_____. "Flight of the Meteor," *The Elite: The Bombers.* Harrisburg, PA: National Historical Society, 1989.

Jankowski, Paul. *Verdun: The Longest Battle of the Great War.* New York: Oxford University Press, 2014.

Jensen, Lt. Oliver, USNR. *Carrier War.* New York: Simon & Schuster, 1945.

Jones, Wing Commander Ira. *Tiger Squadron.* London: White Lion, 1954.

Keskinen, Kalevi, Kari Stenman, and Klaus Niska. *Finnish Fighter Aces.* Espoo, Finland: Tietoeos, 1978.

Khazanov, Dmitriy, and Aleksander Medvedev. *La-5/7 vs. Fw 190: Eastern Front, 1942–45.* Botley, Oxford, UK: Osprey, 2011.

Khazanov, Dmitriy, and Aleksander Medvedev. *MiG-3 Aces of World War 2.* Botley, Oxford, UK: Osprey, 2012.

Kilduff, Peter. *Red Baron: The Life and Death of an Ace.* Cincinnati, OH: David & Charles, 2008.

_____. *The Red Baron Combat Wing: Jagdgeschwader Richthofen in Battle.* London: Arms & Armour, 1997.

_____. *Richthofen: Beyond the Legend of the Red Baron.* London: Arms & Armour, 1993.

_____. *"That's My Bloody Plane."* Chester, CT: Pequot Press, 1975.

Klaeylé, Bernard, and Philippe Osché. *Guynemer: Les avions d'un as (Histoire de l'aviation).* With the assistance of Christophe Cony. Outreau: Éd. Lela presse, 1998. [In French; my translation.]

Krybus, Josef. *The Avia B.534.* Aircraft in Profile No. 152. Leatherhead, Surrey, UK: Profile, 1968.

Kuipers, Lt. J. P. *"Un spectacle fascinant." Icare Revue De L'aviation Française. 1939–40 La Bataille de France.* Vol IX. *L'aviation néer-landaise* (Winter 1976–77): 104–7. [In French; my translation].

Kulikov, Victor. *Russian Aces of World War 1.* Botley, Oxford, UK: Osprey, 2013.

"The Last Swallow of Summer . . . the Extraordinary Story of the Ki.100." *Air International* 11, no. 4 (October 1976): 185–91.

Lawson, Eric, and Jane Lawson. *The First Air Campaign, August 1914–November 1918.* New York: Da Capo Press, 1997.

Lee, Arthur Gould. *No Parachute: A Fighter Pilot in World War I.* New York: Harper & Row, 1968.

Leyvastre, Pierre. "Bloch's Fighters: The Contentious Combatants." *Air Enthusiast Fourteen* (April 1978): 179–189, 204–5.

Lighthall, W. S. "The Royal Air Force in the Palestine Campaign, 1917–1918." *Cross & Cockade Journal (USA)* 11, no. 2 (Summer 1970): 171.

Longoluso, Alfred. *Fiat CR.32 Aces of the Spanish Civil War.* Botley, Oxford, UK: Osprey, 2010.

López, Rafael A. Permuy. *Spanish Republican Aces.* Botley, Oxford, UK: Osprey, 2012.

"Major 'Buffalo' Wong Sun-Shui," Håkans Aviation Page, last modified July 1, 2009, accessed April 22, 2014, http://surfcity.kund.dalnet.se/china_wong2.htm.

Martinez, Luis García. "Los Katiuskas." *Air Enthusiast Thirty-two* (December 1986–April 1987): 45–55.

Maslov, Mikhail. *Polikarpov I-15, I-16 and I-153 Aces.* Botley, Oxford, UK: Osprey, 2010.

Mason, Herbert Molloy. *High Flew the Falcons: French Aces of World War I.* Philadelphia: Lippincott, 1965.

McCudden, Maj. James T. B. *Flying Fury: Five Years in the Royal Flying Corps.* New York: Ace, 1968.

Mellinger, George. *LaGG and Lavochkin of World War 2.* Botley, Oxford, UK: Osprey, 2003.

Mendenhall, Charles A. *Wildcats & Hellcats: Gallant Grummans in World War II.* Osceola, WI: Motorbooks International, 1984.

"Messerschmitt Over Spain." *Air Enthusiast Seven* (July–September 1978): 182–94.

Millman, Nicholas. *Ki-27 'Nate' Aces.* Oxford, UK: Osprey, 2013.

_____. *Ki-44 'Tojo' Aces of World War 2.* Botley, Oxford, UK: Osprey, 2011.

Morgan, Hugh. *Soviet Aces of World War 2.* Botley, Oxford, UK: Osprey, 1997.

Morgan, Hugh, and John Weal. *German Jet Aces of World War 2.* Botley, Oxford, UK: Osprey, 1998.

Morison, Samuel Eliot. *History of U.S. Naval Operations in World War II.* Vol. 3, *The Rising Sun in the Pacific, 1931-April 1942.* Boston: Little Brown, 1988.

Morton, Fred. "5 jours de folles battaille." *Icare Revue De L'aviation Française. 1939–40 La Bataille de France.* Vol IX. *L'aviation néer-landaise* (Winter 1976–77): 72–87. [In French; my translation].

Moyes, Philip J.R. *The De Havilland Mosquito Mks. I–IV.* Leatherhead, Surrey, UK: Profile, 1965.

Mullins, John D. *An Escort of P-38s: The First Fighter Group in World War II*. St. Paul, MN: Phalanx, 1995.

Nohara, Shigeru. *A6M Zero in Action*. Carrollton, TX: Squadron/Signal, 1983.

"An Oriental Swallow." *Air International* 9, no. 2 (August 1975): 75–83.

Ostric, Sime I., and Cedomir J. Janic. *IK Fighters (Yugoslavia: 1930-40s)*. Profile 242. Berkshire, UK: Profile, 1972.

"Pacific Peregrine . . . The Nakajima Ki.43 Hayabusa." *Air International* 18, no. 1 (January 1980): 27–31, 44–46.

Pilawskii, Erik. *Soviet Air Force Fighter Colours, 1941–1945*. Hersham, Surrey, UK: Classic, 2003.

Poiencot, Kelly P. "The Father of Aerial Combat," *Aviation History* 6 (July 1996): 30–36.

Polmar, Norman. *Aircraft Carriers: A History of Carrier Aviation and Its Influence on World Events*. Vol. 1, *1909–1945*. Washington, DC: Potomac Books, 2006.

Price, Alfred. *Spitfire at War*. London: Ian Allen, 1974.

_____. *The Spitfire Story*. New York: Jane's, 1982.

"Raiden: The Asiatic Thunderbolt." *Air Enthusiast* 1, no. 2 (July 1971): 67–73, 103.

Revell, Alex. *British Fighter Units: Western Front, 1914-16*. Botley, Oxford, UK: Osprey, 1978.

_____. *High in the Empty Blue: The History of No. 56 Sqn RFC/RAF, 1916–1919*. Mountain View: Flying Machines Press, 1995.

Ries, Karl, and Hans Ring. *The Legion Condor*. West Chester, PA: Schiffer, 1992.

Rimell, Raymond Laurence. *Zeppelin! A Battle for Air Supremacy in World War I*. London: Conway Maritime Press, 1984.

Rimell, R. L., and Peter M. Grosz. *Pfalz DIII*. Windsock Datafile 7. Berkhamsted, UK: Albatros, 1988.

Robertson, Seona, and Les Wilson. *Scotland's War*. Edinburgh: Mainstream, 1995.

Roscoe, Theodore. *United States Submarine Operations in World War II*. Annapolis: Naval Institute Press, 1988.

Rust, Kenn C. *Twentieth Air Force Story in World War II*. Temple City, CA: Historical Aviation Album, 1979.

Rust, Kenn C., and Dana Bell. *Thirteenth Air Force Story in World War II*. Temple City, CA: Historical Aviation Album, 1981.

Sakai, Saburo. *Samurai!* With the assistance of Martin Caidin and Fred Saito. New York: Bantam, 1978.

Sakaida, Henry. *Imperial Japanese Navy Aces, 1937-1945*. Botley, Oxford, UK: Osprey, 1998.

_____. *Japanese Army Air Force Aces, 1937-45*. Botley, Oxford, UK: Osprey, 1997.

_____. *The Siege of Rabaul*. St. Paul, MN: Phalanx, 1996.

Samson, Jack. *The Flying Tiger: The True Story of General Claire Chennault and the U.S. 14th Air Force in China*. Guilford, CT: Lyons Press, 1987.

Scutts, Jerry. *German Night Fighter Aces of World War 2*. Botley, Oxford, UK: Osprey, 1998.

_____. *Mustang Aces of the Ninth and Fifteenth Air Forces & The RAF*. Botley, Oxford, UK: Osprey, 1995.

Seidl, Hans D. *Stalin's Eagles*. Atglen, PA: Schiffer, 1998.

Shores, Christopher. *Dust Clouds in the Middle East*. London: Grub Street, 1996.

_____. *Spanish Civil War Air Forces*. Botley, Oxford, UK: Osprey, 1977.

Shores, Christopher, and Brian Cull. *Bloody Shambles: The First Comprehensive Account of Air Operations over South-East Asia, December 1941–April 1942*. Vol. 1, *The Drift to War to the Fall of Singapore*. With the assistance of Yasuho Izawa. London: Grub Street, 1993.

Shores, Christopher, and Brian Cull. *Bloody Shambles: The First Comprehensive Account of Air Operations over South-East Asia, December 1941-May 1942*. Vol. 2, *The Defence of Sumatra to the Fall of Burma*. With the assistance of Yasuho Izawa. London: Grub Street, 1993.

Shores, Christopher, and Brian Cull. *Air War for Yugoslavia, Greece and Crete 1940-41*. With the assistance of Nicola Malizia. London: Grub Street, 1987.

Shores, Christopher, and Brian Cull. *Malta: The Hurricane Years, 1940-41*. With the assistance of Nicola Malizia. London: Grub Street, 1987.

Shores, Christopher, Norman Franks, and Russell Guest. *Above the Trenches*. London: Grub Street, 1990.

Shores, Christopher, and Clive Williams. *Aces High*. London: Grub Street, 1994.

Skelton, Marvin L. "Major H.D. Harvey-Kelly, Commanding Officer, No. 19 Squadron." *Cross & Cockade (USA) Journal* 16, no. 4 (Winter 1975): 365–69.

"Slayer of Dragons: The Story of Toryu." *Air Enthusiast* 5, no. 5 (November 1973): 225–29.

Stapfer, Hans-Heiri. *LaGG Fighters in Action*. Carrollton, TX: Squadron/Signal, 1996.

Stark, Rudolf. *Wings of War: An Airman's Diary of the Last Year of the War*. Translated by Claud W. Sykes. London: Arms & Armour Press, 1973. Originally published as *Die Jagdstaffel unsere Heimat: Ein Fliegertagebuch aus dem letzten Kriegsjahr* (London: John Hamilton, 1933).

Stenge, Csaba B. *Baptism of Fire: The First Combat Experiences of the Royal Hungarian Air Force and Slovak Air Force, March 1939*. Solihull, UK: Helion, 2014.

Stenman, Kari. "38 to 1: The Brewster 239 in Finnish Service." *Air Enthusiast*, no. 46 (June–August 1997): 40–42.

Stenman, Kari, and Kalevi Keskinen. *Finnish Aces of World War 2*. Botley, Oxford, UK: Osprey, 1998.

Styling, Mark. *Corsair Aces of World War 2*. Botley, Oxford, UK: Osprey, 1996.

Sullivan, Jim. *F4U Corsair in Action*. Carrollton, TX: Squadron/Signal, 1994.

Swanborough, Gordon. "Beaufighter—Innovative Improvisation by Bristol." *Air Enthusiast International* 6, no. 1 (January 1974): 25–47.

Sweeting, C. G. "Duel in the Clouds." *Aviation History* 23, no. 3 (January 2013): 52–56.

Thomas, Andrew. *Beaufighter Aces of World War 2*. Botley, Oxford, UK: Osprey, 2006.

_____. *Gloster Gladiator Aces*. Botley, Oxford, UK: Osprey, 2002.

Thomas, Chris. *Typhoon and Tempest Aces of World War 2*. Botley, Oxford, UK: Osprey, 1999.

Thompson, Warren. *P-61 Black Widow Units of World War 2*. Botley, Oxford, UK: Osprey, 1998.

Tillman, Barrett. *Corsair: The F4U in World War II and Korea*. Annapolis, MD: Naval Institute Press, 1979.

_____. *Hellcat Aces of World War 2*. Botley, Oxford, UK: Osprey, 1996.

Van Wyngarden, Greg. *Jagdstaffel 2 'Boelcke': Von Richthofen's Mentor*. Botley, Oxford, UK: Osprey, 2007.

Weal, John. *Focke-Wulf Fw 190 Aces of the Western Front*. Botley, Oxford, UK: Osprey, 1996.

_____. *Junkers Ju 87 Stukageschwader, 1937-41*. Botley, Oxford, UK: Osprey, 1977.

Werner, Johannes. *Knight of Germany: Oswald Boelcke, German Ace*. Havertown, PA: Casemate, 2009. First published 1932 by John Hamilton.

"Whirlwind: First of the Four-Cannon Fighters." *Air Enthusiast* 4, no. 7 (July 1973): 30–39.

Wixey, Ken. "Corpulent Feline: Grumman's F4F Wildcat." *Air Enthusiast* 70 (July–August 1997): 50–52.

Wolf, William. *The Fifth Fighter Command in World War II*. Vol. 1, *Pearl Harbor to the Reduction of Rabaul*. Atglen, PA: Schiffer, 2011.

INDEX